Technische Universität München
Lehrstuhl für Hochfrequenztechnik

On Design of Coherent
Multiple-Input-Multiple-Output Radars with
Sparse Antenna Arrays

Andreas Josef Kirschner

Vollständiger Abdruck der von der Fakultät für Elektrotechnik und Informationstechnik der Technischen Universität München zur Erlangung des akademischen Grades eines

– *Doktor-Ingenieurs* –

genehmigten Dissertation.

Vorsitzender: Univ.-Prof. Dr.-Ing., Dr.-Ing. habil.,
                        Dr. h.c. Alexander W. Koch
Prüfer der Dissertation: 1. Univ.-Prof. Dr.-Ing.,
                        Dr.-Ing. habil. Jürgen Detlefsen
                        2. Prof. Dr. Ir Wim Mees
                        (Königliche Militärakademie Brüssel, Belgien)

Die Dissertation wurde am 02.11.2015 bei der Technischen Universität München eingereicht und durch die Fakultät für Elektrotechnik und Informationstechnik am 20.06.2016 angenommen.

Bibliografische Information der Deutschen Nationalbibliothek

Die Deutsche Nationalbibliothek verzeichnet diese Publikation in der
Deutschen Nationalbibliografie; detaillierte bibliografische Daten sind
im Internet über http://dnb.d-nb.de abrufbar.

ISBN 978-3-8325-4410-2

Logos Verlag Berlin GmbH
Comeniushof, Gubener Str. 47,
10243 Berlin
Tel.: +49 (0)30 42 85 10 90
Fax: +49 (0)30 42 85 10 92
INTERNET: http://www.logos-verlag.de

A variety of methods had to be conducted to find model descriptions and processing solutions in order to get completely to the bottom of all problems in this thesis. This could also be quoted in apt manner by:

**Mathematik ist das Alphabet,
mit dessen Hilfe Gott das Universum beschrieben hat.**

**Mathematics is the alphabet,
which God has described the universe with.**

Galileo Galilei ∗ 1564, † 1642

Preface

This thesis is the result of my work as member of scientific staff at the institute *Hochfrequente Felder und Schaltungen - High Frequency Fields and Circuits* at *Technische Universität München*. I would like to take the occasion and thank all students who helped to implement hardware and conducted tests to the benefit of the institute's project. Among the group of more than 30 students I had the privilege to supervise, I would especially thank Lionel Arend, Christian Berger and Chistian Maschall who had been freelancers one could always rely on in busy times.

In the same context, the team of the institute's mechanical workshop headed by Manfred Agerer first and then Josef Fanzisi later on, deserve my deep respect and many thanks for their cooperative manner and professional quality of advice and work. Further, I would like to thank my colleagues Christop Hornsteiner, Sebastian Bertl, Noora Al-Kahachi, Johanna Gütlein and Christian Speck. Many helping hands were needed to get all those systems running. A propos running system: Special thanks are dedicated to Thoma Mittereder and Prof. Dr.-Ing. Uwe Siart, who helped us to keep our IT network running, collaborated in administrative things or gave LATEX advice.

Further, I would like to thank LtCol Prof. Dr.Ir Wim Mees, who had always been a commendable colleague and partner during the SUM project and who has not only escorted but also guided the generation of this thesis. Special thanks are dedicated to my wife, Isabel, who had to raise a lot of patience for many years due to my work on desk.

At the very end, I would like to thank especially Prof. Dr.-Ing. habil. Jürgen Detlefsen, who had not only given the opportunity to work on a PhD thesis to me but also offered the chance to work a broad variety of technical issues in international projects. He always set a tremendous trust in our young team and in our capabilities to mature systems. Besides that, he had always offered technical advice and lent an open ear to all scientific questions. Prof. Detlefsen has been a very well-known name with respect to radar science in Germany for several decades.

Andreas Josef Kirschner, Pittenhart 2015

Dedicated to my wife and family.

Three months after defence of this thesis, Prof. Jürgen Detlefsen died. I will keep him in honored memory.

Andreas Josef Kirschner, Pittenhart 2016

Abstract

This thesis covers so-called *coherent* MIMO radar systems. Among others, the focus is set on the array configuration which is chosen sparsely subject to considerations on minimum redundancy in element spacing occurrence. For a better understanding, the reader is taken from the fundamentals of electromagnetic wave propagation and guided quickly to antenna arrays. The principle of *virtual arrays* is derived explicitly from a general bistatic setup. The signal design and modulation is explained in detail, especially the principle of analytic signals in combination with FMCW principle and the consequences of those for the modulated array signal. Besides that, the covariance matrix for uncorrelated and correlated signal sceneries are briefly examined. The reader is referred to most common angular processing algorithms which are introduced in appendix to assist for better understanding.

Then the reader is made familiar with the topic of MR arrays. A brute-force algorithm is suggested for definition of possible MR configurations. Furthermore, a mathematical description of MR arrays leading to setups with global minima in their redundancy of element spacing is presented. By use of these methods, no greedy optimization is necessary anymore. Consequently these insights are extrapolated to cascaded MR setups by shifting in one or two lateral dimensions. This is the point where connections to coded aperture imaging methods were found.

An important topic is errors and their propagation within MR processing. Here mathematical descriptions are provided. Especially, for angular processing in coherent signal scenario, the reader is guided through methods of compressed sensing.

After this, the systems which this thesis is founded on, are described: The vehicle borne SUM system and the helicopter landing aid radar HeLAR. Both exhibit MIMO configurations, the first as *uniform linear array* (ULA), the latter as *general* MR array. In order to provide high quality data for angular processing, the developed pre-processing, calibration and ramp correction techniques are introduced, performed and proven on real measurement data. Special focus is set on the connection of MR arrays with coherent signal scenarios, which bears challenges with respect to angular processing. The thesis closes with a short conclusion.

Zusammenfassung

Diese Arbeit behandelt sogenannte *kohärente* MIMO Radarsysteme. Unter anderem wird auf ausgedünnte Gruppenantennen im Hinblick auf minimale Redundanz (MR) der Elementabstände Wert gelegt. Zum besseren Verständnis wird der Leser von den Grundlagen der Ausbreitung von elektromagnetischen Wellen schnell zur Thematik von Gruppenantennen geführt. Anhand eines allgemeinen bistatischen Aufbaus wird der Begriff der *virtuellen Gruppenantenne* hergeleitet. Die Modulationsformen im Allgemeinen und das FMCW-Prinzip im Besonderen, sowie die Erzeugung von analytischen Signalen und die Auswirkungen auf das Signal der Gruppenantenne, stehen im Mittelpunkt des folgenden Kapitels. Danach werden kurz die wichtigsten Eigenschaften der Kovarianzmatrizen im korrelierten und unkorrelierten Signalfall erläutert. Der Leser wird danach auf die wichtigsten Prozessierungsmethoden zu Richtungsschätzung verwiesen, da diese als Grundlage für den Entwurf der folgenden Gruppenanordnungen unverzichtbar sind. Zum besseren Verständnis sind diese im Anhang aufgeführt.

Danach wird der Leser mit dem Gebiet der MR-Gruppenantennen vertraut gemacht. Hierzu wird ein Brute-Force-Algorithmus zur automatischen Erzeugung von MR-Gruppenantennen vorgeschlagen. Darüber hinaus wird eine mathematisch geschlossene Beschreibung für beide Sorten von MR-Gruppenantennen vorgestellt. Diese erlaubt es Konfigurationen mit globalen Minima bzgl. der Häufigkeit ihrer Elementabstände zu entwerfen, ohne dass man aufwendige Optimierungen bemühen muss. Konsequenterweise werden die neu gewonnenen Einsichten verwendet, um kaskadierte MR-Gruppenantennen zu entwerfen, die entweder in einer oder sogar zwei lateralen Dimensionen verschoben werden. Hier schließt sich der Kreis zum Themengebiet der *kodierten Aperturen*.

Ein wichtiges Kapitel ist die Vorstellung von möglichen Fehlerquellen und deren Fortpflanzung in der MR-Prozessierung. Speziell für den korrelierten Signalfall, wird der Leser in die wichtigsten Techniken von Compressed Sensing eingeführt.

Danach werden die zwei wesentlichen Systeme, auf denen diese Arbeit fußt, vorgestellt: Das fahrzeuggebunden SUM-System und eine Hubschrauberlandehilfe, hier als HeLAR bezeichnet. Beide sind kohärente MIMO-Systeme. Ersteres mit einer gleichförmigen linearen Gruppe, letzteres mit einer *generellen* MR-Gruppe. Damit eine ausreichende Qualität von Rohdaten für die Winkelprozessierung zur Verfügung gestellt werden kann, müssen Vorprozessierungsschritte, Kalibrierung und Korrektur von Modulationsfehlern vorangestellt werden. Diese Methoden werden an echten Messdaten erprobt und demonstriert. Spezieller Augenmerk wird hierbei auf Verbindung von MR-Gruppenantennen und kohärenter Signalumgebungen gelegt. Diese beinhalten besondere Herausfordungen, was die Winkelauswertung betrifft. Die Arbeit schließt mit kurzen Schlussfolgerungen.

Contents

1 Introduction

1.1 The different MIMO approaches

The multiple-input multiple-output (MIMO) principle is traditionally well-known from communication applications. During the last years, this term has also found its way into radar system theory, signal design and signal processing, resulting in a large variety of experiments and publications. In the overall context, MIMO radar systems fit into the technological trend of *digital beamforming* or *digital radar*, i.e. the signals collected by the radar are sampled at the very first technological level where it is possible. Then all signal processing is performed digitally by digital signal processors (DSP), computers etc. A second aspect is, to reduce the efforts and therefore the costs in hardware, or optimizing the capabilities of existing hardware, by a sophisticated signal design. One of these methods is the principle of *virtual arrays* which is usually used equivalently at the definition of *coherent* MIMO radar. But before going into more detail, a short definition of different terms should precede.

In general, several transmitters and receivers or respectively transceivers are used in MIMO radar systems, where principally two different groups can be distinguished: The *statistical* MIMO and the *coherent* MIMO radar approach [1], [2].

For the first, the radar receivers, transmitters or transceivers are largely spaced to each other [1], [3], [4], [5], [6]. By sharing the information of all system parts, the same objects can be monitored from different directions. The strategy is to countervail the enormously fluctuating backscattering characteristics (therefore *statistical* MIMO) by the increased space diversity in order to increase the probability of detection [1], [3], [4], [5], [6], [7], [8]. This type can be found in radio astronomy, air-traffic surveillance etc., but will not be covered in this thesis in more detail.

The latter group, *coherent* MIMO, has got its origin in the last-mentioned *digital beamforming* topic [9], [10]. Unlike in *statistical* MIMO radars, the receivers, transmitters or transceivers are closely spaced to each other [1], [9], [11]. Due to the technically small distances between the different system parts, not only information can be shared, but even a coherent signal evaluation is possible. Hence, this denoted as *coherent* MIMO radar [1], [11]. But contrary to *phased arrays*, not the same signal is used for the whole radar arrangement [10], [12]. It is of major interest, to identify the different single signal paths from each transmitter to the objects and back to the different receivers in order to improve radar detection by coherent MIMO processing. This corresponds to the estimation of channel properties in communication theory.

General considerations on degrees of freedom, resolution capabilities or realizations of different MIMO approaches can be followed in [2], [13], [14].

1.2 The project background of the thesis

Two radar sensor systems were designed and implemented at the institute *Hochfrequente Felder und Schaltungen*. Those sensors served as fundamental platform for experiments.

The first system was constructed in context of a European scientific program denoted by *Surveillance in an Urban environment using Mobile sensors* (SUM), funded by *European Defence Agency* (EDA). European armed forces had been confronted constantly with *improvised explosive devices* (IEDs) during their abroad peace-keeping missions. Over several decades, these devices had emerged to lethal traps in asymmetric conflicts for attacking e. g. patrol crews. Very often daily-life items are misused to manufacture and/ or camouflage IEDs in scenarios where military personnel usually gets into touch with the local population. To countersteer this development, the objective of SUM program was the enhancement of situational awareness for moving vehicles during patrols or at checkpoint scenarios with respect especially to such IEDs, but also road blocks, land mines etc.

In order to demonstrate the capabilities of a joint sensor system, a Spanish-Belgian-German consortium consisting of Grupo Tecnològico e Industrial GMV S.A. (Spain), Royal Military Academy, RMA (Belgium), Deutsches Zentrum für Luft- und Raumfahrt, DLR (Germany) and Technische Universität München, TUM (Germany) was established.

The second system, is a helicopter landing aid, ordered by *Elektronik und System GmbH* (ESG) Fürstenfeldbruck. Here, the task was, to provide terrain information through raised and swirled dust or snow during the landing phase of a helicopter. Especially at the very last 30 ft altitude, the downwash of the helicopter's main rotor presses aside lax dust or snow. A few meters away of the vehicle, those particles can ascent and are then intaken into the downwash of the main rotor again. Therefore, the same dust/ snow cloud is circulated again and again. The effect for the pilots is a total loss of vision, which is usually denoted as *brown-out* (in case of dust) or *white-out* (in case of snow). Moreover, the standard altimeters exploiting air-pressure are usually too imprecise at altitudes below 30 ft. Therefore, the crew depends on their artificial horizon and possibly a radar altimeter which only provides height information but no details on the terrain like inclination or obstacles like e. g. parking cars, hydrants, street lamps etc. A radar sensor providing 2D/ 3D information should help to improve the actual situation.

1.3 State of the art

After many theoretical examinations on the possibilities of MIMO techniques in radar [15], [16], [17], [18] at the beginning of the new millennium, more and more coherent MIMO radar systems had been implemented for various applications during the years starting from 2002. Among others, there are examples from automotive sector [19], [20] or for airborne systems [21], [22]. Similar to the context of *coherent* MIMO radars are the implementation of satellite services not only providing high spatial resolution due to synthetic aperture SAR techniques but also ground-moving target indication GMTI, which can be achieved by switched apertures [23]. Especially for low-power applications, the progress in solid state devices from 2004 onwards offered new degrees of freedom for very small, power-efficient, highly integrated and finally rather cheap systems [20].

Coherent MIMO systems fit consequently into history and trends of *beam-forming*. First, in *phased array* systems, all antenna elements were fed with the same signal. Phase factors for *beamforming* were implemented as hardware devices in the antenna feed lines [24]. Later, with advanced technology, the phase factors could be incorporated into signal design. Then each antenna element could be fed with its own special signal, altered in amplitude and phase. As further development, the sceneries were illuminated in broad sectors and the beamforming was performed on sampled digital receiver data which is generally denoted as *digital beamforming*. The next logical step was to excite all array elements with a sum of several orthogonal signals, which provides the possibility of forming e. g. several beams at the same time. This technique was e. g. used in smart communication antenna arrays via so-called *Butler matrices* by beginning of the years 2000 [10]. The *coherent* MIMO radar application substituted the *Butler matrix* by *adaptive digital beamforming* shortly afterwards.

Apart from that, the topic of minimum redundancy (MR) arrays had been a completely independent history. Here, the attempts started in late 1950s, beginning 1960s in radio-astronomy [25], [26]. Here the main targets had been reducing the number of receiving dishes due to their tremendous costs by keeping the quality of measured signal acceptable at the same time [25], [26]. In order to combine those approaches which were suited for a single point-like target observation such as a quasar etc., with a multi-target scenario, MR fundamentals were tested for beamforming techniques already at in early 1990s [27]. However, the convolutional character of coherent MIMO systems had beared itself as obstacle for a straight-forward system design with respect to minimum redundancy. A remedy were optimization approaches [28], [29] starting in the new decade of 2000. Nevertheless, those optimization procedures were greedy algorithms and therefore unattractive for system design again.

This thesis shall now give insight in the possibilities of combining both fields of arrays. The convolutional character of coherent MIMO is exploited as fundament for system designs without greedy optimization. The insights of this thesis are therefore at the leading edge of ongoing worldwide research on array setups. Minimum redundancy techniques will help to reduce costs, shorten data acquisition times or decrease data amount to be collected. Maybe during coming years, the combination to MR-MIMO systems might find its way to influence various fields of applications, such as millimetre-wave imaging, person security screening, threat detection, ground penetrating radar, landmine detection and many more.

1.4 Properties of millimetre waves

In terms of radar detection or imaging, bandwidth is of great importance for range resolution. Principally, the higher the center frequency f_0 is chosen, the more bandwidth Δf can be provided physically - although a legal license is then a different kettle of fish - with keeping the (hardware) advantage of handling a narrow-band system and therefore the need of moderate quality factors Q for e. g. filters [30]:

$$\Delta f_{rel} = \frac{\Delta f}{f_0} \approx \frac{1}{Q} \ll 0.2 \quad . \tag{1.1}$$

Furthermore, it becomes possible to implement high gain antennas (G) at moderate mechanical dimensions (area size of aperture A_{ap}) for shorter wavelengths λ [31]:

$$G \sim \frac{A_{ap}}{\lambda^2} \quad . \tag{1.2}$$

However, for shorter wavelengths, the free-space attenuation increases. The so-called *radar equation*, which is a derivate of *Friis' formula*, shows the effect regarding attenuation due to propagation [24], [31], [32]:

$$\frac{P_{rx}}{P_{tx}} = \frac{G_{tx}}{4\pi\, r^2} \cdot \frac{rcs}{4\pi\, r^2} \cdot G_{rx} \cdot \frac{\lambda^2}{4\pi} \ . \tag{1.3}$$

Moreover, the additional absorption rates due to oxygen or water resonances in the atmosphere increases [33]. By the same token, the penetration capabilities decrease.

Generally, millimeter-wave sensors (30 GHz-300 GHz) provide a well-suited compromise between microwave and infra-red technologies. They overwhelm drawbacks of visual/ infra-red cameras in unfavourable environmental conditions like e.g. fog, dust or rain, at relatively low system's dimension or weight. Also remaining penetration capabilities through clothes or awnings can be exploited e. g. for security screening.

2 Electromagnetic wave propagation

Primary[1] radar systems generally evaluate echoed electromagnetic fields, thus attempt to solve a *scattering problem*. However, the *scattering problem* in turn can always be led back to *source problem* [34].

2.1 The source problem and Huygen's principle

A source distribution function $q(\boldsymbol{r}, t)$ shall generate a field $u(\boldsymbol{r}, t)$ in free space. Hereby $u(\boldsymbol{r}, t)$ can represent e.g. an electrical field $\boldsymbol{E}$, a magnetic field $\boldsymbol{H}$, a magnetic vector potential $\boldsymbol{A}$ etc. They all fulfill a *Helmholtz equation* [31], [32], [35], [36], [37]:

$$\left(\Delta - \frac{1}{c^2} \frac{\partial^2}{\partial t^2} \right) u(\boldsymbol{r}, t) = -q(\boldsymbol{r}, t) \quad . \tag{2.1}$$

Now, it shall be assumed that the field propagates spherically with the velocity c through a loss-less, non-dispersive medium [32], [34]. Then all transmitted energy is conserved on those spherical surfaces. Since the surface grows with r^2, the power density must decline by the order of $\mathcal{O}(r^{-2})$, the field respectively with $\mathcal{O}(r^{-1})$ [34]. This simple consideration helps to define a *point pulse response* $s_p(\boldsymbol{r}, t)$ describing spherical waves [34]:

$$s_p(\boldsymbol{r}, t) = \frac{1}{4\pi t} \delta(r - ct) = \frac{c}{4\pi r} \delta(r - ct) = \frac{1}{4\pi r} \delta\left(t - \frac{r}{c} \right) \quad . \tag{2.2}$$

A possible attempt for modelling the source distribution is *separation of variables* [34]. Then, $q(\boldsymbol{r}, t)$ can be considered as a product of the so-called *range domain function* or *object function* $o(\boldsymbol{r})$, only depending on $\boldsymbol{r}$, and a time function $q_t(t)$, only depending on t [34]. In order to ease mathematics, the time harmonic case shall be considered. Hence, $q(\boldsymbol{r}, t)$ can then be written as [34]:

$$q(\boldsymbol{r}, t) = o(\boldsymbol{r}) \cdot q_t(t) = o(\boldsymbol{r}) \cdot s_0 \, e^{+j\omega_0 t} \quad . \tag{2.3}$$

According to the *Huygens' principle*, each point on a wavefront can be considered as source point for a secondary spherical wave [38]. Thus, the

---

[1] non-cooperative, contrary to secondary radars

field solution can be illustrated by the convolution of $s(\boldsymbol{r}, t)$ and the source distribution $q(\boldsymbol{r}, t)$ [34]. Thereby the convolution in $\boldsymbol{r}$ and t can be tackled separately [34]:

$$u(\boldsymbol{r}, t) = q(\boldsymbol{r}, t) \overset{\boldsymbol{r}}{*} \overset{t}{*} s_p(\boldsymbol{r}, t) = o(\boldsymbol{r}) \cdot s_0 \, e^{+j\omega_0 t} \overset{\boldsymbol{r}}{*} \frac{1}{4\pi r} e^{-jk_0 r} \quad , \quad (2.4)$$

$$u(\boldsymbol{r}) = o(\boldsymbol{r}) \overset{\boldsymbol{r}}{*} g(\boldsymbol{r}) \quad \text{at frequency } \omega_0 \, . \tag{2.5}$$

This yields the *Green's function* for wave propagation in free space as spatial point response [31], [32], [36], [37]:

$$g(\boldsymbol{r}) = \frac{1}{4\pi|\boldsymbol{r}|} e^{-j\,\boldsymbol{k}\,\boldsymbol{r}} \, , \text{ with } \left(\Delta - \frac{1}{c^2} \frac{\partial^2}{\partial t^2} \right) g(\boldsymbol{r}) = -\delta(\boldsymbol{r}) \, . \tag{2.6}$$

In the following, equation (2.1) (a *Helmholtz equation*) might be transformed into spectral domain (ω- or $\boldsymbol{k}$-space), hereby usually *harmonic waves* will be considered. Those shall follow the convention $\exp(j\omega t - j\boldsymbol{k}\,\boldsymbol{r})$, hence each time derivative $\partial/\partial t$ becomes a multiplication with $j\omega$ ($\partial/\partial t \to j\omega$) and each Nabla operator ∇ becomes $-j\boldsymbol{k}$ ($\nabla \to -j\boldsymbol{k}$).

2.2 The scattering problem

Now how to get from *source problem* to *scattering problem* [34]? A known incident field, depicted by $u_i(\boldsymbol{r}, t)$, illuminates a scenery, whose scattering centres are described by the *object function* $o(\boldsymbol{r})$. Then the back-scattered field $u_s(\boldsymbol{r})$ shall be examined. The total field involved can be described by the sum: $u(\boldsymbol{r}, t) = u_i + u_s$.

Usually, the relation between source function $q(\boldsymbol{r}, t)$ and total field $u(\boldsymbol{r}, t)$ is non-linear, due to e. g. non-linear behaviour over frequency or inhomogeneous distributions of dielectric materials in space [34]: $\epsilon_r = \epsilon_r(f, \boldsymbol{r})$. However, in this thesis, only partially homogeneous distributions shall be examined. The scattering body shall be assumed as homogeneously dielectric, expressed by the *refraction index* $n_2 = \sqrt{\epsilon_{r,2}\,\mu_{r,2}} = c_0/c_2$, surrounded by a homogeneous medium of $n_1 = \sqrt{\epsilon_{r,1}\,\mu_{r,1}} = c_0/c_1$. Then analogously to *source problem*, a *Helmholtz equation* can be imposed again [34], [39]:

$$\left[\Delta - \frac{1}{c^2(\boldsymbol{r})} \frac{\partial^2}{\partial t^2} \right] u(\boldsymbol{r}) = -q(\boldsymbol{r}) \quad , \text{ with } u(\boldsymbol{r}) = u_i(\boldsymbol{r}) + u_s(\boldsymbol{r}) \quad . \tag{2.7}$$

In case of *harmonic waves*, this leads to a frequency domain representation [34], [39]:

$$\left[-k^2 + \frac{\omega^2}{c^2(\boldsymbol{r})} \right] [u_i(\boldsymbol{r}) + u_s(\boldsymbol{r})] = -q(\boldsymbol{r}) \quad , \tag{2.8}$$

Depending on n_2, the wavenumber within the scattering object is then accordingly: $k_2 = \omega/c_2$. Inserted into equation (2.8), this yields:

$$\left[-k^2 + k_2^2(\boldsymbol{r}) \right] [u_i(\boldsymbol{r}) + u_s(\boldsymbol{r})] = -q(\boldsymbol{r}) \quad . \tag{2.9}$$

The source distribution $q(\boldsymbol{r})$ contains the sources which generate the incident field u_i, propagating through the surrounding medium of n_1 without a scattering object. With $k_1 = \omega/c_1$ and k_0 as free-space wavenumber, it can be confessed that [34], [39]:

$$\left[-k^2 + k_1^2(\boldsymbol{r}) \right] u_i(\boldsymbol{r}) = -q(\boldsymbol{r}) \quad , \text{ with } k_1 = k_0 \cdot n_1(\boldsymbol{r}) . \tag{2.10}$$

Now, the equations (2.9) and (2.10) are set equal and manipulated adequately, so that it can be obtained [34], [39]:

$$\left[-k^2 + k_2^2(\boldsymbol{r}) \right] u_s(\boldsymbol{r}) = - \left[k_2^2(\boldsymbol{r}) - k_1^2(\boldsymbol{r}) \right] u_i(\boldsymbol{r}) \quad . \tag{2.11}$$

Equation (2.11) describes the scattered field within the scattering object, however, the actual interest is the scattered field in the surrounding medium, where it can be measured by a sensor like e. g. a radar. Therefore, the term $- \left[k_2^2(\boldsymbol{r}) - k_1^2 \right] u_s(\boldsymbol{r})$ is added to both sides of equation (2.11). This gives [34], [39]:

$$\left(-k^2 + k_1^2 \right) u_s(\boldsymbol{r}) = - \underbrace{k_0^2 \left[n_2^2(\boldsymbol{r}) - n_1^2(\boldsymbol{r}) \right]}_{=o(\boldsymbol{r})} \cdot [u_i(\boldsymbol{r}) + u_s(\boldsymbol{r})] = \tag{2.12}$$
$$= -o(\boldsymbol{r}) \cdot [u_i(\boldsymbol{r}) + u_s(\boldsymbol{r})] \quad .$$

Equation (2.12) demonstrates, that a change in the *refraction index* causes a scattered field u_s. This is conform to intuitive expectations, since no change in $n(\boldsymbol{r})$ would mean that the complete medium was uniform and equation (2.8) would become a well-known homogeneous *Helmholtz equation* for propagation in *isotropic* media [31], [32], [36], [37]. Besides that, the part $k_0^2 \left[n_2^2(\boldsymbol{r}) - n_1^2 \right]$ is denoted again as *object function* $o(\boldsymbol{r})$. For air or vacuum around the scattering object of $n_2 = n$, this simplifies to [34]:

$$o(\boldsymbol{r}) = k_0^2 \left[n_2^2(\boldsymbol{r}) - n_1^2(\boldsymbol{r}) \right] = k_0^2 \left[n^2(\boldsymbol{r}) - 1 \right] \quad \text{object in free-space.} \tag{2.13}$$

The problem formulated in equation (2.12) is implicit, since the scattered field $u_s(\boldsymbol{r})$, is also part of the source function. Anyway, the solution for equation (2.12) can be formally expressed by a convolution between the *source function* $q(\boldsymbol{r})$ and *Green's function* (compare sections 2.1):

$$u_s(\boldsymbol{r}) = \{o(\boldsymbol{r}) \cdot [u_i(\boldsymbol{r}) + u_s(\boldsymbol{r})]\} * g(\boldsymbol{r}) \quad . \tag{2.14}$$

By convenient simplifications, a solution for equation (2.14) can be approximated or determined iteratively [34], [39].

2.3 The Born approximation

A very well-known method is the so-called *Born approximation* [34], [39], [40]. For the *Born approximation of 1st order*, the scattering objects are supposed to be slightly scattering [34], [39], [40], thus:

$$|u_s(\boldsymbol{r})| \ll |u_i(\boldsymbol{r})| \quad \forall \, \boldsymbol{r} \text{ with } o(\boldsymbol{r}) \neq 0 \quad . \tag{2.15}$$

Therefore, u_s can be neglected in equation (2.14) and the solution for the scattered field $u_s(\boldsymbol{r})$ becomes:

$$u_s(\boldsymbol{r}) \approx u_{s,1}(\boldsymbol{r}) = [o(\boldsymbol{r}) \cdot u_i(\boldsymbol{r})] * g(\boldsymbol{r}) \quad . \tag{2.16}$$

If the object function $o(\boldsymbol{r})$ is supposed to be a set of discrete points of infinitesimal separation to each other, then all scattered spherical waves interact with each other, which is expressed in equation (2.14). Now, by help of *Born approximation*, this interacting is neglected, hence the overall resulting scattered field is approximately a superposition of spherical waves centred at the different scattering points of $o(\boldsymbol{r})$, equation (2.16).

Furthermore, equation (2.16) neglects not only the necessary partially penetration of the object to reach each single scattering volume element [34] but also that the backscattered waves cannot propagate in undisturbed manner out of the distribution [34]. Therefore, there is also a limit for *Born approximation of 1st order* [34]:

$$k_0 \, |n_2 - n_1| \, D \ll \pi \quad ,$$

$$k_0 \, |n - 1| \, D \ll \pi \;\Rightarrow\; |n - 1| \, D \ll \frac{\lambda}{2} \quad , \tag{2.17}$$

with D as maximum dimension of object.

Hence, the difference in *electrical length* between the path for a wave travelling through the surrounding medium and the path through the dielectric body of maximum dimension D shall be much smaller than half of the wavelength [34].

A possibility to encounter the drawbacks of *Born approximation of first order* is to solve equation (2.14) iteratively. Here, the scattered field is first approximated by *Born approximation of 1st order*. The outcome is inserted in equation (2.14), which gives the *Born approximation of 2nd order* for the scattered field u_s [34], [39]:

$$u_s^{(2)}(\boldsymbol{r}) = \left[o(\boldsymbol{r}) \cdot \left(u_i(\boldsymbol{r}) + u_s^{(1)}(\boldsymbol{r}) \right) \right] * g(\boldsymbol{r}) \quad . \tag{2.18}$$

Now by recursively inserting u_s back into equation (2.14) also approximations of higher order can be achieved. The stationary point of this process can be considered as the result for u_s. This procedure can be formally written as [34], [39]:

$$u_s^{(\nu+1)}(\boldsymbol{r}) = \left[o(\boldsymbol{r}) \cdot \left(u_i(\boldsymbol{r}) + u_s^{(\nu)}(\boldsymbol{r}) \right) \right] * g(\boldsymbol{r}) \quad , \text{with } \nu \in \mathbb{N} \quad ,$$
$$u_s^{(\nu)} \to u_s \text{ for } \nu \to +\infty \quad . \tag{2.19}$$

2.4 Rytov approximation

Unlike the Born approximation, the Rytov approximation does not assume an additive scattering contribution to the total field but a multiplicative one [34], [40], [41]:

$$u(\boldsymbol{r}) = u_i(\boldsymbol{r}) \cdot u_s(\boldsymbol{r}) \quad . \tag{2.20}$$

However, this approximation will not be exploited in this thesis but shall be mentioned for the sake of completeness here.

3 Antenna arrays and MIMO radar signal theory

3.1 From vector potentials to array factor

In the following, the fields of general antenna arrays shall be introduced. A general electric field $\boldsymbol{E}(\boldsymbol{r}, t)$ can be represented by the time derivative of a *magnetic vector potential* $\boldsymbol{A}$ and the gradient of a *scalar electric potential* ψ [31], [32], [36]:

$$\boldsymbol{E}(\boldsymbol{r}, t) = -\frac{\partial}{\partial t} \boldsymbol{A}(\boldsymbol{r}, t) - \nabla \psi \quad . \tag{3.1}$$

Further, $\boldsymbol{A}(\boldsymbol{r}, t)$ and $\psi(\boldsymbol{r})$ fulfill a *Helmholtz equation* [32], [35], thus correspond to the term u in section 2. Under *Lorenz condition* [31], [36], this yields [32]:

$$\left(\Delta + \frac{1}{c^2} \frac{\partial^2}{\partial t^2} \right) \boldsymbol{A}(\boldsymbol{r}) = -\mu \boldsymbol{J}(\boldsymbol{r}) \quad \text{, with } \boldsymbol{J}(\boldsymbol{r}) \text{ as current density,}$$
$$\tag{3.2}$$

$$\left(\Delta + \frac{1}{c^2} \frac{\partial^2}{\partial t^2} \right) \psi(\boldsymbol{r}) = -\frac{\varrho(\boldsymbol{r})}{\epsilon} \quad \text{, with } \rho(\boldsymbol{r}) \text{ as charge density.} \tag{3.3}$$

Whereat $\boldsymbol{J}(\boldsymbol{r}, t)$ represents an alternating current density, ρ an electric charge density respectively. As demonstrated in source problem (section 2.1) by separation of variables, the possible formal solutions for the equations (3.2), (3.3) are convolutions of *Green's function* in free space $g(\boldsymbol{r})$ with the corresponding source distribution functions [32]:

$$\boldsymbol{A}(\boldsymbol{r}) = \int\limits_V \mu \boldsymbol{J}(\boldsymbol{r}') \cdot g\left(\boldsymbol{r} - \boldsymbol{r}'\right) dV \quad , \tag{3.4}$$

$$\psi(\boldsymbol{r}) = \int\limits_V \frac{\varrho(\boldsymbol{r}')}{\epsilon} \cdot g\left(\boldsymbol{r} - \boldsymbol{r}'\right) dV \quad . \tag{3.5}$$

The shifted *Green's function* $g\left(\boldsymbol{r} - \boldsymbol{r}'\right)$ can be approximated depending on the distance to the source distributions (see Fig. 3.1). In this thesis, only the *radiating near-field* (*Fresnel*) *region* [31] and the *far-field* (*Fraunhofer*)

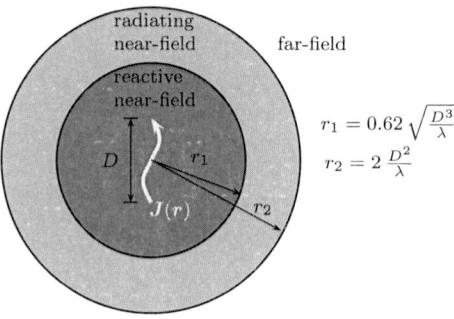

Fig. 3.1 Field regions of a source distribution [31].

region [31] shall be of further interest. In the first case, field parts containing reactive power have already declined rapidly [31] (as well as radial field components), such that the radiating field components dominate [31]. But for the angular pattern, still the curvature of the spherical waves must be taken into account (function of r) [31]. In second case, the pattern is independent of r and the pattern can be determined with the help of *Fourier transformation* [31]. Fig. 3.1 shows the boundaries of all field regions defined by a maximum phase error (of e. g. $\pi/8$) at the bended edges of phase fronts [31], [42]. This classification is supposed to be common sense nowadays, however, a different value of phase error could be chosen [31]. The remaining curvature of the spherical waves can be demonstrated by

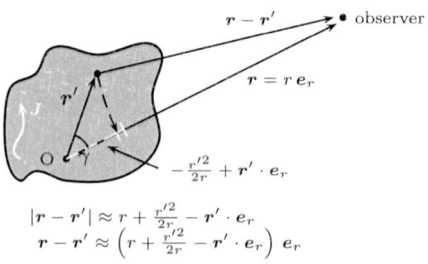

Fig. 3.2 Principal constellation for radiating near-field approximation.

the *law of cosine* [43] imposed on the triangle in Fig. 3.2 [32]:

$$
\begin{aligned}
|\boldsymbol{r} - \boldsymbol{r}'| &= \sqrt{r^2 + r'^2 - 2\,rr'\,\cos\left(\gamma\right)} = \\
&= \sqrt{r^2 + r'^2 - 2\,r\,\boldsymbol{e}_r\left(\vartheta, \varphi\right) \cdot \boldsymbol{r}'} = \\
&= r\,\sqrt{1 + \left(\frac{r'}{r}\right)^2 - \frac{2}{r}\,\boldsymbol{r}' \cdot \boldsymbol{e}_r\left(\vartheta, \varphi\right)}\,.
\end{aligned}
\tag{3.6}
$$

Please note, that the wavenumber shall be expressed as vector along the radial unit vector: $\boldsymbol{k} = k\,\boldsymbol{e}_r$. Furthermore, the Euclidean norm of a vector shall be denoted as $|\boldsymbol{r}| = r$. The squared version is then $r^2 = |\boldsymbol{r}|^2 = \boldsymbol{r} \cdot \boldsymbol{r} = \boldsymbol{r}^2$.

Similar problems with respect to *Fresnel approximation* can be found in *antenna* or *diffration theory* [31], [32], [38]. There it is a very feasible method to substitute the *root* in equation (3.6) by a *Taylor series approximation*, which is aborted after a suited term. In this case, the series shall be aborted after the linear term, thus $\sqrt{1 + x} \approx 1 + x/2$ with $|x| < 1$ [43]. Mapped onto equation (3.6) and with the substitution $\chi = r'/r \in \mathbb{R}$, the following condition must be fulfilled:

$$
|\chi^2 - 2\chi\,\cos(\gamma)| \overset{!}{<} 1\,.
\tag{3.7}
$$

The (vector) space $\mathbb{R}^1$ is metric [43] and therefore, a *triangle inequality* is defined [43], which can be applied to equation (3.7) for a *worst case assessment*. Furthermore, the trigonometric function provides values like $|\cos(\gamma)| \leq 1$, hence it is set equal to one for the assessment. If it can be shown that the evolution of equation (3.7) to worst case is still smaller than one, then equation (3.7) itself is also satisfied. This yields:

$$
\begin{aligned}
|\chi^2 - 2\chi\,\cos(\gamma)| &\leq |\chi^2| + 2|\chi| \overset{!}{<} 1\,, \\
|\chi| &< \sqrt{2} - 1\,, \quad \Rightarrow \quad \frac{r'}{r} \in \left[0;\, \sqrt{2} - 1\right]\,.
\end{aligned}
\tag{3.8}
$$

The special value $r'/r \to 0$ can only be found by $r \to +\infty$, which would be the *far-field* case, where all waves are assumed to be plane [31], or by $r' = 0$, which would lead to a source distribution of maximum dimension zero which would not radiate in praxis [32]. If the condition in equation (3.8) is satisfied, equation (3.6) can be approximated by:

$$
|\boldsymbol{r} - \boldsymbol{r}'| \approx r + \frac{r'^2}{2r} - \boldsymbol{r}' \cdot \boldsymbol{e}_r(\vartheta, \varphi)\,, \text{ Fresnel approx.}
\tag{3.9}
$$

Setting $2r' \approx D$ and by use of far-field boundary condition $r < 2D^2/\lambda$, the useful limits for the approximation in equation (3.9) are:

$$\frac{\lambda}{4D} < \frac{D}{2r} < \sqrt{2} - 1 \text{ with } \lambda \text{ as wavelength.} \tag{3.10}$$

Further, it can be assumed that $|\boldsymbol{r}'| \ll |\boldsymbol{r}|$, the overall attenuation due to the propagation is dominated by $\boldsymbol{r}$. If equation (3.9) is exploited for the spatially shifted *Green's function* in addition, it can be obtained:

$$g\left(\boldsymbol{r} - \boldsymbol{r}'\right) \approx \frac{\mathrm{e}^{-jkr}}{4\pi r} \, \mathrm{e}^{-jkr'^2/2r} \, \mathrm{e}^{+jk\, \boldsymbol{e}_r \cdot \boldsymbol{r}'} = g(\boldsymbol{r}) \cdot \mathrm{e}^{-jkr'^2/2r} \, \mathrm{e}^{+j\boldsymbol{k}\cdot\boldsymbol{r}'} \ . \tag{3.11}$$

Inserted into equations (3.4), (3.5) gives then:

$$\boldsymbol{A}(\boldsymbol{r}) = \mu \underbrace{\frac{\mathrm{e}^{-jkr}}{4\pi r}}_{g(\boldsymbol{r})} \int_V \boldsymbol{J}(\boldsymbol{r}') \, \mathrm{e}^{-jk\frac{r'^2}{2r}} \, \mathrm{e}^{+j\,\boldsymbol{k}\cdot\boldsymbol{r}'} \, dV \quad , \text{ in radiating near-field,}$$

$$\tag{3.12}$$

$$\psi(\boldsymbol{r}) = \frac{1}{\epsilon} \frac{\mathrm{e}^{-jkr}}{4\pi r} \int_V \varrho(\boldsymbol{r}') \, \mathrm{e}^{-jk\frac{r'^2}{2r}} \, \mathrm{e}^{+j\,\boldsymbol{k}\cdot\boldsymbol{r}'} \, dV \quad , \text{ in radiating near-field.}$$

$$\tag{3.13}$$

The term $\exp^{-jkr'^2/2r}$ imposes a quadratic phase term depending on the lateral dimensions, which classifies the equations (3.12), (3.13) as *Fresnel integrals* [32], appendix A.2. By increasing distance, the effect of the quadratic phase term declines, such that in *far-field* case, the expression $(\boldsymbol{r} - \boldsymbol{r}')$ (3.5), can be approximated by [31], [32], [34]:

$$|\boldsymbol{r} - \boldsymbol{r}'| \approx r - \boldsymbol{r}' \cdot \boldsymbol{e}_r \ , \text{ Fraunhofer approx.} \tag{3.14}$$

This projection onto the line of sight in equation 3.14 gives then:

$$g\left(\boldsymbol{r} - \boldsymbol{r}'\right) \approx \frac{\mathrm{e}^{-jkr}}{4\pi r} \, \mathrm{e}^{+jk\, \boldsymbol{e}_r \cdot \boldsymbol{r}'} = g(\boldsymbol{r}) \cdot \mathrm{e}^{+j\boldsymbol{k}\cdot\boldsymbol{r}'} \ . \tag{3.15}$$

Then $\boldsymbol{A}(\boldsymbol{r})$ can be calculated from the product of *Green's function* in free-space and the *Fourier transformation* respectively *inverse Fourier transformation*[1] $\boldsymbol{F}(\boldsymbol{k})$ of $\boldsymbol{J}(\boldsymbol{r})$. In the following, the term $\exp(+jkr)$ shall be

---

[1]depending on the convention $\exp(\pm jkr)$

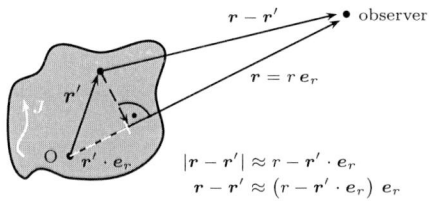

Fig. 3.3 Principal constellation for far-field approximation.

specific for the *inverse Fourier transformation*. A similar approach can be made for the scalar potential ψ [32]:

$$A(r) = \mu\, g(r) \underbrace{\int_V J(r')\, \mathrm{e}^{+j\, k\cdot r'}\, dV}_{=F(k)} \quad \text{, in far-field,} \tag{3.16}$$

$$\psi(r) = \frac{1}{\epsilon}\, g(r) \underbrace{\int_V \varrho(r')\, \mathrm{e}^{+j\, k\cdot r'}\, dV}_{=Q(k)} \quad \text{, in far-field.} \tag{3.17}$$

The term $F(k)$ is also often denoted as *radiation vector*, $Q(k)$ as *charge form-factor* [32]. Equation (3.16) stands for the far-field idea, that all source filaments are projected onto the line of sight between object on observer with a correction phase. This is represented by the *radiation vector* $F(k)$, whereas the general phase is determined by the centre of gravity depicted by $g(r)$ [34]. With a closer investigation of $F(k)$, it can be shown that all radial terms decline by orders of at least $\mathcal{O}(r^{-2})$, whereas the tangential components in ϑ- and φ-direction are of order $\mathcal{O}(r^{-1})$ and decline less rapidly [32]. Hence these determine the *far-field* [31], [32], [36], [37]. Fig. 3.3 shows the principal setup for this far-field approximation.

For the sake of completeness, a general electromagnetic field can always be described by a superposition of a magnetic vector potential A with a scalar electrical potential ψ and an electrical vector potential A_m with a scalar magnetic potential ψ_m [31], [32], [36]. The preceding considerations can be made for $\boldsymbol{A}_m$, ψ_m analogously [31], [32], [36]. For isotropic media, this yields [32]:

$$\boldsymbol{E} = -\frac{\partial}{\partial t}\boldsymbol{A} - \nabla \psi - \frac{1}{\epsilon}\nabla \times \boldsymbol{A}_m \text{ , electric field,} \qquad (3.18)$$

$$\boldsymbol{H} = -\frac{\partial}{\partial t}\boldsymbol{A}_m - \nabla \psi_m + \frac{1}{\mu}\nabla \times \boldsymbol{A} \text{ , magnetic field.} \qquad (3.19)$$

However, more details would lead to far. Therefore, the reader shall be relegated to different standard literature like [31], [32], [36] or [37].

3.2 The array factor

Now, this knowledge on distributed filaments shall be applied to a set of discrete sources like this is the case for *antenna arrays*: N antennas shall be distributed arbitrarily in space. The very first antenna element serves

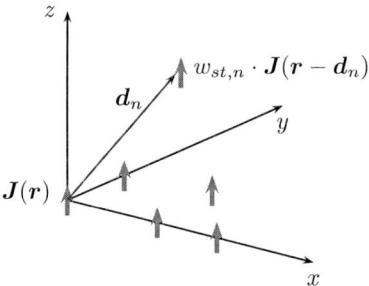

Fig. 3.4 Example for an arbitrary array

as reference element and without loss of generality, its position is also the origin of the coordinate system (figure 3.4). This reference element can be described by a current density $\boldsymbol{J}(\boldsymbol{r})$. According to that, the *radiation vector* of the reference element is denoted by $\boldsymbol{F}_0(\boldsymbol{k})$. All other antennas are shifted geometrically by arbitrary vectors $\boldsymbol{d}_n$ and are described by a weighted and shifted copy of the reference current density $w_{st,n}\,\boldsymbol{J}(\boldsymbol{r}-\boldsymbol{d}_n)$, with $w_{st,n} \in \mathbb{C}$ as complex weights. According to *Born approximation of 1st order* [34], it shall be assumed that the filaments hardly interact with each other, thus the resulting field is a superposition of the single fields. Then, by the rules of *Fourier transformation* and the substitution $\boldsymbol{\xi} = \boldsymbol{r} - \boldsymbol{d}_n$, the *radiation vectors* of the antenna elements can then be determined far-field case as [32]:

$$
\begin{aligned}
\boldsymbol{F_{dn}(k)} &= \int\limits_V w_{st,n}\, \boldsymbol{J}(\boldsymbol{r} - \boldsymbol{d_n})\, \mathrm{e}^{+j\boldsymbol{k}\cdot\boldsymbol{r}}\, dV = \\
&= w_{st,n}\, \mathrm{e}^{+j\boldsymbol{k}\cdot\boldsymbol{d_n}} \int\limits_V \boldsymbol{J}(\boldsymbol{\xi})\, \mathrm{e}^{+j\boldsymbol{k}\cdot\boldsymbol{\xi}}\, dV = \\
&= w_{st,n}\, \mathrm{e}^{+j\boldsymbol{k}\cdot\boldsymbol{d_n}}\, \boldsymbol{F_0(k)} \quad , n \in [1;\ N]\ .
\end{aligned}
\tag{3.20}
$$

The overall *radiation vector* of the array $\boldsymbol{F_{ar}(k)}$, is then the sum over all single elements:

$$
\boldsymbol{F_{ar}} = \boldsymbol{F_0} \cdot \sum_{n=1}^{N} w_{st,n}\, \mathrm{e}^{+j\,\boldsymbol{k}\cdot\boldsymbol{d_n}} = \boldsymbol{F_0} \cdot N \cdot AF \quad ,
\tag{3.21}
$$

$$
\Rightarrow AF = \frac{1}{N} \sum_{n=1}^{N} w_{st,n}, \mathrm{e}^{+j\boldsymbol{k}\cdot\boldsymbol{d_n}} \quad ,
\tag{3.22}
$$

$$
\Rightarrow \boldsymbol{E_{ar}} = \boldsymbol{E_0} \cdot N \cdot AF\,, \quad \boldsymbol{H_{ar}} = \boldsymbol{H_0} \cdot N \cdot AF\,, \text{ overall fields,}
\tag{3.23}
$$

$$
\Rightarrow \boldsymbol{G_{ar}} = \boldsymbol{G_0} \cdot N^2\, |AF|^2\,, \text{ overall array gain.}
\tag{3.24}
$$

The term AF is the so-called *array factor* [31], [32], [37]. Please note, that AF shall usually be normalized by the number of involved elements N in this thesis. Please note that for equation (3.24), all antennas were supposed to head into the same direction.

In plain words, the array factor is a spatial *Fourier series* with the weights $w_{st,n}$ as coefficients (or a *discrete spatial inverse Fourier transformation* of the weights $w_{st,n}$) [31], [32], [37]. The complete field of the array can be described by the single element's field multiplied by the array factor [31], [32], [37]. The terms $w_{st,n}$ does not only a weighting e. g. for windowing purposes, but also an additional phase shift, so that the maximum of directivity can be altered in direction [31], [32], [37]. Whereas the alternation is denoted as *steering* (indices of $w_{st,n}$), the overall topic is denoted as *beamforming* [31], [32], [37]. Nevertheless, the focus will be set on *digital beamforming* methods in this thesis, i. e. the scenery is illuminated in large sectors and *steering* and *beamforming* is performed on sampled digital data by post-processing actions.

3.3 Phase relations and steering vector

We introduced the array factor [31], [32], [37]:

$$AF = \frac{1}{N} \sum_{n=1}^{N} w_{st,n}\, e^{+j\boldsymbol{k}\cdot\boldsymbol{d}_n} \quad . \tag{3.25}$$

The phase terms $(\boldsymbol{k} \cdot \boldsymbol{d}_n)$ are the projection of the elements' position vectors $\boldsymbol{d}_n$ onto the propagation vector $\boldsymbol{k}$ (or vice versa). Depending on the direction of propagation, $\boldsymbol{k}$ can be expressed by the radial unity vector in spherical parameterisation [43]:

$$\boldsymbol{k} = \pm k \cdot \boldsymbol{e}_r = \pm \frac{2\pi}{\lambda} \cdot \boldsymbol{e}_r \,, \quad \text{with } \boldsymbol{e}_r = \begin{bmatrix} \cos(\varphi)\sin(\vartheta) \\ \sin(\varphi)\sin(\vartheta) \\ \cos(\vartheta) \end{bmatrix} \quad . \tag{3.26}$$

The array elements can be distributed arbitrarily along different coordinate axes, however w. l. o. g., in discrete steps η of a fundamental distance d_0:

$$\boldsymbol{d}_n = d_n \cdot \boldsymbol{e}_\gamma = \eta \cdot d_0 \cdot \boldsymbol{e}_\gamma \,, \quad \text{with } \eta \in \mathbb{Z}\,,\, n \in [1\,;\,N]\,. \tag{3.27}$$

Therefore, the phase terms always contain trigonometric functions, the fundamental distance d_0 and, as a degree of freedom, the parameter η. Usually, all those terms except η are represented together by the so-called *digital wavenumber* Ψ [32]:

$$\Psi(\vartheta, \varphi, d_0, k) = k\, d_0\, (\boldsymbol{e}_\gamma \cdot \boldsymbol{e}_r) \text{ with } n \in [1\,;\,N]\,. \tag{3.28}$$

A widely used array configuration is the so-called *uniform linear array* (ULA) [32]: There, all array elements are distributed along a straight line (optionally along one of the system's coordinate axes) by integer multiples $\eta = n$ of d_0, thus the digital wavenumber Ψ is the same for all elements. The normalized array factor AF is then [32]:

$$AF(\Psi) = \frac{1}{N} \sum_{n=1}^{N} w_{st,n}\, e^{+j\,\Psi_n \cdot \eta} = \frac{1}{N} \sum_{n=1}^{N} w_{st,n}\, e^{+j\,\Psi \cdot n} \text{ for a ULA.} \tag{3.29}$$

For later considerations, the so-called steering weights $w_{st,n}$ are aligned into a vector $\boldsymbol{w}_{st} = {}^{1}\!/\!\sqrt{N}\, [\ldots , w_{st,n},\, \ldots]^{\mathrm{T}}$, which is also known as *steering vector* [31], [32]. The array factor can be multiplied by $\left(e^{j\Psi} - 1\right)$ and then manipulated to [31]:

$$AF(\Psi)\left(e^{+j\Psi} - 1\right) = \cdots = e^{+j\Psi \cdot N} - 1 \quad ,$$

$$AF(\Psi) = \cdots = \frac{1}{N} \cdot e^{+j\frac{\Psi}{2}(N-1)} \cdot \frac{\sin\left(N \cdot \Psi/2\right)}{\sin\left(\Psi/2\right)} \quad . \tag{3.30}$$

For a symmetric array around the origin and for small Ψ, a non-normalized *sinus cardinalis*[2] can be obtained [31]:

$$AF(\Psi) \approx \frac{\sin\left(N \cdot \Psi/2\right)}{N \cdot \Psi/2} = \mathrm{si}\left(N \cdot \Psi/2\right) \quad . \tag{3.31}$$

Now, some relations to *complex analysis* can be imposed. By use of the substitute $e^{j\,\Psi} = z$, equation (3.29) can be rewritten to [32]:

$$AF(z) = \frac{1}{N} \sum_{n=0}^{N-1} w_{st,n} \cdot z^n \quad \textit{finite Laurent series.} \tag{3.32}$$

The similarity to a finite *Laurent series* are inherent [32], [43], [44]. All steering weights beyond the interval $[0, N-1]$ shall be zero, since the corresponding elements do not exist. Furthermore, it can be stated that in case of inverse approach, hence starting from a known array factor $AF(z)$ searching the steering weights, $w_{st,n}$ can be determined analogously by *Cauchy's integration formula* [43], [44].

$$w_{st,n} = \frac{1}{2\pi j} \oint_{\partial C} \frac{AF(z)}{z^{n+1}} \, dz \quad . \tag{3.33}$$

With help from *residual theorem* [43], [44], it can be shown that equation (3.32) can be simplified to the following expression (appendix D.1) [32]:

$$w_{st,n} = \frac{1}{2\pi} \int_{-\pi}^{+\pi} AF(\Psi)\, e^{-j\,\Psi\,n} \, d\Psi \quad . \tag{3.34}$$

At the beginning of this section, the array factor was stated to be the spatial *Fourier series* with the steering weights as coefficients. Now, equa-

---

[2]More specific compared to international convention, it is distinguished between $\mathrm{si}(x) = \frac{\sin x}{x}$ and its normalized version $\mathrm{sinc}(x) = \frac{\sin(\pi\,x)}{\pi\,x}$ in German use [44].

tion (3.34) can directly be recognized as the analysis formula for its *Fourier series coefficients* [43]. Further equation (3.34) demonstrates that the steering weights $w_{st,n}$ can be determined by use of *Fourier transformation* over the interval $[-\pi, +\pi]$. Besides that, the complete issue shows the innate relation to discrete filter design, such as *finite impulse response filter* (FIR-filter) [32]. This feature is exploited e.g. at *linear prediction method* for estimation of direction of arrival in array processing [45].

3.3.1 Visible region, grating lobes and angular resolution

In preceding section, the *digital wavenumber* of an ULA was introduced. In case of a distribution along e.g. x-axis, Ψ is [32]:

$$\Psi = \nu \, k \, d_0 \, \cos\varphi \, \sin\vartheta \quad , \text{with } \nu \in \{1, 2\} \; . \tag{3.35}$$

Since the trigonometric functions only deliver values between -1 and $+1$, Ψ is between: $-\nu \, k \, d_0 \leq \Psi \leq +\nu \, k \, d_0$. This is also denoted as *visible region* of Ψ [32]. Please note that $\nu = 1$ in case of arrays assigned for transmit or receive purpose only (thus a bistatic or multistatic radar setup). However, $\nu = 2$ for monostatic arrays caused by the fact that the phase shift due to element spacing must be run through twice, once for transmitting a wave and then for receiving it again (see sections 3.4, 3.5).

For a choice of $\nu \, d_0 = \lambda/2$, it can be obtained: $-\pi \leq \Psi \leq +\pi$, which corresponds exactly to the *principal value* of exponential functions in *complex analysis* [44]. Within the complete azimuth half-space $\varphi \in [0°; 180°]$ (with $\vartheta = const.$), hence a plane in azimuth, Ψ has got a unique complex value. The same is valid for a plane along the polar angle $\vartheta \in [0°; 180°]$ with $\varphi = const.$ However, without loss of generality, the following effects shall be demonstrated with a cutting plane of $\vartheta = 90°$, hence the $xy-$plane.

If the array is considered as a discrete spatial sampling, the condition $\nu \, d_0 = \lambda/2$ could be regarded as fulfilled *Nyquist criterion* for a complete azimuth or polar half-space $[0°; 180°]$ [32]. For element spacings $\nu \, d_0 < \lambda/2$, the *visible region* shrinks to a smaller interval $W_{vr} = [-\pi; +\pi]$. Actually, the values in the cutting planes $\varphi = const.$ and $\vartheta = const.$ stay unique. This case can be compared to spatial oversampling [32] in angular direction. However, for element spacings $\nu \, d_0 > \lambda/2$, the possible interval for Ψ becomes larger than $[-\pi; +\pi]$, thus, with periodicity of 2π, the same value for the array factor is possible for several azimuth or polar angles. This corresponds to spatial undersampling in angular direction, which generates *aliasing* effects, denoted as *grating lobes* [32]. These *grating lobes* are

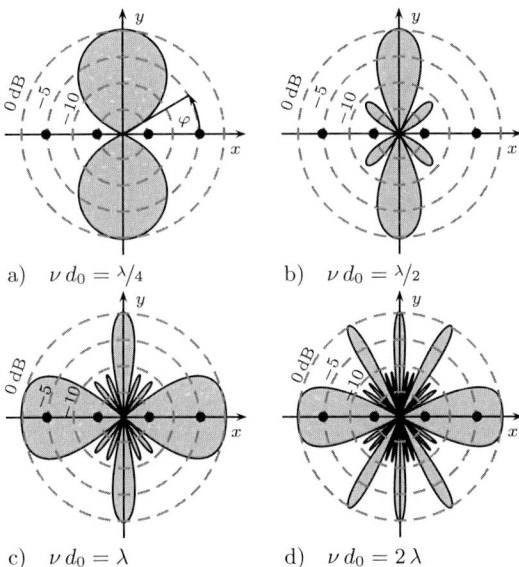

a) $\nu\,d_0 = \lambda/4$ b) $\nu\,d_0 = \lambda/2$

c) $\nu\,d_0 = \lambda$ d) $\nu\,d_0 = 2\,\lambda$

Fig. 3.5 Examples for normalized broadside array gain pattern of a four element unifrom linear array along x-axis (black dots), variations in element spacings d_0, normalized to 1, $\vartheta = 90\,^\circ$.

no sidelobes, but fully equivalent main lobes [32]. Therefore, an angular assignment is no longer possible uniquely for the complete half-space.

Figure 3.5 shows an example of a uniform linear array (ULA) consisting of four elements sited along the positive $x-$axis starting at zero. Thereby the element spacing $\nu\,d_0$ is altered from $\lambda/4$ and $\lambda/2$ over λ to $2\,\lambda$ in a), b), c) and d). It can be seen that especially in d) grating lobes occur next to sidelobes, indicating the spatial aliasing. Furthermore, it can be seen that the main looking direction for a ULA along $x-$axis is at $\varphi = 90\,^\circ$ and $\vartheta = 90\,^\circ$, due to the fact that the steering weights $w_{st,n}$ were set to one. This mode is also denoted as *broadside array* [31], [32].

It can be stated that the element spacing determines the (solid) angle without angular ambiguities. Whereas, the total length of the array determines the width of the main lobe and therefore the angular resolution

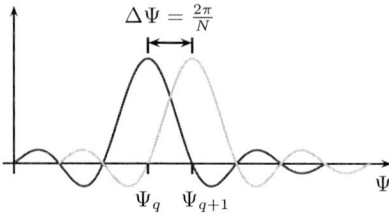

Fig. 3.6 Principle angular resolution by array factor with rectangular window.

capabilities of the array [31], [32]. This is analog to dimensions of antenna apertures.

The resolution, thus the capability to detect two si-functions separately (Fig. 3.6), is defined by the first null of equation (3.31) [46]: $N \cdot \Psi/2 = \pi$, which gives $\Delta\Psi = \frac{2\pi}{N}$. In general, two si-functions are orthogonal to each other by integer shifts of π [43].

The steering weights $w_{st,n}$ can also differ from the absolute value of one, in order to apply a window function e. g. a *Hamming* window, onto the array factor. In general, this is a possibility drop the sidelobe level, which gives a higher relation between the mainlobe level and the first sidelobe, the so-called *peak sidelobe level ratio* (PSLR) [31]. The drawback usually is a broader main beam [31].

3.3.2 Array steering

In case of a *broadside array*, the main direction corresponds to the digital wavenumber $\Psi = 0$. But now, the direction of the main beam shall be altered from the broadside to an arbitrary digital wavenumber Ψ_0. This can be achieved by acting upon the phases of the steering weights $w_{st,n}$. First, the array factor must be manipulated formally from $AF(\Psi)$ to $AF[\Psi_1 = (\Psi - \Psi_0)]$ [32]. This leads to an altered array factor:

$$AF(\Psi_1) = AF(\Psi - \Psi_0) =$$
$$= \frac{1}{N} \sum_{n=0}^{N-1} w_{st,n} \cdot e^{+j\,(\Psi-\Psi_0)\cdot n} = \frac{1}{N} \sum_{n=0}^{N-1} \tilde{w}_{st,n} \cdot e^{+j\,\Psi\cdot n} \;. \tag{3.36}$$

The steering weights $w_{st,n}$ now not only contain a window function, but also a phase factor.

$$\tilde{w}_{st,n} = w_{st,n} \cdot \mathrm{e}^{-j\,\Psi_0 \cdot n} \quad . \tag{3.37}$$

If all phase values of the elements are collected in a vector $\boldsymbol{a}$ of Euclidean norm one, denoted as *directional vector* in the following [32], and all steering weights collected in the so-called *steering vector* $\boldsymbol{w}_{st}$ analogously, the normalized array factor is simply the inner product of two vectors [32]:

$$\boldsymbol{a} = \frac{1}{\sqrt{N}} \left[1, \quad \mathrm{e}^{+j\,\Psi\cdot 1}, \quad \ldots, \quad \mathrm{e}^{+j\,\Psi\cdot(N-1)} \right]^{\mathrm{T}} , \tag{3.38}$$

$$\boldsymbol{w}_{st} = \frac{1}{\sqrt{N}} \left[1, \quad \mathrm{e}^{-j\,\Psi_0\cdot 1}, \quad \ldots, \quad \mathrm{e}^{-j\,\Psi_0\cdot(N-1)} \right]^{\mathrm{T}} , \tag{3.39}$$

$$AF = \boldsymbol{w}_{st}^{\mathrm{T}}(\Psi_0)\,\boldsymbol{a}(\Psi) = \boldsymbol{w}^{\mathrm{H}}(\Psi_0)\,\boldsymbol{a}(\Psi) . \tag{3.40}$$

From equation (3.40), it can be conducted that the wanted *steering vector* can be derived from the complex conjugate of an intended *directional vector* [45]. Figure 3.7 shows an example for an steered array gain of the before-mentioned array example. This fact will be exploited for the *conventional beamformer* method in later sections [45]. For the sake of completeness, in so-called *end-fire arrays*, the main beam is aligned to the antenna distribution (unlike to *broadside arrays* where it is perpendicular to it) [31], [32].

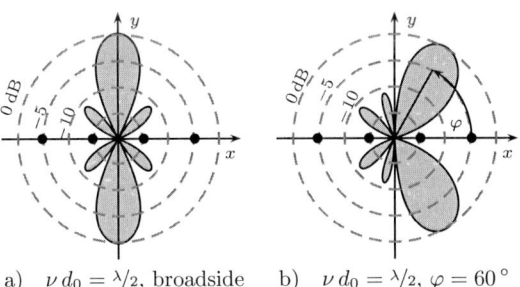

a) $\nu\,d_0 = \lambda/2$, broadside   b) $\nu\,d_0 = \lambda/2$, $\varphi = 60°$

Fig. 3.7 Example for normalized array gain pattern of a four element uniform linear array along x-axis (black dots), a) broadside, b) steered to $60°$ azimuth.

3.4 The general bistatic antenna array

For the following considerations, the principle ideas from *source* and *scattering problem* , as well as the *Born approximation* of *1st order* (compare to section 2.3, [34]) shall be applied. As a consequence, all object functions and large targets are considered to be a set of point scatterers with complex reflection coefficients. The echoes of those are superimposed linearly and form approximately the total received echo.

Now, this knowledge shall be applied to antenna configurations (Fig. 3.8). For the sake of completeness, please compare also to [1], [9], [11], [12]. The expression $k_0 = 2\pi/\lambda_0 \, e_r$ is the wavenumber corresponding to the center frequency of the exploited signal bandwidth, $\nu = 1$ for the digital wavenumber Ψ. It is assumed that all N transmitters are fed with the same signal $s(t) = s_0 \, e^{+j\omega_0 t}$, of power $P_{s0} = \mathcal{E}\left\{|s_0|^2\right\}$. Due to the fact that sources with $f = 0$ do not radiate [32], $s(t)$ becomes automatically zero mean by the transmit antenna. Each transmit element's output can be altered in amplitude and phase by the steering weights $w_{st,n}$. The same is valid for the receiver element weights $w_{st,m}$ respectively.

The number of receivers shall be M. The single antenna characteristics can be expressed by the antenna gain (pattern) G_{Tn} and G_{Rm} respectively. The scattering behaviour of each object can be described by the help of the so-called *radar cross-section* (RCS) [24], [47] depicted as rcs_q. The Q objects are illuminated under slightly different angular directions due to the array setup. It shall be assumed that the RCS is approximately constant under small angular changes, which is often not the case for real targets. However, the scintillation effects of RCSs shall be left to *statistical* MIMO radar. Furthermore, it shall be assumed that the RCS is approximately constant over the exploited frequency bandwidth. The path lengths from the transmitter n to the object q shall be denoted by r_{nq}, the receive path length from the object q to the antenna m is then r_{qm}. For further considerations, it shall be assumed that direct cross-talk between antenna elements can be neglected and the single antenna pattern remain undisturbed by their neighbour elements.

First, a power contemplation shall be engaged by use of *Friis equation* modified for the radar case, which is also denoted as *radar equation* [24], [47]. It describes the received power of a single scattering object without considering multi-path or multi-scattering scenarios:

$$P_{nqm} = \frac{P_{s0}}{4\pi \, r_{nq}^2} \, G_{Tn}(\vartheta_{nq}, \varphi_{nq}) \cdot \frac{rcs_q}{4\pi \, r_{qm}^2} \cdot \frac{\lambda_0^2}{4\pi} \, G_{Rm}(\vartheta_{qm}, \varphi_{qm}) \, . \quad (3.41)$$

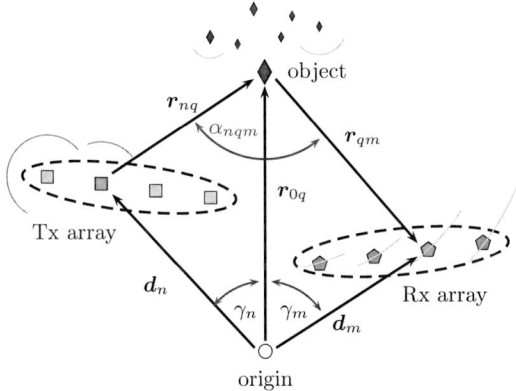

Fig. 3.8 A general array setup with several objects and sketched spherical wave propagation.

Hereby, the values of the antenna patterns $G_{Tn}(\vartheta_{nq}, \varphi_{nq})$, $G_{Rm}(\vartheta_{qm}, \varphi_{qm})$ depend on the *directional cosines* (thus inner products) between the heading of the antennas ($e_{h,Tx}$, $e_{h,Rx}$) and the direction to the object (r_{nq}, r_{qm}). Now it shall be assumed that all transmit antennas have the same gain G_{Tx} and the same heading $e_{h,Tx}$, all receiver antennas G_{Rx} and $e_{h,Rx}$ respectively. Then, the *range domain functions* $o_{Tx}(r)$ and $o_{Rx}(r)$ describe the discrete antenna distributions:

$$o_{Tx}(\boldsymbol{r}) = \sqrt{G_{Tx}} \overset{\boldsymbol{r}}{*} \sum_{n=1}^{N} w_{st,n} \cdot \delta(\boldsymbol{r} - \boldsymbol{d}_n) =$$
$$= \sum_{n=1}^{N} w_{st,n} \cdot \sqrt{G_{Tx}(\boldsymbol{r} - \boldsymbol{d}_n)} \quad , \tag{3.42}$$

$$o_{Rx}(\boldsymbol{r}) = \sqrt{G_{Rx}} \sqrt{\frac{\lambda_0^2}{4\pi}} \overset{\boldsymbol{r}}{*} \sum_{m=1}^{M} w_{st,m} \cdot \delta(\boldsymbol{r} - \boldsymbol{d}_m) =$$
$$= \sqrt{\frac{\lambda_0^2}{4\pi}} \cdot \sum_{m=1}^{M} w_{st,m} \cdot \sqrt{G_{Rx}(\boldsymbol{r} - \boldsymbol{d}_m)} \quad . \tag{3.43}$$

$$G_{Tn} = G_{Tx} \,, \forall n \in [1; N] \ \wedge \ G_{Rm} = G_{Rx} \,, \forall m \in [1; M] \quad .$$

A comparison of coefficients between the *radar equation* (3.41) and the final power calculation determines the coefficients in equation (3.42) and (3.43). According to equation (2.4), the illuminating field $u_i(\boldsymbol{r}, t)$ is:

$$u_i(\boldsymbol{r}, t) = o_{Tx}(\boldsymbol{r}) \cdot s_0 \, \mathrm{e}^{+j\omega_0 t} \overset{\boldsymbol{r},t}{*} s_p(\boldsymbol{r}, t) \, . \tag{3.44}$$

According to equation (2.5), the ansatz is reduced by the time dependence:

$$
\begin{aligned}
u_i(\boldsymbol{r}) &= o_{Tx}(\boldsymbol{r}) \overset{\boldsymbol{r}}{*} g(\boldsymbol{r}) = \\
&= \frac{1}{4\pi r} \, \mathrm{e}^{-j\,k_0 r} \overset{\boldsymbol{r}}{*} \sum_{n=1}^{N} w_{st,n} \cdot \sqrt{G_{Tx}(\boldsymbol{r} - \boldsymbol{d}_n)} = \\
&= \sum_{n=1}^{N} w_{st,n} \cdot g(\boldsymbol{r}) \overset{\boldsymbol{r}}{*} \sqrt{G_{Tx}(\boldsymbol{r} - \boldsymbol{d}_n)} \, .
\end{aligned}
\tag{3.45}
$$

Phase shifts not only occur due to signal propagation and different antenna positions but also due to the kind of object reflection (dielectric, metallic), which is expressed by $\mathrm{e}^{+j\Delta\Phi_q}$. In case of a point scattering model, the object function can be described by a discrete distribution, too:

$$o(\boldsymbol{r}) = \sum_{q=1}^{Q} \zeta_q \cdot \delta(\boldsymbol{r} - \boldsymbol{r}_{0q}) \quad , \text{with } \zeta_q = 4\pi \cdot \sqrt{rcs_q} \cdot \mathrm{e}^{+j\Delta\Phi_q} \, . \tag{3.46}$$

The objects serve as spatial sampling function for the illuminating field $u_i(\boldsymbol{r}, t)$. After their complex weights ζ_q are applied they appear again as source points for spherical waves. With the *Born approximation* of *1st order* (section 2.3, [34]), the scattered field $u_s(\boldsymbol{r}, t)$ can be written as:

$$u_s(\boldsymbol{r}, t) = o(\boldsymbol{r}) \cdot u_i(\boldsymbol{r}, t) \overset{\boldsymbol{r},t}{*} s_p(\boldsymbol{r}, t) \, ,$$

$$u_s(\boldsymbol{r}) = o(\boldsymbol{r}) \cdot u_i(\boldsymbol{r}) \overset{\boldsymbol{r}}{*} g(\boldsymbol{r}) =$$

$$= \sum_{n=1}^{N} \sum_{q=1}^{Q} \left[w_{st,n} \, \zeta_q \cdot \sqrt{G_{Tx}(\boldsymbol{r} - \boldsymbol{d}_n, \boldsymbol{r}_{nq})} \cdot g(\boldsymbol{r}_{0q} - \boldsymbol{d}_n) \cdot \delta(\boldsymbol{r} - \boldsymbol{r}_{0q}) \overset{\boldsymbol{r}}{*} g(\boldsymbol{r}) \right]$$

$$= \sum_{n=1}^{N} \sum_{q=1}^{Q} \left[w_{st,n} \, \zeta_q \cdot \sqrt{G_{Tx}(\boldsymbol{r} - \boldsymbol{d}_n, \boldsymbol{r}_{nq})} \cdot g(\boldsymbol{r}_{0q} - \boldsymbol{d}_n) \cdot g(\boldsymbol{r} - \boldsymbol{r}_{0q}) \right] \, .$$

$$\tag{3.47}$$

The receiver array in turn samples the incoming field and applies its weights due to array steering $w_{st,m}$ and antenna gain G_{Rx}:

$$u_{Rx}(\boldsymbol{r}, t) = u_s(\boldsymbol{r}, t) \cdot o_{Rx}(\boldsymbol{r}) \ ,$$

$$u_{Rx}(\boldsymbol{r}) = u_s(\boldsymbol{r}) \cdot o_{Rx}(\boldsymbol{r}) =$$

$$= \sqrt{\frac{\lambda_0^2}{4\pi}} \cdot \sum_{n=1}^{N} \sum_{q=1}^{Q} \sum_{m=1}^{M} \left[w_{st,n} \, w_{st,m} \, \zeta_q \cdot \sqrt{G_{Tx}\left(\boldsymbol{r} - \boldsymbol{d}_n, \boldsymbol{r}_{nq}\right)} \cdot \right. \tag{3.48}$$

$$\left. \cdot \sqrt{G_{Rx}\left(\boldsymbol{r} - \boldsymbol{d}_m, \boldsymbol{r}_{qm}\right)} \cdot g(\boldsymbol{r}_{0q} - \boldsymbol{d}_n) \cdot g(\boldsymbol{d}_m - \boldsymbol{r}_{0q}) \right] \ .$$

The terms $g(\boldsymbol{r}_{nq}) = g(\boldsymbol{r}_{0q} - \boldsymbol{d}_n)$ and $g(\boldsymbol{r}_{qm}) = g(\boldsymbol{d}_m - \boldsymbol{r}_{0q})$ in equation (3.48) are well-known from section 3.1. According to section 3.1 and $d_i^2 = |\boldsymbol{d}_i|^2 = \boldsymbol{d}_i^2$, the following approximations can be imposed for the *radiating near-field* [32], [34], [38]:

$$|\boldsymbol{r}_{0q} - \boldsymbol{d}_n| \approx r_{0q} + \frac{1}{2}\frac{d_n^2}{r_{0q}} - \boldsymbol{d}_n \cdot \boldsymbol{e}_r(\vartheta_q, \varphi_q) \ , \text{ Fresnel approx. } , \tag{3.49}$$

$$|\boldsymbol{d}_m - \boldsymbol{r}_{0q}| \approx r_{0q} + \frac{1}{2}\frac{d_m^2}{r_{0q}} - \boldsymbol{d}_m \cdot \boldsymbol{e}_r(\vartheta_q, \varphi_q) \ , \text{ Fresnel approx. } \tag{3.50}$$

Or for the *far-field* [32], [34], [38]:

$$|\boldsymbol{r}_{0q} - \boldsymbol{d}_n| \approx r_{0q} - \boldsymbol{d}_n \cdot \boldsymbol{e}_r(\vartheta_q, \varphi_q) \ , \text{ Fraunhofer approx.,} \tag{3.51}$$

$$|\boldsymbol{d}_m - \boldsymbol{r}_{0q}| \approx r_{0q} - \boldsymbol{d}_m \cdot \boldsymbol{e}_r(\vartheta_q, \varphi_q) \ , \text{ Fraunhofer approx.} \tag{3.52}$$

For a more general perspective, the term $g(\boldsymbol{r}_{0q} - \boldsymbol{d}_n) \cdot g(\boldsymbol{d}_m - \boldsymbol{r}_{0q})$ in equation (3.48) shall be manipulated by the near-field approximation (equations (3.49), (3.50)). By use of equation (3.11) and the recall that r_{0q} dominates the attenuation, the two *Green's functions* in equation (3.48) can be approximated by:

$$g(\boldsymbol{r}_{0q} - \boldsymbol{d}_n) \approx \frac{1}{4\pi \, r_{0q}} \cdot e^{-jk_0 \, r_{0q}} \cdot e^{-jk_0 \, d_n^2/(2\, r_{0q})} \cdot e^{+jk_0 \, \boldsymbol{e}_r \, \boldsymbol{d}_n} \ , \tag{3.53}$$

$$g(\boldsymbol{d}_m - \boldsymbol{r}_{0q}) \approx \frac{1}{4\pi \, r_{0q}} \cdot e^{-jk_0 \, r_{0q}} \cdot e^{-jk_0 \, d_m^2/(2\, r_{0q})} \cdot e^{+jk_0 \, \boldsymbol{e}_r \, \boldsymbol{d}_m} \ . \tag{3.54}$$

The product of those yields:

$$g(\boldsymbol{r}_{0q} - \boldsymbol{d}_n) \cdot g(\boldsymbol{d}_m - \boldsymbol{r}_{0q}) = \cdots =$$
$$= g^2(\boldsymbol{r}_{0q}) \cdot e^{-j2k_0 \frac{1}{4r_{0q}}\left(d_n^2 + d_m^2\right)} \cdot e^{+j2k_0 \frac{1}{2}\left(\boldsymbol{d}_n + \boldsymbol{d}_m\right) \cdot \boldsymbol{e}_r(\vartheta, \varphi)} \ . \tag{3.55}$$

Since it was assumed that all transmit antennas, as well as the receive antennas, are oriented identically and due to large r_{0q}, which in turn lead to small bistatic angles α_{nqm}, the individual angles between the antenna elements and the object become more and more similar to each other. Therefore, the individual values for the antenna gains can be approximated by a single one, depending on the main direction measured from the origin to the objects, thus:

$$
\begin{aligned}
G_{Tx}(\boldsymbol{e}_{h,Tx}, \boldsymbol{r}_{nq}) &\approx G_{Tx}(\boldsymbol{e}_{h,Tx}, \boldsymbol{r}_{0q}) = G_{Tx}(\boldsymbol{r}_{0q}) \quad, \\
G_{Rx}(\boldsymbol{e}_{h,Rx}, \boldsymbol{r}_{qm}) &\approx G_{Rx}(\boldsymbol{e}_{h,Rx}, \boldsymbol{r}_{0q}) = G_{Rx}(\boldsymbol{r}_{0q}) \quad.
\end{aligned}
\tag{3.56}
$$

Thereby, the mathematical operation also alters from a (discrete) convolution with two antenna patterns, to a single multiplication with a central joint antenna pattern. With that knowledge and the approximation from equation (3.55), equation (3.48) can be manipulated in two different ways. The first one yields:

$$
\begin{aligned}
u_{Rx}(\boldsymbol{r}, t) &\approx s_0 \, \mathrm{e}^{+j\omega_0 t} \cdot \sqrt{\frac{\lambda_0^2}{4\pi}} \cdot \sum_{q=1}^{Q} \left\{ \zeta_q \cdot g^2\left(\boldsymbol{r}_{0q}\right) \cdot \right. \\
&\quad \cdot \sqrt{G_{Tx}(\boldsymbol{r}_{0q})} \sqrt{G_{Rx}(\boldsymbol{r}_{0q})} \cdot \mathrm{e}^{-j2k_0 \frac{1}{4r_{0q}}\left(d_n^2 + d_m^2\right)} \\
&\quad \cdot \left(\sum_{n=1}^{N} w_{st,n} \cdot \mathrm{e}^{+jk_0 \, \boldsymbol{d}_n \cdot \boldsymbol{e}_r(\vartheta_q, \varphi_q)} \right) \cdot \left(\sum_{m=1}^{M} w_{st,m} \cdot \mathrm{e}^{+jk_0 \, \boldsymbol{d}_m \cdot \boldsymbol{e}_r(\vartheta_q, \varphi_q)} \right) \right\} = \\
&= s_0 \, \mathrm{e}^{+j\omega_0 t} \cdot \sqrt{\frac{\lambda_0^2}{4\pi}} \cdot \sum_{q=1}^{Q} \left\{ \zeta_q \cdot g^2\left(\boldsymbol{r}_{0q}\right) \cdot \mathrm{e}^{-j2k_0 \frac{1}{4r_{0q}}\left(d_n^2 + d_m^2\right)} \right. \\
&\quad \left. \cdot \sqrt{G_{Tx}(\boldsymbol{r}_{0q})} \sqrt{G_{Rx}(\boldsymbol{r}_{0q})} \cdot N \, M \cdot AF_{Tx}\left(\vartheta_q, \varphi_q\right) \cdot AF_{Rx}\left(\vartheta_q, \varphi_q\right) \right\} \quad.
\end{aligned}
\tag{3.57}
$$

The array factors for the transmit and receive array can be extracted from the equation (3.57), as introduced in section 3.2.

$$
AF_{Tx}(\vartheta, \varphi) = \frac{1}{N} \cdot \sum_{n=1}^{N} w_{st,n} \cdot \mathrm{e}^{+jk_0 \, \boldsymbol{d}_n \cdot \boldsymbol{e}_r(\vartheta_q, \varphi_q)} \quad,
\tag{3.58}
$$

$$
AF_{Rx}(\vartheta, \varphi) = \frac{1}{M} \cdot \sum_{m=1}^{M} w_{st,m} \cdot \mathrm{e}^{+jk_0 \, \boldsymbol{d}_m \cdot \boldsymbol{e}_r(\vartheta_q, \varphi_q)} \quad.
\tag{3.59}
$$

The second possibility gives:

$$u_{Rx}(\boldsymbol{r},t) \approx s_0\, e^{+j\omega_0 t} \cdot \sqrt{\frac{\lambda_0^2}{4\pi}} \cdot \sum_{q=1}^{Q} \Big\{ \zeta_q \cdot g^2\,(\boldsymbol{r}_{0q}) \cdot$$

$$\cdot \sqrt{G_{Tx}(\boldsymbol{r}_{0q})}\, \sqrt{G_{Rx}(\boldsymbol{r}_{0q})} \cdot e^{-j2k_0\frac{1}{4r_{0q}}\left(\boldsymbol{d}_n^2 + \boldsymbol{d}_m^2\right)}$$

$$\cdot \left(\sum_{n=1}^{N} \sum_{m=1}^{M} w_{st,n},\, w_{st,m} \cdot e^{+j2k_0\frac{1}{2}\,(\boldsymbol{d}_n + \boldsymbol{d}_m)\cdot \boldsymbol{e}_r(\vartheta_q,\varphi_q)} \right) \Big\} = \qquad (3.60)$$

$$= s_0\, e^{+j\omega_0 t} \cdot \sqrt{\frac{\lambda_0^2}{4\pi}} \cdot \sum_{q=1}^{Q} \Big\{ \zeta_q \cdot g^2\,(\boldsymbol{r}_{0q}) \cdot e^{-j2k_0\frac{1}{4r_{0q}}\left(\boldsymbol{d}_n^2 + \boldsymbol{d}_m^2\right)}$$

$$\cdot G_{TRx}(\boldsymbol{r}_{0q}) \cdot N\, M \cdot AF_{TRx}\,(\vartheta_q,\,\varphi_q) \Big\}\ .$$

Hereby, the antenna pattern have been substituted by a single joint antenna pattern:

$$G_{TRx}(\boldsymbol{r}_{0q}) = \sqrt{G_{Tx}(\boldsymbol{r}_{0q})}\, \sqrt{G_{Rx}(\boldsymbol{r}_{0q})}\ . \qquad (3.61)$$

Additionally, the effective array factor:

$$AF_{TRx}\,(\vartheta_q,\,\varphi_q) =$$

$$= \frac{1}{NM} \left(\sum_{n=1}^{N} \sum_{m=1}^{M} w_{st,n},\, w_{st,m} \cdot e^{+j2k_0\frac{1}{2}\,(\boldsymbol{d}_n + \boldsymbol{d}_m)\cdot \boldsymbol{e}_r(\vartheta_q,\varphi_q)} \right)\ , \qquad (3.62)$$

corresponds to a monostatic array with the effective element positions:

$$\boldsymbol{d}_i = \frac{1}{2}\,(\boldsymbol{d}_n + \boldsymbol{d}_m)\ ,\ i \in [1; NM]\ . \qquad (3.63)$$

These effective element positions lie on the centre of gravity of the connecting lines between the actual transmit and receive elements and are also denoted as *virtual elements* [21].

3.5 The virtual array principle

The *coherent* MIMO approach is a consecutive development of the *phased array* approach. In *coherent* MIMO systems, the distances between the different elements are supposed to be much smaller than the distances to the target scenery [1], [21]. Then, the length of the bistatic propagation paths from the transmitters to the observed objects and again back to the receivers can be approximated by the doubled monostatic paths from *virtual array* elements to the objects [1], [21]. These *virtual elements* are then sited at the centre of gravity between the transmitting and receiving elements (equation (3.63)) [1], [21]. Hence, the *virtual array* principle is a monostatic approximation for a bistatic antenna setup.

3.5.1 The position of virtual elements by convolution

Equation (3.57) demonstrated that the incoming signal of an arbitrary *bistatic array* is directly proportional to the product of the *array factors*:

$$u_{Rx}(\boldsymbol{r}, t) \sim AF_{Tx} \cdot AF_{Rx} \sim \frac{1}{N} \sum_{n=1}^{N} w_{st,n} \cdot \mathrm{e}^{+j\boldsymbol{k}_0 \, \boldsymbol{d}_n} \cdot \frac{1}{M} \sum_{m=1}^{M} w_{st,m} \cdot \mathrm{e}^{+j\boldsymbol{k}_0 \, \boldsymbol{d}_m} \quad .$$
$$(3.64)$$

The result of the same approach for a monostatic setup was presented in section 3.4, now with $\Gamma = N\,M$:

$$u_{Rx}(\boldsymbol{r}, t) \sim AF_{TRx} \sim \frac{1}{NM} \sum_{i=1}^{N\,M} w_{st,i}^2 \cdot \mathrm{e}^{+j\boldsymbol{k}_0 \, 2\,\boldsymbol{d}_i} \quad . \qquad (3.65)$$

The array factors can be considered as discrete versions of the *Fourier* integral over the sources in e. g. equation (3.16) or equation (3.20). Hence, they can be regarded as a spectral transformation of the steering weights[3]. Their back-transformation gives the positions of the antenna elements, whereby the multiplication becomes a convolution due the rules of *Fourier transformation* [43]:

$$\frac{1}{N} \sum_{n=1}^{N} w_{st,n} \cdot \mathrm{e}^{+j\boldsymbol{k}_0 \, \boldsymbol{d}_n} \cdot \frac{1}{M} \sum_{m=1}^{M} w_{st,m} \cdot \mathrm{e}^{+j\boldsymbol{k}_0 \, \boldsymbol{d}_m}$$

$$\updownarrow$$

$$\frac{1}{N} \sum_{n=1}^{N} w_{st,n} \cdot \delta\left(\boldsymbol{r} - \boldsymbol{d}_n\right) * \frac{1}{M} \sum_{m=1}^{M} w_{st,m} \cdot \delta\left(\boldsymbol{r} - \boldsymbol{d}_m\right) = \qquad (3.66)$$

$$= \frac{1}{NM} \sum_{n=1}^{N} \sum_{m=1}^{M} w_{st,n} \cdot w_{st,m} \cdot \delta\left[\boldsymbol{r} - \left(\boldsymbol{d}_n + \boldsymbol{d}_m\right)\right] \quad .$$

---

[3]depending on the convention $\exp(\pm jkr)$

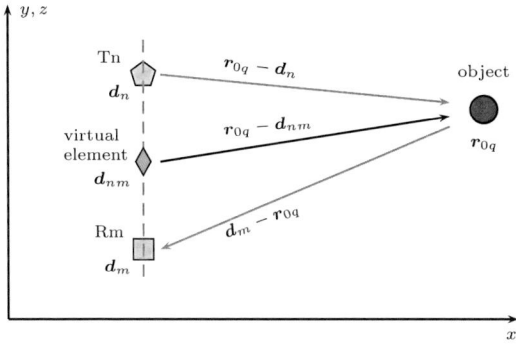

Fig. 3.9 The principle of a virtual antenna element.

For the monostatic virtual array:

$$\frac{1}{NM} \sum_{i=1}^{N\,M} w_{st,i}^2 \cdot \mathrm{e}^{+jk_0\,2\,d_i} \quad\bullet\!-\!\!\circ\quad \frac{1}{NM} \sum_{i=1}^{N\,M} w_{st,i}^2 \cdot \delta\left(r - 2\,d_i\right) \ . \tag{3.67}$$

A comparison of equation (3.66) with (3.67) gives:

$$\begin{aligned} w_{st,i}^2 \cdot \delta(r - 2\,d_i) &= w_{st,n} \cdot w_{st,m} \cdot \delta\left[r - (d_n + d_m)\right] \\ \Rightarrow \quad d_i &= \frac{1}{2} \cdot (d_n + d_m) \ , \quad w_{st,i}^2 = w_{st,n} \cdot w_{st,m} \ . \end{aligned} \tag{3.68}$$

It can be seen, that those virtual elements have to be placed at the centre of gravity between the transmit and receive elements, independently from the centre frequency f_0. For later use, the term d_i shall be depicted by d_{nm}. This represents automatically the Tx and Rx element which had created the virtual element.

The formulations incorporating the convolutional character of virtual arrays such as equations (3.57), (3.60) or (3.66) can be expressed by a *Kronecker product* [48]. In advance, the steering weights, as well as the phasor

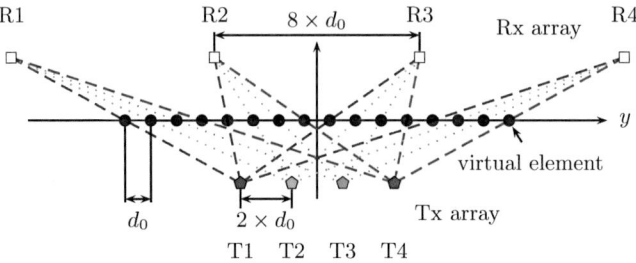

Fig. 3.10 A combination of four transmitters (Tx) and four receivers (Rx) elements delivers sixteen virtual elementss

terms are collected in normalized steering vectors and a directional vectors for Tx and Rx elements respectively:

$$
\begin{aligned}
\boldsymbol{w}_{st,Tx} &= \frac{1}{\sqrt{N}} \left[\ldots, \quad w_{st,n}, \quad \ldots \right]^{\mathrm{T}}, \\
\boldsymbol{w}_{st,Rx} &= \frac{1}{\sqrt{M}} \left[\ldots, \quad w_{st,m}, \quad \ldots \right]^{\mathrm{T}}, \\
\boldsymbol{a}_{Tx,q} &= \frac{1}{\sqrt{N}} \left[\ldots, \quad \mathrm{e}^{+j\,\boldsymbol{k}_0(\vartheta_q,\varphi_q)\cdot\boldsymbol{d}_n}, \quad \ldots \right]^{\mathrm{T}}, \\
\boldsymbol{a}_{Rx,q} &= \frac{1}{\sqrt{M}} \left[\ldots, \quad \mathrm{e}^{+j\,\boldsymbol{k}_0(\vartheta_q,\varphi_q)\cdot\boldsymbol{d}_m}, \quad \ldots \right]^{\mathrm{T}}.
\end{aligned}
\tag{3.69}
$$

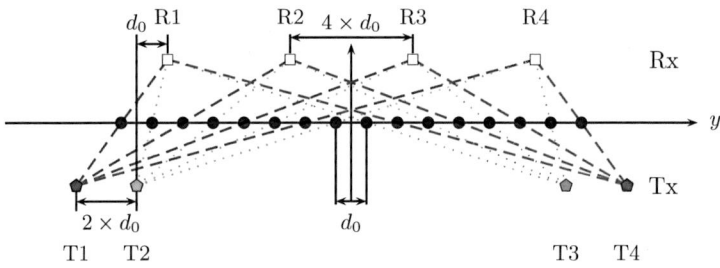

Fig. 3.11 Alternative setup creates same virtual array.

Then equation (3.64) can be rewritten as:

$$u_{Rx}(\boldsymbol{r},t) \sim AF_{Tx} \cdot AF_{Rx} \sim$$
$$\sim \left(\boldsymbol{w}_{st,Rx} \otimes \boldsymbol{w}_{st,Tx}\right)^{\mathrm{T}} \left(\boldsymbol{a}_{Rx,q} \otimes \boldsymbol{a}_{Tx,q}\right) = \qquad (3.70)$$
$$= \left(\boldsymbol{w}_{st,Rx}^{\mathrm{T}} \otimes \boldsymbol{w}_{st,Tx}^{\mathrm{T}}\right) \left(\boldsymbol{a}_{Rx,q} \otimes \boldsymbol{a}_{Tx,q}\right) \quad .$$

In this special case, the *Kronecker products* are even *commutative* due to the reciprocity theorem of antennas [31], [32].

3.5.2 Description by grid vectors

The composition of virtual arrays is not necessarily unique: Fig. 3.10 and Fig. 3.11 demonstrate that two different bistatic antenna configurations can result in the same *virtual array* [21], [49]. Furthermore, virtual arrays are not limited to one dimension. Fig. 3.12 demonstrates the principle with a general example: Three transmitters and three receivers, each arranged in uniform linear arrays (ULA), form a two-dimensional virtual array, thus a grid of virtual elements. This allows signal detection in two directions like e. g. in azimuth and elevation angle. Furthermore, from Fig. 3.12, it can be seen, that the virtual array can be described by two fundamental grid vectors $\boldsymbol{e}_1$, $\boldsymbol{e}_2$.

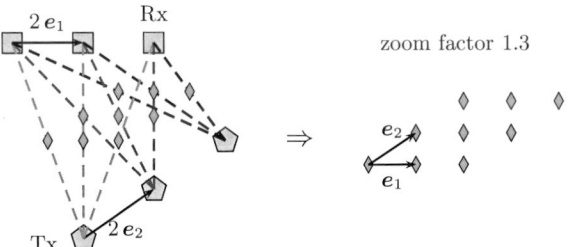

Fig. 3.12 Two uniform linear arrays of transmitters (Tx) and receivers (Rx) form a two-dimensional virtual array.

3.6 The MIMO radar signal structure

Each virtual element stands for a Tx-Rx combination. Thus, there is a must to separate the different Tx contributions at each Rx element, which can be achieved by the signal structure: In case of orthogonal signals for each transmitter, all signal echoes could be separated by matched filters or correlation receivers at each receiver [10], [21]. Then for orthogonal signals the correlation coefficients between the different transmit signals $s_{Ti}(t)$ have to satisfy:

$$\rho_{ij} = \frac{\text{cov}\left\{s_{Ti}(t),\, s_{Tj}(t)\right\}}{\sqrt{\text{var}\left\{s_{Ti}(t)\right\}} \cdot \sqrt{\text{var}\left\{s_{Tj}(t)\right\}}} = \delta_{ij} =$$

$$= \begin{cases} \pm\ 1 & ,\ \text{for } i = j\,, \\ 0 & ,\ \text{for } i \neq j\,, \end{cases} \quad ,\ \text{with } i,\, j \in [1; N]\,. \tag{3.71}$$

Whereby s_{Ti} and s_{Tj} denote the different transmit signals. The covariance and the variance are generally defined as:

$$\text{cov}\left\{s_{Ti}(t),\, s_{Tj}(t)\right\} = \mathcal{E}\left\{(s_{Ti}(t) - \mu_{Ti}) \cdot (s_{Tj}(t) - \mu_{Tj})\right\} \quad , \tag{3.72}$$

$$\text{var}\left\{s_{Ti}(t)\right\} = \mathcal{E}\left\{(s_{Ti}(t) - \mu_{Ti})^2\right\} = \mathcal{E}\left\{s_{Ti}^2(t)\right\} - \mu_{Ti}^2 \quad . \tag{3.73}$$

The mean value is represented by μ_{Ti}. For the following, it shall be assumed that the scattering processes are *stationary*, i.e. the mean-values, as well as the variances, do not change over time [50]. If the processes are additionally *ergodic*, the expected value can be substituted by the mean over time (stationarity is a requirement for ergodicity) [50]. This means, that the numerator's covariance in equation (3.71) becomes a cross-correlation function ϕ_{ij}, whereas the variances in the denominator become auto-correlation functions ϕ_{ii}, ϕ_{jj}. Applying these conditions, it can be concluded for power-limited radar signals, such as *continuous wave* (CW) signals [50], [51], [52]:

$$\phi_{ij}(t) = \lim_{T \to \infty} \frac{1}{2T} \int\limits_{-T}^{+T} s_{T,i}(\tau) \cdot s_{T,j}^*(t + \tau)\, d\tau = 0 \quad ,\ \text{for } i \neq j\,. \tag{3.74}$$

Whereas for energy limited signals, like for a single radar pulse, this simplifies to [50], [51], [52]:

$$\phi_{ij}(t) = \int\limits_{-\infty}^{+\infty} s_{T,i}(\tau) \cdot s_{T,j}^*(t + \tau)\, d\tau = 0 \quad ,\ \text{for } i \neq j\,. \tag{3.75}$$

Principally, the signal orthogonality can be achieved by three methods: First by *frequency-division-multiplex* FDM [12]: Signals in different frequency bands are orthogonal due to the following property of harmonic waves [52], [53]:

$$\frac{1}{T} \int_T e^{+j(l-m)\omega\,t}\,dt = \delta_{lm} = \begin{cases} 1 & \text{, for } l = m \\ 0 & \text{, for } l \neq m \end{cases} \text{, with } l, m \in \mathbb{Z} \ . \ (3.76)$$

For non-narrowband transmit signals with a real bandwidth B_n, it must be secured that those have no overlap in their spectra [52].

The second possibility is *code-division-multiplex* CDM Here, the signal phases are modulated by codes which are a-priori orthogonal to each other (even after different round trip delays), like e. g. Pseudo-Noise-, Hadamard- or Gold-Codes [54]. This principle is well-known from communication and radar applications [24], [47], [55]. The big advantage is the re-use of the same signal band and the same bandwidth for all MIMO channels.

The orthogonality of the codes leads to cross-correlation functions between two different MIMO channels which are ideally zero. However, in real systems, the different cross-combinations add a non-Gaussian and coloured noise contribution to all other channels. This gives a slightly lower SNR than for the *phased array* case. Again a big advantage is the fact, that the virtual array can be filled by one measurement.

The third possibility is *time-division-multiplex* (TDM). Here, a switch ensures that only one transmit signal is active while the others are blanked out. Therefore, always one function in (3.74) is equal to zero over the complete integration for $i \neq j$ and the condition of orthogonality is fulfilled independently from the signal modulation. Hence, the same signal scheme with the same bandwidth, as well as the same frequency band can be exploited for all transmitters.

A drawback is that TDM requires the coherent evaluation of N consecutive time slots to provide the same SNR as for a *phased array*. However, for moving platforms, the longer period data acquisition can cause problems in terms of violating the *Nyquist's sampling theorem* [21] (here in spatial case). Regarding this, the switching sequence of the transmitters exhibits a degree of freedom.

The big advantage of this method is, that there are no matched filters or correlation receivers necessary. By knowledge of the time slots, it is possible to distinguish the different transmit signal contributions. Then these signals can be mapped to transmitter-receiver combinations and therefore to *virtual elements*. In comparison to the two methods before, the time division can be considered as sequential *single-input multiple-output* SIMO which will be processed to a MIMO system by means of post-processing.

Figure 3.10 shows a switching example of the MIMO configuration in figure 3.10, thus which virtual element is padded at one time step. The chosen default switching sequence of the transmitters is straight forward $T_1 - T_2 - T_3 - T_4$. There, at each time step, the coloured pentagons indicate which transmitter is active and which virtual element the four received signals are mapped onto.

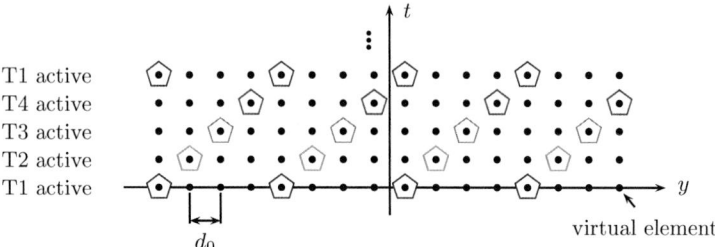

Fig. 3.13 Switching sequence mapped onto virtual elements.

3.7 Output power and SNR

The output power is equal to the expected absolute value of equation (3.57). If the different factors are statistically independent from each other, the expected value over a product decomposes into: $\mathcal{E}\{a \cdot b\} = \mathcal{E}\{a\} \cdot \mathcal{E}\{b\}$ [43]. The output power of a bistatic array P_b is then:

$$
P_\text{b} = \mathcal{E}\left\{u_{Rx}(t)\, u_{Rx}^*(t)\right\} = \mathcal{E}\left\{|u_{Rx}(t)|^2\right\} =
$$
$$
= \mathcal{E}\left\{|s_0|^2 \cdot \frac{\lambda_0^2}{4\pi}\right\} \cdot \mathcal{E}\left\{\left|\sum_{q=1}^{Q} 4\pi \cdot \frac{\sqrt{rcs_q}}{(4\pi\, r_{0q})^2}\cdot\right.\right.
$$
$$
\left.\left.\cdot\sqrt{G_{Tx}(\boldsymbol{r}, \boldsymbol{r}_{0q})}\sqrt{G_{Rx}(\boldsymbol{r}, \boldsymbol{r}_{0q})}\cdot N\,M \cdot AF_{Tx,q}\, AF_{Rx,q}\right|^2\right\}. \tag{3.77}
$$

The very last term in equation (3.78) needs a closer inspection: W. l. o. g. shall be assumed that all weights $w_{st,n}$, $w_{st,m}$ are equal to one, then the array factors become Dirac-delta function shifted by ϑ_q, φ_q for large N, M (ideally ∞). Therefore, mixed terms containing $AF_{Tx,i} \cdot AF_{Tx,j}$ automatically become equal to zero for $i \neq j$, thus all q echo contributions are orthogonal to each other. However, for smaller N, M, the array factors are si-functions. Hence, the mixed terms only become zero if the different maxima a placed at the nulls of the other array factors. Since this cannot be guaranteed in general, a residual mixed term P_{res} will remain:

$$
\begin{aligned}
P_{\mathrm{b}} = {} & P_{s0} \cdot N^2 \, M^2 \cdot \frac{\lambda_0^2}{4\pi} \cdot \sum_{q=1}^{Q} \left[\frac{1}{4\pi \, r_{0q}^2} \cdot \frac{rcs_q}{4\pi \, r_{0q}^2} \cdot \right. \\
& \left. \cdot G_{Tx}(\boldsymbol{r}, \boldsymbol{r}_{0q}) \, G_{Rx}(\boldsymbol{r}, \boldsymbol{r}_{0q}) |AF_{Tx,q}|^2 \, |AF_{Rx,q}|^2 \right] + \\
& + P_{res} \; .
\end{aligned}
\tag{3.78}
$$

If the residual term is neglected at first glance and the sum-term in equation (3.78) is represented by Υ_Q, then the P_{b} can be re-written briefly by:

$$
P_{\mathrm{b}} = P_{s0} \cdot N^2 \, M^2 \cdot \Upsilon_Q \quad .
\tag{3.79}
$$

If it is assumed that only uncorrelated thermal noise is present at each receiver (with bandwidth B_0), it can be obtained for the signal-to-noise ratio (SNR):

$$
P_{\mathrm{b},n} = M \cdot k \, T \, B_0 \quad ,
\tag{3.80}
$$

$$
SNR_{\mathrm{b}} = N^2 \, M \cdot \Upsilon_Q \cdot \frac{P_{s0}}{k \, T \, B_0} \quad .
\tag{3.81}
$$

By a closer look on the equations (3.57), (3.60) and (3.81), there is no principal difference in signal-to-noise ratio between a *virtual array* and a classical *phased array*. Although they may differ in antenna gain pattern, the basic difference is that the classical phased array approach is a physical beamforming, whereas the virtual array is usually processed coherently by digital beamforming. Nevertheless, the reader might get tempted to say, *coherent* MIMO is always equal or even superior to *phased array* techniques. However, especially for far-distant targets, it is important to project significant illumination power onto the target, in order to gain relevant echoes. In *coherent* MIMO systems in turn, usually large sectors are illuminated which reduces the power density at the target and therefore the level of potential echoes. If the reflected energy ends well below thermal noise limit, hardware meets its limits. Especially for search and track radars like e.g. missile early-warning systems, it can be more power efficient to stick to classical *phased arrays*.

4 Modulation, demodulation, signal design

4.1 Generation of complex signals

A completely real signal $s(t)$ consists of an even part $s_{r,e}(t)$ and an odd part $s_{r,o}(t)$: $s(t) = s_{r,e}(t) + s_{r,o}(t)$ [56]. According to the rule of assignment of Fourier analysis, the real even part $s_e(t)$ is assigned to the real and even part of the spectrum $S_{r,e}(f)$, whereas the real but odd signal part $s_{r,o}(t)$ is assigned to the odd but imaginary part of the spectrum $S_{i,o}(f)$ [51], [54], [56], [57]:

$$s(t) = s_{r,e}(t) + s_{r,o}(t) + j\, s_{i,e}(t) + j\, s_{i,o}(t) \quad ,$$

$$S(f) = S_{r,e}(f) + j\, S_{i,o}(f) + j\, S_{i,e}(f) + S_{r,o}(f) \quad . \tag{4.1}$$

Furthermore, by a closer look on discrete Fourier transformation, time discrete, thus sampled, real signals of finite lengths L, have a Fourier transform with only $L/2$ linear independent coefficients [56]. The second half of the coefficients are the complex conjugated of the first half, hence they are conjugated symmetric: $S(f) = S^*(-f)$ [56]. This can also be derived from the rule of assignment and the periodicity [56] of discrete Fourier transformation.

4.1.1 Complex signal generation by IQ-demodulation

A general signal, e.g. a received radar signal shifted by a phase difference ϕ, can be written as [47], [51], [56], [58], [59]:

$$
\begin{aligned}
s_{Rx}(t) &= U_0(t) \cdot \cos\left(\omega_0\, t - \Phi\right) = \\
&= U_0(t) \cdot \left[\cos\Phi\,\cos\left(\omega_0\, t\right) + \sin\Phi\,\sin\left(\omega_0\, t\right)\right] \quad .
\end{aligned} \tag{4.2}
$$

During standard *inphase-quadrature* (IQ) conversion, the received signal in equation (4.2) is once multiplied by $2\cos\left(\omega_0\, t\right)$ for the I-channel and once by $-2\sin\left(\omega_0\, t\right)$ for the Q-channel, respectively. All higher conversion products are suppressed by low-pass filters. This yields for baseband:

$$s_I(t) = +U_0(t)\cos\Phi \quad \text{, for I-channel,} \tag{4.3}$$

$$s_Q(t) = -U_0(t)\sin\Phi \quad \text{, for Q-channel.} \tag{4.4}$$

The complex envelope $s_{IQ}(t)$ is then:

$$s_{IQ}(t) = s_I(t) + j \cdot s_Q(t) = U_0(t) \cdot \mathrm{e}^{-j \cdot \Phi} \quad . \tag{4.5}$$

By this method, the full spectrum (positive and negative frequencies) is conserved, whereby the negative part is the complex conjugate of the positive one [56]. Furthermore, the sign of the exponential term in equation (4.5), thus the direction of rotation of the complex phasor, is clearly determined.

4.1.2 Analytic signal generation by Hilbert transformation

The second possibility to generate a complex signal from a real signal is by *Hilbert transformation* [54], [56], [59]:

$$s_Q(t) = \mathcal{H}\{s_I(t)\} \quad , \tag{4.6}$$

$$s(t) = s_I(t) + j \cdot \mathcal{H}\{s_I(t)\} \quad . \tag{4.7}$$

Hereby, the *Hilbert transformator* is defined as [54], [56], [59]:

$$\mathcal{H}\{s_I(t)\} = s_I(t) * \frac{1}{\pi t} = \frac{1}{\pi} \int\limits_{-\infty}^{+\infty} \frac{s_I(\xi)}{t - \xi} \, d\xi \quad . \tag{4.8}$$

Transformed into frequency domain, this gives [54], [56], [59]:

$$\mathcal{H}\{s_I(t)\} \overset{\mathcal{F}}{\circ\!\!-\!\!\bullet} -j \cdot \mathrm{sgn}\,(f)\, \mathcal{F}\{s_I(t)\} = -j \cdot S_I(f) \cdot \mathrm{sgn}\,(f) \quad . \tag{4.9}$$

The complex spectrum is then [54], [56], [59]:

$$S_+(f) = S_I(f) \cdot [1 + \mathrm{sgn}\,(f)] \quad . \tag{4.10}$$

Usually an arbitrary signal is centred around a carrier frequency f_0, thus the spectrum is centred around $\exp\,(+j\,2\pi f_0 t)$. If the signal is downconverted to centre frequency $f_{\mathrm{IF}} = 0$ or by means of mathematics multiplied by $\exp\,(-j\,2\pi f_0 t)$, then the complex signal can be considered as a complex envelope signal and is then denoted as *analytical signal* [54], [56], [59].

The advantage is a reduced need of hardware like a second mixer or an IQ-mixer. Nevertheless, this also holds a drawback: The only knowledge is $s_I(t)$, which can be considered as the real part, thus the projection of a complex signal onto the $\Re$-axis. The orthogonal component is chosen to be $90°$ ahead ($+j$ in equation (4.7) instead of $-j$) by convention. However, the

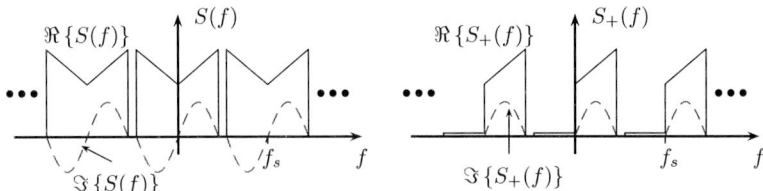

Fig. 4.1 Complex spectra of real signals after time-discrete sampling: generated by IQ-conversion (left), by Hilbert transformation (right).

phase shift in equation (4.2) can be positive or negative, thus two complex conjugated signals deliver the same projection due to their same real part [56]. The creation of analytical signals is therefore only unique due to convention [56]. Since the *Hilbert transform* of $\cos\phi$ is $+\sin\phi$ [43], the complex envelope according to equation (4.5) would be:

$$s_+(t) = U_0(t)\,[\cos\Phi + j \cdot \mathcal{H}\{\cos\phi\}] = U_0(t) \cdot \mathrm{e}^{+j\,\Phi} \quad . \tag{4.11}$$

Additionally, the spectral part with negative frequency are set to zero, the periodic spectrum (periodic due to time-discrete sampling) is blanked out periodically to zero, too. Hence, the complete signal (and its spectrum) can still be represented with half of the samples. Therefore, only half the sampling frequency, as required for the IQ-conversion case, would be needed to fulfil the *sampling theorem* [56] (Fig. 4.2). However, the negative frequency part of the spectrum can still have influence by *spill-over* effects due to finite length of signal windows [56]. In order to reduce such *spill-overs* in

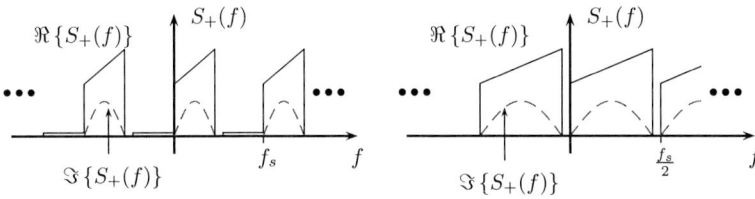

Fig. 4.2 Complex spectra of complex time-discrete signals generated using Hilbert transformation: sampling rate as required for IQ-conversion (left), possible subsampling (right).

this thesis, usually a window function is applied on the real signals before *Hilbert transformation*. This attenuates the values close to the edges.

4.2 Fundamental parameters

For the following sections, a few terms must be introduced. First of all, the distance between the transmitter Tn, an object enumerated by q and the receiver Rm, thus r_{nqm} or accordingly the *round trip delay* or *flight delay* τ_q shall be described by [24], [47], [55], [59]:

$$\tau_q = \frac{1}{c_0} \left(r_{nqm} + 2\, v_{r,q} \cdot t \right) = \tau_{0q} + \tau_{d,q} \ . \tag{4.12}$$

In general, the *round trip delay* shall consist of a static part τ_{0q} due to the radial distance of the object r_{nqm} and a dynamic part $\tau_{d,q}$ due to the relative radial velocity between object q and radar $v_{r,q}$. The term $v_{r,q}$ which is assumed to be constant for duration of measurement, resuls in a *Doppler* shift of the signal echo. Hereby, transmitter and receiver shall be collocated close to each other (compare to section 3.4). The *Doppler* frequency $f_{d,q}$ is then accordingly [24], [47], [55], [59]:

$$f_{d,q} = 2 \cdot f_0 \, \frac{v_{r,q}}{c_0} \quad . \tag{4.13}$$

The impulse response $h_{\mathrm{mf}}(t)$ of a *matched filter* to a signal $s(t)$ is generally [24], [47], [55], [59]:

$$h_{\mathrm{mf}}(t) = s^*(-t) \ \overset{\mathcal{F}}{\circ\!\!-\!\!\bullet} \ H_{\mathrm{mf}}(f) = S^*(f) \ . \tag{4.14}$$

4.3 Pulse radar systems

A well-known type of radar systems are pulse radars. There, a continuous wave signal is altered in amplitude by a pulse modulation or *on-off keying*. This shall be represented by a window function $\mathcal{U}(t)$ which shall be symmetric without loss of generality. As an example, the single pulse envelope $\mathcal{U}(t)$ shall be rectangular in time domain and normalized by the square-root of pulse duration T_p [55], [59]:

$$s(t) = \mathcal{U}(t) \cdot \mathrm{e}^{+j\,\omega_0 t} \ , \text{e.\,g.} \ \mathcal{U}(t) = \frac{1}{T_p} \cdot \mathrm{rect}\left(\frac{t}{T_p} \right) \ , \tag{4.15}$$

which yields a *sinc*-function in spectral domain. With equations (4.12) and (4.13), the time shifted echo $s(t - \tau_q)$ of a single target (again normalized by amplitude and square root of pulse duration) can be written as:

$$s(t - \tau_q) = \mathcal{U}(t - \tau_q) \cdot e^{+j\,\omega_0(t - \tau_q)} = s(t - \tau_{0q}) \cdot e^{-j\,2\pi f_{d,q} t} \,. \qquad (4.16)$$

The *matched filter* response $s_{\mathrm{mf}}(t)$ to the time shifted echo $s(t - \tau_q)$ gives then:

$$s_{\mathrm{mf}}(t) = s(t - \tau_q) * h_{\mathrm{mf}}(t) = s(t - \tau_q) * s^*(-t) =$$

$$= \int\limits_{-\infty}^{+\infty} s(\xi - \tau_{0q}) \cdot s^*(\xi + t') \cdot e^{-j\,2\pi f_{d,q}\xi}\,d\xi = \text{(subst. } t = \xi - \tau_{0q}\,,\, t' = 0)$$

$$= e^{-j\,2\pi f_d \tau_{0q}} \int\limits_{-\infty}^{+\infty} s(t) \cdot s^*(t + \tau_{0q}) \cdot e^{-j\,2\pi f_{d,q} t}\,dt =$$

$$= e^{-j\,2\pi f_d \tau_{0q}} \cdot \chi(\tau_{0q},\, f_{d,q}) \,.$$

$$(4.17)$$

The complex modulation term $\chi(\tau_{0q},\, f_{d,q})$ is denoted as *ambiguity function* [24], [55], [59], [60]. Depending on conventions, the signs for τ_{0q} and $f_{d,q}$

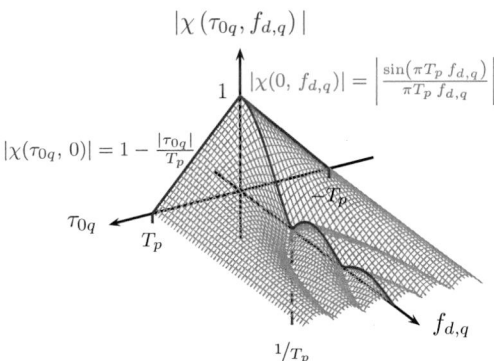

Fig. 4.3 Ambiguity function of a single rectangular radar pulse [55].

45

can differ [55], [59]. Therefore, the most common convention shall be used in the following:

$$\chi(\tau_{0q}, f_{d,q}) = \int\limits_{-\infty}^{+\infty} s(t) \cdot s^*(t + \tau_{0q}) \cdot e^{+j\,2\pi f_{d,q}t} \, dt \, . \tag{4.18}$$

By a closer look, equation (4.18) can be identified as parametric integral, thus a so-called *Radon transform* [34]. The principle behaviour of $\chi(\tau_{0q}, f_{d,q})$ for a single pulse signal of rectangular shape is sketched in Fig. 4.3. Here, only the absolute value of $\chi(\tau_{0q}, f_{d,q})$ shall be of further interest. In this case, equation (4.18) can be altered and evaluated to:

$$|\chi(\tau_{0q}, f_{d,q})| = \left| \left(1 - \frac{|\tau_{0q}|}{T_p} \right) \cdot \frac{\sin(\pi T_p f_{d,q})}{\pi T_p f_{d,q}} \right| \, , \tag{4.19}$$

providing us with the central slices through Fig. 4.3 [55], [59]:

$$|\chi(\tau_{0q}, 0)| = 1 - \frac{|\tau_{0q}|}{T_p} \, , \text{ auto-corr. fctn. of envelope fctn.,} \tag{4.20}$$

$$|\chi(0, f_{d,q})| = |\text{si}(\pi T_p f_{d,q})| \, , \text{ along Doppler axis.} \tag{4.21}$$

The first nulls of the cuts determine approximately the resolutions [55], [59]:

$$\Delta\tau_{0q} = T_p \, , \text{ delay resolution,} \tag{4.22}$$

$$\Delta f_{d,q} = \frac{1}{T_p} \, , \text{ Doppler resolution.} \tag{4.23}$$

the more interesting range resolution Δr can be derived from the delay resolution [24], [46], [47]:

$$\Delta r = \frac{c_0}{2} \cdot \Delta\tau_0 = \frac{c_0 \, T_p}{2} = \frac{c_0}{2\,\Delta F} \, . \tag{4.24}$$

Please note, that the spectrum of the rectangular pulse shape is *sinc*-function. The first zero of the aforesaid can be found at $^1/_{T_p}$ which is approximated by the signal bandwidth ΔF of the pulse [24], [46], [47]. Therefore the *time-bandwidth product* (TBP) is:

$$TBP = T_p \cdot \frac{1}{T_p} = 1 = PCR \, . \tag{4.25}$$

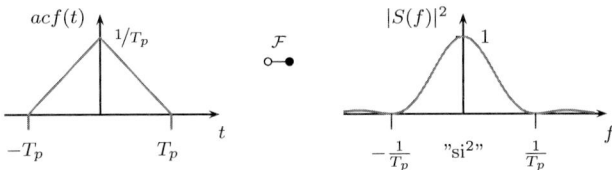

Fig. 4.4 Triangular auto-correlation function and its spectral power density [55].

For a pulse signal, the *time-bandwidth product* is equal to the so-called *pulse compression ratio* (PCR) or *dispersion factor*, which is the ratio of transmitted pulse length T_p to compressed pulse length (here T_p again) [47], [55], [59].

Equation (4.25) indicates the fundamental conflict about T_p: Bandwidth and pulse duration cannot be chosen independently. A large value for T_p provides a high energy per pulse, thus a high *signal-to-noise ratio* (SNR) and therefore high detection rate as well as a high *range accuracy* [24], [46], [47]. Whereas a short T_p yields a fine range resolution [24], [46], [47].

Besides that, the shorter the pulse T_p is, the larger the bandwidth ΔF becomes. However, the final step in today's radar sensors is usually the analog-to-digital conversion (ADC). In order to meet the *sampling theorem*, the ADC must perform the conversion with a sampling frequency at least twice the maximum frequency ($f_s > 2(f_0 + 1/T_p)$). A high bandwidth thus leads to complex and expensive hardware. Possible remedies of this conflicts are trains of pulse [55] or, like in this thesis, the use of signal modulation [55].

4.4 Modulated radars

4.4.1 Linear frequency modulation

A popular modulation scheme is the linear frequency modulation (LFM) which shall be covered in more detail here. The sweep bandwidth shall be symbolized by ΔF, T_c stands for the *sweep time* (or *chirp length*). The linearly modulated transmit frequency f_t shall then be [47]:

$$f_{Tx}(t) = f_0 \pm \frac{\Delta F}{T_c} \cdot (t - t_i) \text{ , for: } t_i - \frac{T_c}{2} \le t < t_i + \frac{T_c}{2} . \quad (4.26)$$

In the following the modulation rate shall be represented by $K = \Delta F/T_c$ [61]. A so-called *up-chirp* gives a positive sign for the slope in equation (4.26),

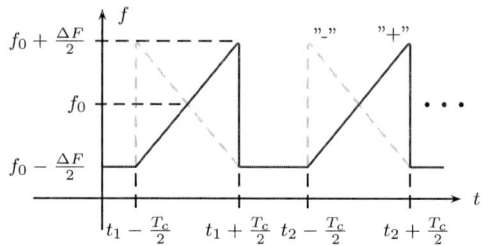

Fig. 4.5 Chirps with linear frequency modulation and their representation in equation (4.26): *up-chirp* ("+") and *down-chirp* ("-").

whereas a *down-chirp* must impose a negative sign in equation (4.26). Figure 4.5 shows an example. However, w.l.o.g. the instantaneous transmit frequency can also be related to the starting frequency of each chirp. Therefore the substitution $\tilde{t} = t + T_c/2$ has to be applied to equation (4.26), which gives us:

$$f_{Tx}(t) = \cdots = f_0 \mp \frac{\Delta F}{2} \pm K \cdot (\tilde{t} - t_i) = f_{start} \pm K \cdot (\tilde{t} - t_i) \, ,$$
$$\text{for: } t_i \leq \tilde{t} < t_i + T_c \, . \tag{4.27}$$

The drawback of this ansatz is, that the starting frequency f_{start} has to be adapted for the up-chirp or down-chirp case. Anyway, since both presentations are considered to be equal, we chose the version of equation (4.26) for the following parts The corresponding phase $\varphi_{Tx}(t)$ to equation (4.26) is then [47]:

$$\varphi_{Tx}(t) = 2\pi \cdot \int_{t_i}^{t} f_{Tx}(\xi) \, d\xi = 2\pi \cdot \left[\pm \frac{K}{2} t^2 + (f_0 \mp K t_i) \, t \right] + \varphi_0 \quad . \tag{4.28}$$

For a time slot $t_i = 0$ this simplifies to [46]:

$$\varphi_{Tx}(t) = 2\pi \cdot \left[\pm \frac{1}{2} K t^2 + f_0 t \right] + \varphi_0 \quad . \tag{4.29}$$

The transmit signal is then:

$$u_{Tx}(t) = \mathcal{U}(t) \cdot \cos\left[\varphi_{Tx}(t)\right] \quad . \tag{4.30}$$

In the following, we only concentrate on the *up-chirp* case, hence the positive sign in equation (4.29). Furthermore, the real signals shall be complemented to an *analytical* signal by *Hilbert transformation*:

$$u_{Tx}(t) = \mathcal{U}(t) \cdot e^{+j\,\varphi_{Tx}(t)} \quad . \tag{4.31}$$

4.4.2 Compressing chirps by matched filter

As for pulse radar signals, chirps can be compressed by a *matched filter* [55]. In terms of hardware, this can be achieved by dispersive filters like e. g. *surface-acoustic-wave* (SAW) filters. Later we will focus on a direct mixing method combined with a *Fourier transformation* [62], [63]. Especially for systems where all τ_q are much larger than T_c, thus the transmit chirp and its echoes do not overlap in time-domain anymore, like e. g. in space-born systems, direct mixing techniques become less feasible [61]. Alternatively, the echoes have to be down-converted ($\cdot \exp(-j\,2\pi f_0 t)$) and compressed by a *matched filter* [61]. The uncompressed echo of a single target in baseband ($f_{\text{IF}} = 0$) is proportional to [61]:

$$
\begin{aligned}
u(t) &\sim u_{Tx}(t - \tau_q) \cdot e^{-j\,2\pi f_0 t} = \\
&= \mathcal{U}(t - \tau_q)\, e^{+j\,\pi K (t - \tau_q)^2} \cdot e^{-j\,2\pi f_0 \tau_q} \quad .
\end{aligned}
\tag{4.32}
$$

The impulse response of the corresponding *matched filter* shall be [55], [61]:

$$h_{\text{mf}}(t) = \mathcal{U}^*(-t) \cdot e^{-j\,\pi K t^2} = \mathcal{U}^*(t) \cdot e^{-j\,\pi K t^2} \quad . \tag{4.33}$$

The *matched filter* output with equations (4.12), (4.13) is then [55], [61]:

$$
\begin{aligned}
s_{\text{mf}}(t) &= u(t) * h_{\text{mf}}(t) = \ \ldots \ = \\
&= \int_{-\infty}^{+\infty} \mathcal{U}(t) \cdot \mathcal{U}^*(t - \tau_q) \cdot e^{-j\,2\pi K \tau_q t} \cdot e^{+j\,\pi K \tau_q^2} \cdot e^{-j\,2\pi f_0 \tau_q}\, dt \quad .
\end{aligned}
\tag{4.34}
$$

In order to simplify the integral in equation (4.34), it shall be assumed, that $\mathcal{U}(t - \tau_q) \approx \mathcal{U}(t - \tau_{0q})$. The complex term $\exp\left(+j\,\pi K \tau_q^2\right)$ is often depicted as *residual video phase* (RVP) [64]. Please note, that the use of the term RVP can differ in literature: Alternatively, the remaining phase distortion after the attempt of compensating $\exp\left(+j\,\pi K \tau_q^2\right)$ is denoted as RVP [65]. However, all have in common that any use of RVP describes small phase deviations due to dependence on (quadratic) round trip delays. The RVP effect is that two targets with constant radial distance appear with larger separation at larger distances [64]. This is especially a problem for systems with large *round-trip delays*. Please note, that by applying $\tau_q^2 \approx \tau_{0q}^2$ (small *Doppler* effect) on equation (4.34), the RVP term can be written outside the integral:

$$s_{\mathrm{mf}}(\tau_{0q}, v_{r,q}) \approx \mathrm{e}^{-j\,\omega_0\,\tau_{0q}} \cdot \mathrm{e}^{+j\,\pi K\,\tau_{0q}^2}$$

$$\cdot \int_{-\infty}^{+\infty} \mathcal{U}(t) \cdot \mathcal{U}^*(t - \tau_{0q}) \cdot \mathrm{e}^{-j\,2\pi K\,\tau_{0q}\,t} \cdot \mathrm{e}^{-j\,2\pi K\,\frac{2\,v_{r,q}}{c_0}\,t^2} \cdot \mathrm{e}^{-j\,2\pi f_{d,q}(v_{r,q})\,t}\,dt\;.$$

$$(4.35)$$

Furthermore, there is a second quadratic phase term $\exp\left(-j\,2\pi K\,2\,v_{r,q}/c_0\,t^2\right)$ in equation (4.35) caused by the relative velocity. Its effect is a defocus or broadening of the compressed chirp [42], which can be demonstrated by interpreting the integral in equation (4.34) as convolution :

$$s_{\mathrm{mf}}(\tau_{0q}, v_{r,q}) = \mathrm{e}^{-j\,\omega_0\,\tau_{0q}} \cdot \mathrm{e}^{+j\,\pi K\,\tau_{0q}^2}\cdot$$

$$\cdot \int_{-\infty}^{+\infty} \mathcal{U}(t) \cdot \mathcal{U}^*(t - \tau_{0q}) \cdot \mathrm{e}^{-j\,2\pi(K\,\tau_{0q}+f_{d,q})\,t}\,dt \; * \int_{-T_c/2}^{+T_c/2} \mathrm{e}^{-j2\pi\,K\frac{2v_{r,q}}{c_0}\,t^2}\,dt\;.$$

$$(4.36)$$

This evolution provides the so-called *ambiguity function* of a LFM radar signal which is the first integral in equation (4.36) [55], [59], [66]. Again the signs of τ_{0q}, $f_{d,q}$ can differ due to conventions [55], [59]. The difference to pulse radar signals is the term $(K\tau_{0q} + f_{d,q})$ in the exponent, causing a *shearing effect* [55], [59] which in turn will lead to a higher *pulse compression rate* [55], [59]. Besides that, after applying a substitution, the second integral in equation (4.36) can be expressed by a *Fresnel diffraction integral* [32], [47]:

$$t = \frac{T_c}{2}\,\xi \quad \Rightarrow \quad dt = \frac{T_c}{2}\,d\xi\;, \tag{4.37}$$

$$\varsigma_q^2 = 2K\,\frac{v_{r,q}}{c_0} \cdot T_c^2 = 2\,\frac{v_{r,q}}{c_0} \cdot TBP\;, \tag{4.38}$$

$$\kappa_q = 0\;, \tag{4.39}$$

resulting in:

$$\int\limits_{-T_c/2}^{+T_c/2} e^{-j2\pi \, K \frac{2v_{r,q}}{c_0} \, t^2} \, dt = \frac{T_c}{2} \int\limits_{-1}^{+1} e^{+j \, \pi\kappa_q\xi} \cdot e^{-j \frac{\pi}{2}\varsigma_q^2\xi^2} \, d\xi = \frac{T_c}{2} \, \mathcal{F}_0\left(\kappa_q, \, \varsigma_q\right) \; .$$
(4.40)

The *matched filter* output can then be expressed by the following expression:

$$s_{\mathrm{mf}}(t) \approx \frac{T_c}{2} \, e^{-j\,\omega_0\,\tau_{0q}} \cdot e^{+j\,\pi K\,\tau_{0q}^2} \cdot \chi(\tau_{0q}, \, K\tau_{0q}+f_{d,q}) \; * \; \mathcal{F}_0\left(0, \, \varsigma_q\right) \; . \quad (4.41)$$

The convolution with the *Fresnel diffraction integral* results in the before mentioned broadening effect [42]. Analogously to equation (4.18), the absolute value of *ambiguity function* $|\chi(\tau_{0q}, \, K\tau_{0q}+f_{d,q})|$ for a single LFM pulse with rectangular envelope is [55], [59]:

$$|\chi(\tau_{0q}, \, K\tau_{0q}+f_{d,q})| = \left| \left(1 - \frac{|\tau_{0q}|}{T_c}\right) \cdot \frac{\sin\left[\pi T_c \, (K\tau_{0q} + f_{d,q}) \cdot \left(1 - \frac{|\tau_{0q}|}{T_c}\right)\right]}{\left[\pi T_c \, (K\tau_{0q} + f_{d,q}) \cdot \left(1 - \frac{|\tau_{0q}|}{T_c}\right)\right]} \right| \; .$$
(4.42)

The first null of $|\chi(\tau_{0q}, \, 0|$ can be found for the sinus argument equal to π,

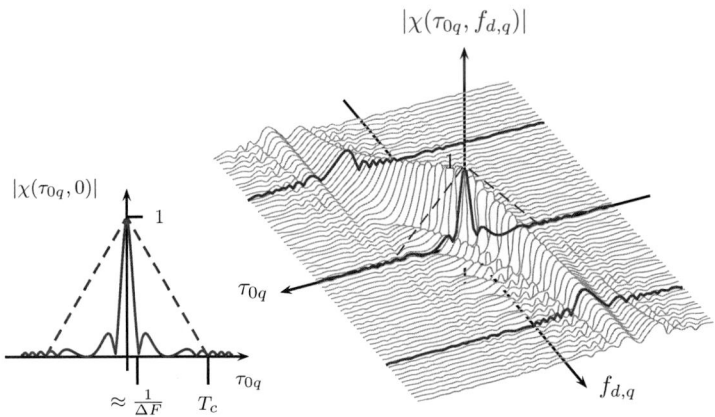

Fig. 4.6 Ambiguity function of a single LFM radar pulse [55].

which yields a quadratic function in τ_{0q}:

$$\pi T_c \left(K\tau_{0q} - f_{d,q} \right) \cdot \left(1 - \frac{|\tau_{0q}|}{T_c} \right) \overset{!}{=} 1 \cdot \pi \tag{4.43}$$

For $\tau_{0q} > 0$ and $f_{d,q} = 0$, equation (4.43) can be altered to:

$$K\tau_{0q}^2 - \Delta F\tau_{0q} + 1 \overset{!}{=} 0 . \tag{4.44}$$

The solutions of equation (4.44) can be approximated by $\sqrt{1-x} \approx (1 - x/2)$ for $|x| < 1$, [43], [55], [59]:

$$\tau_{0q,1,2} = \frac{T_c}{2} \pm \frac{T_c}{2} \sqrt{1 - \frac{4}{TBP}} \approx \frac{T_c}{2} \pm \frac{T_c}{2} \left(1 - \frac{2}{TBP} \right) , \text{ for } TBP > 4 . \tag{4.45}$$

The very fist null is then $\tau_{0q,1} \approx 1/\Delta F$. Therefore, the *pulse compression ratio* (PCR) is again approximately equal to the *time-bandwidth product* (TBP) [55], [59]:

$$PCR = \frac{T_c}{\tau_{0q,1}} \approx \Delta F T_c = TBP . \tag{4.46}$$

The *delay resolution*, thus *range resolution* is now independent from the pulse duration T_c and is only determined by the bandwidth ΔF:

$$\Delta r = \frac{c_0}{2} \cdot \Delta \tau_0 = \frac{c_0}{2\Delta F} . \tag{4.47}$$

However, the drawback is that a *Doppler* shift is immediately mapped onto a *run time* shift [55]: This can be encountered e. g. by evaluation of chirp trains [55] or other modulation forms like a triangular shaped one [19], [47], [67], [68], [69]. More details on ambiguity functions of monostatic, bistatic or MIMO configurations can be found in [15], [16], [17], [18].

4.4.3 The FMCW principle

A well approved technique is *frequency modulated continuous wave* (FMCW) radar [62]. Here, the pulse compression is usually not achieved by a *matched filter*: First, the echo is converted with a copy of the transmit signal [46], [62] where the mixer serves as *phase discriminator* [46], [62]. Then, the output signal of this *homodyne* ansatz ($f_{\text{IF}} = 0$) is compressed by *Fourier*

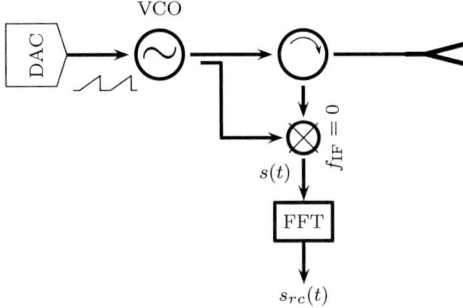

Fig. 4.7 Block diagram of FMCW radar principle [62].

transformation [46], [62], [63]. According to equation 3.57 and 3.60 respectively, the general signal which is back-scattered by the object function $o(t = {^2r}/c)$ consisting of Q point scatterers can be expressed by convolution:

$$u_{Rx}(t) = \mathcal{U}(t) \cdot \mathrm{e}^{+j\,\varphi_{Tx}(t)} \overset{t}{*} o\left(t = \frac{2\,r}{c}\right) =$$

$$= \mathcal{U}(t) \cdot \mathrm{e}^{+j\,\varphi_{Tx}(t)} \overset{t}{*} \sum_{q=1}^{Q} \sigma_q \cdot \delta(t - \tau_q) \quad .$$

(4.48)

The complex term σ_q shall incorporate the complex reflectivity ζ_q of the object, the attenuation $\sim 1/r^4$, the joint antenna pattern $\sqrt{G_{Tx} \cdot G_{Rx}}$, but not the basic phase shift caused by range or array factor (compare equation 3.57). The down-conversion can be expressed by the product of $u_{Rx}(t)$ and a complex conjugate copy of the transmit signal $\mathcal{U}^*(t) \cdot \mathrm{e}^{-j\,\varphi_{Tx}(t)}$. Please note, that the complex signals shall be generated by *Hilbert transformation*, thus they are *analytic* (see section 4.1.2). With additional low-pass filtering, the baseband signal (provided by the mixer) can then be represented by:

$$s_{mx}(t) = \sum_{q=1}^{Q} \sigma_q \cdot \mathcal{U}(t) \cdot \mathcal{U}^*(t - \tau_q) \cdot \mathrm{e}^{+j\,[\varphi_{Tx}(t) - \varphi_{Tx}(t - \tau_q)]} =$$

$$= \sum_{q=1}^{Q} \sigma_q \cdot \mathcal{U}(t) \cdot \mathcal{U}^*(t - \tau_q) \cdot \mathrm{e}^{+j\,2\pi\left[\pm\frac{1}{2}\,K\cdot\left(2t\,\tau_q - \tau_q^2\right) + (f_0 - \Delta F)\tau_q\right]} \quad .$$

(4.49)

A closer inspection of equation (4.49) shows that there is again a *residual video phase* (RVP) [64] term $\exp\left(\mp j\,\pi K\tau_q^2\right)$, which can be neglected for short ranges. Furthermore, $\Delta F \ll f_0$, thus $(f_0 - \Delta F) \approx f_0$ shall be valid. With the additional substitution $\sigma_q \cdot \mathcal{U}(t) \cdot \mathcal{U}^*(t - \tau_q) = U_q$, equation (4.49) (in *up-chirp* case) can be altered to:

$$
\begin{aligned}
s_{mx,nm}(t) &= \sum_{q=1}^{Q} U_q \cdot \mathrm{e}^{-j\,\pi K \tau_q^2} \cdot \mathrm{e}^{+j\,\varphi_q(t)} \approx \\
&\approx \sum_{q=1}^{Q} U_q \cdot \mathrm{e}^{+j\,\varphi_q(t)} = \sum_{q=1}^{Q} U_q \cdot \mathrm{e}^{+j\,2\pi\,(K\,t+f_0)\,\tau_q} \ .
\end{aligned}
\tag{4.50}
$$

With equation (4.12), the phase terms $\varphi_q(t)$ delivered by the mixer can be written explicitly as:

$$
\varphi_q(t) = \frac{2\pi}{c_0}\left(K\,r_{nqm}\cdot t + 2\,f_0\,v_{r,q}\cdot t + 2\,K\,v_{r,q}\cdot t^2 + f_0\,r_{nqm}\right) \ . \tag{4.51}
$$

Again, the object's radial velocity brings in a quadratic phase term $\exp\left(-j\,2\pi K\,2\,v_{r,q}/c_0\,t^2\right)$ which serves as additional linear frequency modulation and defocuses or broadens the compressed chirp [42]. Each object q can be assigned to an instantaneous frequency [24], [46], [47], [55], [59]:

$$
f_q(t) = \frac{1}{2\pi}\frac{\partial\,\varphi_q(t)}{\partial\,t} = \frac{1}{c_0}\,K\,r_{nqm} + 2\frac{v_{r,q}}{c_0}\,(f_0 + 2K\,t) \ . \tag{4.52}
$$

The pulse broadening shall be neglected at first glance by choosing the approximation $2Kt \ll f_0$. This leads to:

$$
f_q(t = 0) = \frac{1}{c_0}\,K\,r_{nqm} + 2f_0\,\frac{v_{r,q}}{c_0} = f_{r,q} + f_{d,q} \ . \tag{4.53}
$$

Thereby each frequency contribution consists of two components: The distance frequency $f_{r,q}$ and the *Doppler* frequency $f_{d,q}$ [46], [47], [59]. In the general case of equation (4.52), it can be seen that the distance frequency $f_{r,q}$ changes over time with $v_{r,q}$. Therefore the time derivative of f_q, thus the second derivative of the phase terms in equation (4.50), gives a hint on the radial velocity $v_{r,q}$:

$$
\frac{\partial}{\partial t}\,f_q(t) = K\,\frac{4}{c_0}\,v_{r,q} + \frac{2}{c_0}\,(f_0 + 2K)\,\frac{\partial\,v_{r,q}}{\partial\,t} \ , \tag{4.54}
$$

which reduces to:

$$\frac{\partial}{\partial t} f_q(t) = K \frac{4}{c_0} v_{r,q} \ , \quad \text{for } v_{r,q} = \text{const.} \tag{4.55}$$

Equation (4.53) repeats the insights of the *ambiguity function* of a single LFM chirp, hence a *Doppler* shift cannot be distinguished from a range shift (with equation (4.47)):

$$
\begin{aligned}
f_{d,b} &\overset{!}{=} \Delta f_{r,q} \ , \\
f_{d,b} &= K \frac{2 \Delta r}{c_0} = \frac{1}{T_c} \quad \Rightarrow \quad v_{r,b} = \frac{\lambda_0}{2 T_c} \ .
\end{aligned}
\tag{4.56}
$$

Already a *Doppler* shift of $1/T_c$, or the movement of half the wavelength during the chirp, causes a shift of one range cell Δr [63]. This coincidence can also be considered as the first *blind speed* of a single LFM chirp [63].

Chirp compression

For showing parallels to the *matched filter* output, the mixer output signal shall be *Fourier* transformed [46], [47]. Hereby, the RVP shall be enlisted for the sake of completeness, however with the approximation $\tau_q^2 \approx \tau_{0q}^2$ (up-chirp case). Since integral and sum operator are linear operators [43], [44], [70] they can be swapped. Then with $\omega_0 \tau_{0q} = k_0 r_{nqm}$ (compare to equation (4.12)), this gives a range compressed signal $s_{rc,nm}(f)$ for an array element:

$$
\begin{aligned}
s_{rc,nm}(f) = \sum_{q=1}^{Q} \sigma_q \cdot \mathrm{e}^{+j\,k_0 r_{nqm}} \cdot \mathrm{e}^{-j\,\pi K \tau_{0q}^2}. \\[2mm]
\cdot \int_{-\infty}^{+\infty} \mathcal{U}(t) \cdot \mathcal{U}^*(t - \tau_{0q}) \cdot \mathrm{e}^{-j\,2\pi[f-(K\,\tau_{0q}+f_{d,q})]t} \, dt. \\[2mm]
* \int_{-T_c/2}^{+T_c/2} \mathrm{e}^{+j2\pi\left(2K\frac{v_{r,q}}{c_0} t^2 - ft\right)} \, dt \ ,
\end{aligned}
\tag{4.57}
$$

which again contains a form of LFM *ambiguity function* as for the *matched filter* case [55], [59]. A separation of the exponential terms in equation (4.57) led to a convolution in spectral domain [43]. By applying several substitutions, again a *Fresnel diffraction integral* can be recognized [32], [47]:

$$t = \frac{T_c}{2} \xi \, , \quad \Rightarrow \quad \frac{\partial t}{\partial \xi} = \frac{T_c}{2} \, , \tag{4.58}$$

$$\kappa_q = -T_c f \, , \tag{4.59}$$

$$\varsigma_q^2 = -2\Delta F T_c \cdot \frac{v_{r,q}}{c_0} = -2 \cdot \frac{v_{r,q}}{c_0} \cdot TBP \, , \tag{4.60}$$

$$f_{r,q} = K\tau_{0q} = K \frac{r_{nqm}}{c_0} \, , \tag{4.61}$$

equation (4.57) can be altered to:

$$s_{rc,nm}(f) = \sum_{q=1}^{Q} \sigma_q \cdot \mathrm{e}^{+j\,k_0 r_{nqm}} \cdot \mathrm{e}^{-j\,\pi K \tau_{0q}^2} \cdot \chi\left[\tau_{0q}, \, f - (K\,\tau_{0q} + f_{d,q})\right]$$

$$* \, \frac{T_c}{2} \int_{-1}^{+1} \mathrm{e}^{+j\,\pi \kappa_q \cdot \xi} \cdot \mathrm{e}^{-j\,\frac{\pi}{2} \varsigma_q^2 \cdot \xi^2} \, d\xi =$$

$$= \sum_{q=1}^{Q} \sigma_q \frac{T_c}{2} \cdot \mathrm{e}^{+j\,k_0 r_{nqm}} \cdot \mathrm{e}^{-j\,\pi K \tau_{0q}^2} \cdot \chi\left[\tau_{0q}, \, f - (f_{r,q} + f_{d,q})\right] \, * \, \mathcal{F}_0(\kappa_q, \varsigma_q) =$$

$$= \sum_{q=1}^{Q} \tilde{\sigma}_q \cdot \mathrm{e}^{+j\,k_0 r_{nqm}} \cdot \mathrm{e}^{-j\,\pi K \tau_{0q}^2} \cdot psf\left[f - (f_{r,q} + f_{d,q})\right] \, .$$

$$\tag{4.62}$$

The convolution operation of the wanted *ambiguity function* and the *Fresnel integral* results again in the already mentioned broadening effect, thus into an effective *point-spread function* $psf(f)$ [42], [71]. Hereby, the complex term $\tilde{\sigma}_q$ shall incorporate all additional factors due to the substitutions. In case of zero *Doppler* shift, thus $\varsigma_q = 0$, $\mathcal{F}_0(\kappa_q, 0)$ simplifies to a *si*-function [32]. The *si*-function in turn narrows with increasing chirp duration T_c, hence equation (4.57) advances to a convolution with a *Dirac delta function*:

$$
s_{rc,nm}(f) = \sum_{q=1}^{Q} \tilde{\sigma}_q \cdot e^{+j\, k_0 r_{nqm}} \cdot e^{-j\,\pi K \tau_{0q}^2} \cdot \chi\left(\tau_{0q},\, f - f_{r,q}\right)
$$

$$
* \; 2\,\mathrm{si}\left[\pi T_c \left(f - f_{r,q}\right)\right] \approx
$$

$$
\approx \sum_{q=1}^{Q} 2\,\tilde{\sigma}_q \cdot e^{+j\, k_0 r_{nqm}} \cdot e^{-j\,\pi K \tau_{0q}^2} \cdot \chi\left(\tau_{0q},\, f - f_{r,q}\right) \quad , \text{ for } \varsigma_q = 0\,,\, T_c \gg 0\,.
$$

$$(4.63)$$

An alternative method of approximating the *Fourier* transformation in equation (4.57) is the *method of stationary phase* [32], [41], [47], [71], appendix A.3:

$$
\int_{-\infty}^{+\infty} h(t)\, e^{+j\,\Phi(x)}\, dt \approx \sqrt{\frac{j\, 2\pi}{\Phi''(t_*)}}\; h(t_*)\, e^{+j\,\Phi(t_*)} \quad ,
$$

$$(4.64)$$

where t_* shall denote the *stationary point* with $\Phi'(t_*) = 0$. The derivatives of the phase term $\Phi_q(t)$ can be derived from equation (4.51):

$$
\Phi_q'(t) = \frac{\partial}{\partial t} \Phi_q(t) = 2\pi \left[4K \frac{v_{r,q}}{c_0} \cdot t - \left(f - f_{r,q} - f_{d,q}\right) \right] \,,
$$

$$(4.65)$$

$$
\Phi_q''(t) = \frac{\partial^2}{\partial t^2} \Phi_q(t) = 2\pi \cdot 4K \frac{v_{r,q}}{c_0} \,.
$$

$$(4.66)$$

From $\Phi_q'(t_*) = 0$, t_* can be derived [32]:

$$
t_* = \frac{1}{4K} \cdot \frac{v_{r,q}}{c_0} \cdot \left(f - f_{r,q} - f_{d,q}\right) \,.
$$

$$(4.67)$$

The phase $\Phi_q(t_*)$ at the stationary point is then:

$$
\Phi_q(t_*) = 2\pi \frac{-1}{8K} \cdot \left(f - f_{r,q} - f_{d,q}\right)^2 + k_0\, r_{nqm} \,.
$$

$$(4.68)$$

The range compressed signal $s_{rc}(f)$ is then approximately:

$$s_{rc,nm}(f) \approx$$

$$\approx \sum_{q=1}^{Q} \tilde{\sigma}_q \cdot e^{+j\,k_0 r_{nqm}} \cdot e^{-j\,\pi K \tau_{0q}^2} \cdot psf\left(f - f_{r,q} - f_{d,q}\right) \ . \tag{4.69}$$

The *points-spread function* is then approximated by;

$$psf\left(f - f_{r,q} - f_{d,q}\right) \approx$$

$$\approx \chi\left[\tau_{0q},\, f - (f_{r,q} + f_{d,q})\right] \cdot e^{-j\,\frac{\pi}{4K}\,\frac{c_0}{v_{r,q}}\cdot(f-f_{r,q}-f_{d,q})^2} \ . \tag{4.70}$$

The FMCW array signal

The last term which must be inspected in more detail is $\exp\left(+j\,k_0 r_{nqm}\right)$:
An approximation for r_{nqm} can be derived from equations 3.49, 3.50:

$$r_{nqm} \approx 2\left[r_{0q} + \frac{1}{4r_{0q}}\left(\boldsymbol{d}_n^2 + \boldsymbol{d}_m^2\right) - \frac{1}{2}\left(\boldsymbol{d}_n + \boldsymbol{d}_m\right) \cdot \boldsymbol{e}_r\left(\vartheta_q,\,\varphi_q\right)\right] =$$

$$= 2\left[r_{0q} + \delta_i^2\left(r_{0q}\right) - \boldsymbol{d}_i \cdot \boldsymbol{e}_r\left(\vartheta_q,\,\varphi_q\right)\right] \ . \tag{4.71}$$

Again r_{0q} depicts the distance from origin to object, $\boldsymbol{d}_i$ stands for the virtual element position, whereas δ_i^2 represents the quadratic phase terms due to array dimensions:

$$\delta_i^2\left(r_{0q}\right) = \frac{1}{4r_{0q}}\left(\boldsymbol{d}_n^2 + \boldsymbol{d}_m^2\right) \ . \tag{4.72}$$

After inserting equation (4.71) into equation (4.50), the baseband signal for a virtual element can be derived:

$$s_i(t) = \sum_{q=1}^{Q} U_q \cdot e^{+j\,\frac{4\pi}{c_0}\,(Kt+f_0)\cdot\left[r_{0q}+\delta_i^2(r_{0q})-\boldsymbol{d}_i\cdot\boldsymbol{e}_r(\vartheta_q,\,\varphi_q)\right]}.$$

$$\cdot e^{+j\,2\pi f_{d,q}t} \cdot e^{+j\,4\pi\frac{v_{r,q}}{c_0}t^2} \cdot e^{-j\,\pi\frac{K}{c_0^2}r_{0q}^2} \ . \tag{4.73}$$

When all $s_i(t)$ are lined to a vector and normalized to length one, this yields $\boldsymbol{s}(t)$:

$$
\boldsymbol{s}(t) = \frac{1}{\sqrt{NM}} \sum_{q=1}^{Q} U_q \cdot \mathrm{e}^{+j\,\frac{4\pi}{c_0}\,(Kt+f_0)\cdot r_{0q}} \cdot \mathrm{e}^{+j\,2\pi f_{d,q}t}.
$$

$$
\cdot\, \mathrm{e}^{+j\,4\pi\,\frac{v_{r,q}}{c_0}\,t^2} \cdot \mathrm{e}^{-j\,\pi\,\frac{K}{c_0^2}\,r_{0q}^2} \cdot \operatorname{diag}\left\{ \mathrm{e}^{-j\,\frac{4\pi}{c_0}\,(Kt)\,\boldsymbol{d}_i\cdot\boldsymbol{e}_r(\vartheta_q,\varphi_q)} \right\} \cdot
$$

$$
\cdot \operatorname{diag}\left\{ \mathrm{e}^{+j\,\frac{4\pi}{c_0}\,(Kt+f_0)\,\delta_i^2(r_{0q})} \right\} \cdot \begin{bmatrix} \vdots \\ \mathrm{e}^{-j\,\frac{4\pi}{c_0}\,(f_0)\,\boldsymbol{d}_i\cdot\boldsymbol{e}_r(\vartheta_q,\varphi_q)} \\ \vdots \end{bmatrix} . \tag{4.74}
$$

The time dependent azimuth shifts were collected in a diagonal matrix:

$$
\operatorname{diag}\left\{ \mathrm{e}^{+j\,\frac{4\pi}{c_0}\,(Kt)\left[\delta_i^2 - \boldsymbol{d}_i\cdot\boldsymbol{e}_r(\vartheta_q,\varphi_q)\right]} \right\} = \begin{bmatrix} \ddots & & 0 \\ & \mathrm{e}^{+j\,\frac{4\pi}{c_0}\,(Kt)\left[\delta_i^2 - \boldsymbol{d}_i\cdot\boldsymbol{e}_r(\vartheta_q,\varphi_q)\right]} & \\ 0 & & \ddots \end{bmatrix} =
$$

$$
= \boldsymbol{\Phi}_q\left(\vartheta_q, \varphi_q, \delta_i^2, t\right) \quad .
$$

$$
\tag{4.75}
$$

However, the fixed phases due to phase front curvature are collected in:

$$
\operatorname{diag}\left\{ \mathrm{e}^{+j\,2k_0\delta_i^2(r_{0q})} \right\} = \begin{bmatrix} \ddots & & 0 \\ & \mathrm{e}^{+j\,2k_0\delta_i^2(r_{0q})} & \\ 0 & & \ddots \end{bmatrix} = \boldsymbol{\Lambda}_\delta\left(r_{0q}\right) \quad, \tag{4.76}
$$

With increasing r_{0q}, the exponents of the complex terms in $\boldsymbol{\Lambda}_\delta$ tend to zero, thus $\boldsymbol{\Lambda}_\delta$ tends to unity matrix for the far-field. The matrix $\boldsymbol{\Phi}_q$ however, collects all phase terms leading to range-cell migration. There, the term of curvature δ_i^2 also tends to zero for far-field.

The so-called *directional vector* is normalized to length one and can also be expressed by the digital wavenumber Ψ_q (compare to equation (3.28)):

$$
\boldsymbol{a}_q\left(\vartheta_q,\,\varphi_q\right) = \frac{1}{\sqrt{NM}}\left[\ldots,\,\mathrm{e}^{-j\,2k_0\,\boldsymbol{d}_i\cdot\boldsymbol{e}_r(\vartheta_q,\varphi_q)},\,\ldots\right]^{\mathrm{T}} =
$$

$$
\boldsymbol{a}_q\left(\Psi_q\right) = \frac{1}{\sqrt{\Gamma_u}}\left[\ldots,\,\mathrm{e}^{-j\,\Psi_q\cdot\eta},\,\ldots\right]^{\mathrm{T}} \quad . \tag{4.77}
$$

According to equation (3.70), the directional vector a_q can also be expressed by the Kronecker product of two directional vectors $a_{Tx,q}$, $a_{Rx,q}$, representing the bistatic origin.

$$
\begin{aligned}
a_q \left(\vartheta_q, \varphi_q \right) &= a_{Rx,q} \left(\vartheta_q, \varphi_q \right) \otimes a_{Tx,q} \left(\vartheta_q, \varphi_q \right) = \\
&= a_{Tx,q} \left(\vartheta_q, \varphi_q \right) \otimes a_{Rx,q} \left(\vartheta_q, \varphi_q \right) .
\end{aligned}
\tag{4.78}
$$

Equation (4.74) can be written as:

$$
\begin{aligned}
s(t) = \sum_{q=1}^{Q} U_q &\cdot e^{+j \frac{4\pi}{c_0} (Kt + f_0) \cdot r_{0q}} \cdot e^{+j \, 2\pi f_{d,q} t} \cdot e^{+j \, 4\pi \frac{v_{r,q}}{c_0} t^2} \cdot \\
&\cdot e^{-j \, \pi \frac{K}{c_0^2} r_{0q}^2} \cdot \Phi_q \left(\vartheta_q, \varphi_q, \delta_i^2, t \right) \Lambda_\delta \left(r_{0q} \right) a_q \left(\vartheta_q, \varphi_q \right) .
\end{aligned}
\tag{4.79}
$$

According to equation (4.69), the range compressed version of equation (4.74) can be expressed by:

$$
\begin{aligned}
s_{rc}(f) = \sum_{q=1}^{Q} \tilde{\sigma}_q &\cdot e^{-j \, \pi \frac{K}{c_0^2} r_{0q}^2} \cdot psf \left(f - f_{r,q} - f_{d,q} \right) \cdot \\
&\cdot e^{+j \, 2k_0 r_{0q}} \cdot \Lambda_\delta \left(r_{0q} \right) \cdot a_q \left(\vartheta_q, \varphi_q \right) .
\end{aligned}
\tag{4.80}
$$

Please note: It was assumed that the range cell migration due to azimuth or phase front curvature phase shifts can be neglected (Φ_q becomes a unity matrix). If this was not the case anymore, the point spread function would be additionally shifted to:

$$
\begin{aligned}
&psf \left(f - f_{r,q} - f_{d,q} \right) \rightarrow \\
&psf \left[f - \left(f_{r,q} + f_{d,q} + \frac{2K}{c_0} \delta_i^2 (r_{0q}) - \frac{2K}{c_0} d_i \cdot e_r (\vartheta_q, \varphi_q) \right) \right] .
\end{aligned}
\tag{4.81}
$$

4.4.4 The SFCW radar vs. sampled FMCW radar

For digital signal processing, the signals are examined at time discrete steps. Therefore, several terms shall be defined:

$$
t = \nu T_s , \quad \frac{T_c}{T_s} \in \mathbb{N} \text{ (even)}, \quad \nu \in \left[-\frac{T_c}{2T_s} ; +\frac{T_c}{2T_s} \right]
\tag{4.82}
$$

$$
\Delta f = \frac{\Delta F \cdot T_s}{T_c} \quad \Rightarrow \quad K \cdot t = \Delta f \cdot \nu .
\tag{4.83}
$$

The sampled signal can be expressed by the multiplication of equation (4.74) with a series of *Dirac delta* functions, resulting in a periodic spectrum [56]:

$$s(\nu) = s(t) \cdot T_s \sum_{l=-\infty}^{+\infty} \delta(t - l \cdot T_s) \overset{\mathcal{F}}{\circ\!\!-\!\!\bullet} S(f) = s(f) * \sum_{p=-\infty}^{+\infty} \delta\left(f - \frac{p}{T_s}\right).$$

$$(4.84)$$

Applied to equation (4.74), this yields:

$$s(\nu) = \frac{T_s}{\sqrt{NM}} \sum_{q=1}^{Q} U_q \cdot e^{+j\frac{4\pi}{c_0}(\Delta f\nu + f_0) \cdot r_{0q}} \cdot e^{+j\,2\pi f_{d,q} T_s \cdot \nu}.$$

$$\text{diag}\left\{e^{+j\frac{4\pi}{c_0}(\Delta f\nu + f_0)\,\delta_i^2(r_{0q})}\right\} \cdot \begin{bmatrix} \vdots \\ e^{-j\frac{4\pi}{c_0}(\Delta f\nu + f_0)\,d_i \cdot e_r(\vartheta_q,\varphi_q)} \\ \vdots \end{bmatrix}. \quad (4.85)$$

$$\cdot\, e^{+j\,4\pi\frac{v_{r,q}}{c_0}T_s^2 \cdot \nu^2} \cdot e^{-j\,\pi\frac{K}{c_0^2}r_{0q}^2}.$$

The range compressed signal, thus the *Fourier* transform, is then periodic:

$$s_{rc}(f) = \sum_{p=-\infty}^{+\infty} \sum_{q=1}^{Q} \tilde{\sigma}_q \cdot e^{-j\,\pi\frac{K}{c_0^2}r_{0q}^2} \cdot psf\left(f - f_{r,q} - f_{d,q} - \frac{p}{T_s}\right). $$

$$\cdot\, e^{+j\,2k_0 r_{0q}} \cdot \Lambda_\delta(r_{0q}) \cdot a_q(\vartheta_q,\varphi_q). \quad (4.86)$$

Additionally, the *sampling theorem* must be fulfilled in order to avoid *aliasing effects* [51], [56], thus the sampling frequency must be at least twice the maximum *distance frequency* $K\tau_{\max}$ which can occur:

$$f_s = \frac{1}{T_s} \geq 2 \cdot K\tau_{\max} \quad \Rightarrow \quad \frac{T_c}{T_s} \geq 2 \cdot \Delta F\tau_{\max}. \quad (4.87)$$

Equation (4.85) demonstrates that the gathered sampled FMCW radar signal cannot be distinguished from a *Stepped Frequency Continuous Wave* (SFCW) radar signal [72]. Fig. 4.8 shows the SFCW principle. The instan-

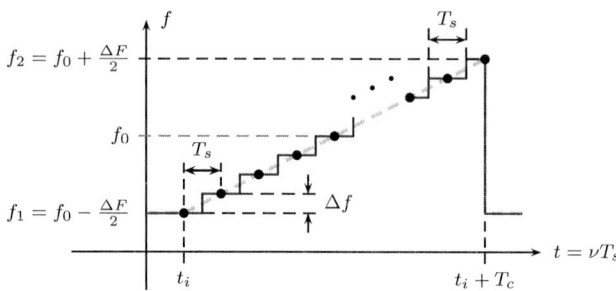

Fig. 4.8 Stepped frequency modulation compared to a sampled linear frequency modulation.

taneous transmit frequency $f_{Tx}(\nu)$ is increased by discrete steps here [59]. The resulting phase is in this case:

$$f_{Tx}(\nu) = f_1 + \nu\,\Delta f \quad,\ t \in \left[t_i + \nu\,T_s\ ;\ t_i + (\nu+1)\,T_s\right[, \qquad (4.88)$$

$$\varphi_{Tx}(t) = 2\pi\,(f_1 + \nu\,\Delta f)\,t + \varphi_{\nu-1} \quad, \qquad (4.89)$$

with $\varphi_{\nu-1} = 2\pi\,T_s\,[\nu\,f_1 + (\nu-1)\,\Delta f] =$

$$= 2\pi\,T_s\,\left[\nu\left(f_0 - \frac{\Delta F}{2}\right) + (\nu-1)\,\Delta f\right] . \qquad (4.90)$$

Equation (4.88) corresponds principally to equation (4.27) but in time discrete manner. If this signal and the received echoes serve as inputs for a mixer, the output phases, hence difference phases will be equal to equation (4.85).

4.4.5 Different ramp signals

At the mixer output, the distance and the *Doppler* frequency cannot be separated for each object. A remedy for this can be a different tuning slope containing an up- and a down-chirp (triangular modulation scheme, duration $2\,T_c$). For the up-chirp K has got a positive sign, whereas it becomes negative for a down-chirp. The *Doppler* contribution f_d however, stays approximately the same in the short time slot [46]:

$$\begin{aligned} f_q(t) &= +f_{r,q}(t) + f_{d,q} \quad, \\ f_q(t + T_c) &= -f_{r,q}(t + T_c) + f_{d,q} \quad, \\ \Rightarrow\ f_q(t + T_c) &= +f_{r,q}(t + T_c) - f_{d,q} \quad. \end{aligned} \qquad (4.91)$$

The third line is due to the fact, that $f_{r,q}$ is usually larger than $f_{d,q}$, but only positive frequencies can be measured at the mixer output. By simple arithmetic, the *Doppler* and the distance frequency can be separated for a single object [19], [46], [68]:

$$f_{r,q}(t) \approx f_{r,q}(t + T_c) = \frac{1}{2} |f_q(t) + f_q(t + T_c)| \quad ,$$

$$f_{d,q} = \frac{1}{2} |f_q(t) - f_q(t + T_c)| \quad .$$

(4.92)

The combination between distance and *Doppler* frequency is unique for each object due to its distance and different velocity. However, in a multi-target scenario, very quickly the appointment becomes ambiguous [46]. Here, a series of up- and down-chirps with different slopes can help to gain independent information on the targets: The measured mixer output frequencies $f_{m,q}(t)$ can be displayed as straight lines in a velocity over range plot with their individual slope [19], [46]. The targets can be identified by the points of highest number of intersections in this plot. Practical analysis has shown that already eight pairs of up- and down-chirps with an additional continuous wave transmission (flat ramp) provides sufficient results for detection in e. g. an automotive traffic scenario [19].

Alternatively, there are methods with so-called *intertwined ramps*, which principally combines *frequency shift keying* (FSK) with *linear frequency modulation* (LFM) [68], [69]. Here, range and velocity can be derived from an up- or down-chirp by evaluating phase and frequency of the mixer output signal [68], [69].

4.5 Sidelobe reduction in range compression

After range compression, the *point-spread function* or more explicitly $\chi(\tau_{0q}, 0)$ not only exhibits a single mainlobe but also a pair (or even several couples) of sidelobes over compressed range, hence over frequency in FMCW radars [47],[56], [59], [67]. There are two possibilities to reduce these sidelobes [47],[56], [59], [67]: *Mismatched filters* or nonlinear frequency modulation NLFM [47],[56], [59], [67]. Nevertheless, *mismatched filters* result in a reduction of sidelobes, on condition that the mainlobe broadens (and even sinks in level) [47],[56], [59], [67].

In case of pulse compression with *matched filters*, an additional filter applies amplitude weights onto the signal by multiplication in time domain, hence convolution in spectral domain [43], [47], [59], [67]. The overall filter response is then maladapted to the original signal, which gave the additional filter the title *mismatched filter* [47], [67].

In case of (NLFM), the signal's amplitude envelope stays rectangular, whereas the *matched filter* response exhibits a windowed behaviour due to the nonlinear phase response of the filter [47], [59], [67].

In terms of window shape, the optimal weighting would be a *Dolph-Chebyshev filter/weighting/window* [47], [56], [59] which however cannot be implemented properly [47], [56], [59]. Feasible approximations for implementations are *Taylor windows* [47], [56], [59] or *Hamming windows* [56], [67]. Hereby, the *Taylor window* gives the minimum mainlobe broadening to a required sidelobe suppression (e. g. 40 dB) [47], [56], [59], [67].

Another type, the *Kaiser window*, features the advantage of defining the stopband attenuation by a single shape parameter [56]. For details please see also appendix C.3.

4.5.1 Windowing and mismatched filters

Since the FMCW technique is equivalent to a *matched filter* response of a LFM signal, however performed in two steps (section 4.4.3), this bears the possibility to apply amplitude weighting on the mixer output signal before *Fourier* transformation [47], [56], [67]. The window function's *Fourier transform* is then convoluted with the target's spectrum in frequency domain by the rules of *Fourier transformation* [43], [44]. Due to processing simplicity, this will be the method of choice for the later resampling process (section 11.3.1).

4.5.2 Nonlinear frequency modulation

The nonlinear frequency modulation attenuates the absolute value of $\chi(\tau_{q,i})$ faster than a linear frequency modulation with larger differences from the target's $\tau_{q,i}$, which leads to a reduction of sidelobe level [59], [67]. The increased relation between mainlobe and sidelobes in turn leads to smooth amplitudes in time domain. Therefore, an *amplitude weighting* in time domain becomes obsolete by this method [47], [59], [67]. As already mentioned, a *Taylor window* provides the minimum mainlobe broadening to a required sidelobe suppression, closely followed by a *Hamming weighting* [47], [56], [59], [67].

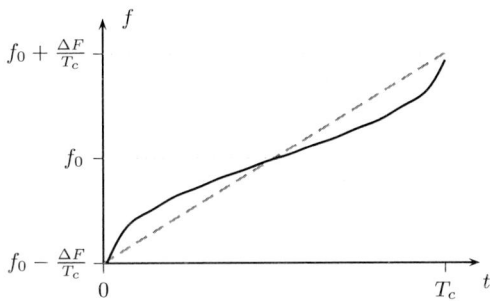

Fig. 4.9 Nonlinear modulation scheme for Taylor window effect with 40 dB sidelobe suppression [67], [59].

The linear modulation from equation (4.26) must be adapted to [47], [59]:

$$f(t) = f_0 + \Delta F \cdot \left[\frac{t}{T_c} + \sum_{\nu} \gamma_\nu \cdot \sin\left(2\pi \nu \frac{t}{T_c}\right) \right] ,$$

$$\text{for: } t_i - \frac{T_c}{2} \leq t < t_i + \frac{T_c}{2} . \tag{4.93}$$

For both types with approximately 40 dB sidelobe suppression, the coefficients γ_ν are listed in Tab. 4.1, 4.2. There is also the possibility to

γ_1	γ_2	γ_3	γ_4	γ_5	γ_6	γ_7
-0.1145	0.0396	-0.0202	0.0118	-0.0082	0.0055	-0.0040

Tab. 4.1 Coefficients for 40dB-*Taylor window* [47], [59].

γ_1	γ_2	γ_3	γ_4	γ_5	γ_6	γ_7	γ_8
-0.1145	0.0389	-0.0169	0.0084	-0.0044	0.0024	-0.0014	0.0008

Tab. 4.2 Coefficients for 40dB-*Hamming window* [47], [67].

increase the number of coefficients [73]. Analogously to equation (4.28), the transmit phase is then:

$$\varphi_{Tx}(t) = 2\pi \left[f_0 t + \frac{K}{2} t^2 - TBP \sum_{\nu} \frac{\gamma_\nu}{2\pi \cdot \nu} \cos\left(2\pi \nu \frac{t}{T_c}\right) \right] + \varphi_0 \,, \quad (4.94)$$

with $TBP = \Delta F T_c$ time-bandwidth product.

With the help of trigonometric addition theorems, the phase difference between transmit and received phase at the mixer output could then look like:

$$\varphi_{mx,i}(t) = \varphi_{Tn}(t) - \varphi_{Rm}(t) =$$

$$= 2\pi \left\{ (Kt + f_0)\, \tau_{q,i} - \frac{K}{2}\, \tau_{q,i}^2 + \right.$$

$$\left. + TBP \sum_{\nu} \frac{\gamma_\nu}{2\pi \cdot \nu} \left[\cos\left(2\pi \nu \frac{t - \tau_{q,i}}{T_c}\right) - \cos\left(2\pi \nu \frac{t}{T_c}\right) \right] \right\} =$$

$$= \cdots = 2\pi \left\{ (Kt + f_0)\, \tau_{q,i} - \frac{K}{2}\, \tau_{q,i}^2 + \right.$$

$$+ TBP \sum_{\nu} \frac{\gamma_\nu}{2\pi \cdot \nu} \left\{ \left[\cos\left(2\pi \nu \frac{\tau_{q,i}}{T_c}\right) - 1 \right] \cos\left(2\pi \nu \frac{t}{T_c}\right) + \right.$$

$$\left. \left. + \sin\left(2\pi \nu \frac{\tau_{q,i}}{T_c}\right) \sin\left(2\pi \nu \frac{t}{T_c}\right) \right\} \right\} \,. \quad (4.95)$$

This can be simplified further for small $\tau_{q,i}$, hence $\cos\left(2\pi \nu \frac{t}{T_c}\right) \approx 1$, $\sin\left(2\pi \nu \frac{t}{T_c}\right) \approx \left(2\pi \nu \frac{t}{T_c}\right)$, $\tau_{q,i}^2 \approx 0$:

$$\varphi_{mx,i}(t) \approx 2\pi \left[(Kt + f_0)\, \tau_{q,i} + \Delta F \cdot \tau_{q,i} \sum_{\nu} \gamma_\nu \cdot \sin\left(2\pi \nu \frac{t}{T_c}\right) \right] \,. \quad (4.96)$$

The ambiguity function is then:

$$\chi(\tau_{q,i}) = \int\limits_{-\infty}^{+\infty} \mathcal{U}(t) \cdot \mathcal{U}^*(t - \tau_{q,i}) \cdot e^{+j\, 2\pi \left[Kt + \Delta F \sum_{\nu} \gamma_\nu \cdot \sin\left(2\pi \nu \frac{t}{T_c}\right) \right] \tau_{q,i}} \, dt \,. \quad (4.97)$$

With respect to the FMCW method, the standard *Fourier* transformation of the mixer output signal is not feasible anymore. The basis functions for an integral transformation need to be altered from $\exp\left(j\,\omega t\right) = \exp\left(j\,2\pi K\tau t\right)$ to:

$$\mathrm{e}^{+j\,2\pi\left[Kt+\Delta F\sum_\nu\,\gamma_\nu\cdot\sin\left(2\pi\,\nu\frac{t}{T_c}\right)\right]\tau}\ . \tag{4.98}$$

Please note, that the coefficients in Tab. 4.1, 4.2 are designed for a phase function. They do not necessarily match the coefficients for amplitude *Taylor* or *Hamming windows* appendix C.3.1, [56], [59].

Intensive academic work has been performed on the field of nonlinear frequency modulation. Adaptations or modifications of this section can also be found in adapted manner in [73], [74], [75], [76], [77], [78], [79], [80].

4.6 Doppler compensation

The effects of Doppler shift which are the kernel topic of motion compensated radar systems are covered in more detail in [81], [82], [83], [84]. General aspects on MIMO radar beamformer processing under Doppler shifts can be found in [85].

4.7 Constant false alarm rate algorithms

Constant false-alarm rate (CFAR) algorithms, in particular *ordered statistics* CFAR or OS-CFAR respectively, were applied to range compressed data in both implemented systems. However, those shall be not part of this thesis. Therefore, the reader shall be remitted to literature [67], [86] for the sake of completeness.

5 Angular signal processing

The signal modulation provides possibilities of range compression which was presented in section 4. This section in turn, shall cover the angular processing methods.

5.1 Array windows and spatial correlation

In section 3.2, the array weights $w_{st,i}$ were introduced as coefficients of a spatial *Fourier series* or respectively a spatial window function over the array put into a discrete (*inverse*) *Fourier transformation* (section 3.3, equations (3.25) (3.58), (3.59), (4.74), (4.77)). The choice of *Fourier transformation* or its inverse depends on the signal conventions: Sign of exponents of basis functions ($\exp(\pm j\omega t)$) or complex signal generation by *IQ-demodulation* or *Hilbert transformation*, which is the case here. Hence, a rectangular window (all $w_{st,i} = 1$) results in a si-function for the radiated and received fields u_{Tx}, u_{Rx}, Fig. 5.1 (due to *reciprocity theorem* [31], [32]). The antenna pattern shall be of rather large beamwidth, so that the overall spatial characteristic is dominated by the array factor (see section 10). Leaning on equation (3.24), the total array gain depends on the squared

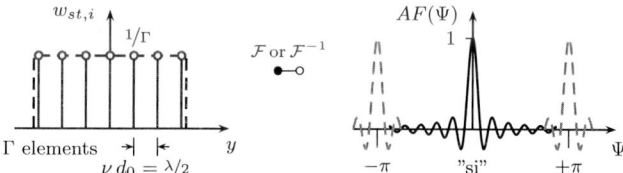

Fig. 5.1 Rectangular window over array elements results in a si-function for radiated/ received fields.

absolute value of the array factor AF, hence a product of AF and AF^*. In case of Fig. 5.1 the squared si-function is directly proportional to the radiated/ received power density over solid angle. Back in spatial domain, this leads to an auto-correlation along the array dimension which gives a triangular function then. The spatial auto-correlation is also depicted as *spatial sensivity function* [25]. Its discrete values represent the number of

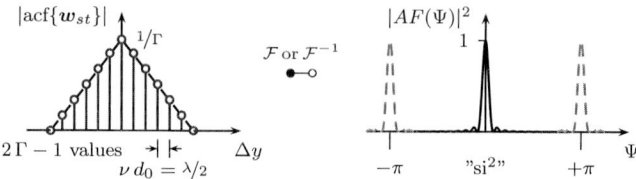

Fig. 5.2 Triangular auto-correlation over array elements results in a squared si-function for radiated/ received power.

combinatorial possibilities to form the corresponding interferometric couple, thus at position e. g. $\Delta y = \xi\, d_0$, the amount of signal combinations of elements which are separated by $\xi\, d_0$ [25].

5.2 Range dependent curvature filter

The signal $s(t)$ provided by radar with FMCW principle (equation (4.74)) still incorporates the curvature of phase fronts, collected in Λ_δ. This can be compensated by a range dependent filter. The range compressed signal $s_{rc}(f)$ from equation (4.86) is altered to:

$$s_{rc,i,p}(r) = s_{rc,i}(r) \cdot e^{-j\, 2k_0 \delta_i^2(r)} \,. \tag{5.1}$$

The term $\delta_i^2(r)$ has been introduced in equation (4.72). By the substitution:

$$f = K\tau = K\,\frac{2r}{c_0} \,, \text{ for monostatic radar,} \tag{5.2}$$

it can be easily switched between frequency and range representation. The output of equation (5.1) is a plane wave representation of $s_{rc}(r)$ in spectral domain. By use of *inverse Fourier transformation*, a plane wave representation of $s(t)$, denoted by $s_p(t)$ with $\Lambda_\delta \approx E$ (unity matrix) can be obtained for time domain. Please note, that the effects of Φ_q (e. g. range cell migration) will be compensated in section 11.

5.3 Analysis of FMCW-signal

The start signal shall be $\boldsymbol{s}_p(t) \in \mathbb{C}^{\Gamma \times 1}$, hence the plane wave representation of equation (4.74) (with all quadratic parasitic signal parts for the sake of completeness):

$$\boldsymbol{s}_p(t) = \sum_{q=1}^{Q} U_q \cdot \mathrm{e}^{+j\frac{4\pi}{c_0}(Kt+f_0)\cdot r_{0q}} \cdot \mathrm{e}^{+j\,2\pi f_{d,q}t}.$$

$$\cdot\, \mathrm{e}^{+j\,4\pi\frac{v_{r,q}}{c_0}t^2} \cdot \mathrm{e}^{-j\,\pi\frac{K}{c_0^2}r_{0q}^2} \cdot \boldsymbol{\Phi}_q\left(\vartheta_q, \varphi_q, \delta_i^2, t\right)\, \boldsymbol{a}_q\left(\vartheta_q,\, \varphi_q\right) = \qquad (5.3)$$

$$= \sum_{q=1}^{Q} s_q(r_{0q}, f_{d,q}, t) \cdot \boldsymbol{\Phi}_q\left(\vartheta_q, \varphi_q, \delta_i^2, t\right)\, \boldsymbol{a}_q\left(\vartheta_q,\, \varphi_q\right)\; .$$

Now, the following matrix shall be imposed:

$$\mathcal{E}\left\{\boldsymbol{s}_p\, \boldsymbol{s}_p^{\mathrm{H}}\right\}\; . \qquad (5.4)$$

Similar to section 3.6, it shall be assumed that the process is *stationary* and *ergodic*, thus the expected value can be determined by the mean over time [50]. Then the matrix in equation (5.5) can be represented by the dyadic product of the mean vectors $\boldsymbol{m}$ over time and a covariance matrix [48].

$$\mathcal{E}\left\{\boldsymbol{s}_p\, \boldsymbol{s}_p^{\mathrm{H}}\right\} = \boldsymbol{m}\, \boldsymbol{m}^{\mathrm{H}} + \mathcal{E}\left\{(\boldsymbol{s}_p - \boldsymbol{m})\, (\boldsymbol{s}_p - \boldsymbol{m})^{\mathrm{H}}\right\}\; . \qquad (5.5)$$

However, the signals in $\boldsymbol{s}_p(t)$ were designed to zero mean. Anyway, the illumination signal has been radiated by an antenna. Since source currents of $f = 0$ do not radiate [31], [32] and the radar signals are of sinusoidal shape, all entries of $\boldsymbol{s}_p(t)$ become automatically zero mean. Therefore the mean vector $\boldsymbol{m}$ is equal to zero vector and the covariance matrix remains:

$$\boldsymbol{R} = \mathcal{E}\left\{\boldsymbol{s}_p\, \boldsymbol{s}_p^{\mathrm{H}}\right\}\; . \qquad (5.6)$$

5.3.1 The covariance matrix of general FMCW signal

If equation (5.5) is written explicitly, Q^2 signal combinations must occur. Herein, Q quadratic terms and $Q(Q-1)$ mixed terms:

$$
\boldsymbol{R} = \mathcal{E}\left\{ \sum_{q=1}^{Q} \left| s_q(r_{0q}, f_{d,q}, t) \right|^2 \cdot \boldsymbol{\Phi}_q\, \boldsymbol{a}_q \boldsymbol{a}_q^H\, \boldsymbol{\Phi}_q^H \right\} +
$$
$$
+ \mathcal{E}\left\{ \sum_{\nu \neq \xi}^{Q(Q-1)} s_\nu(r_{0\nu}, f_{d,\nu}, t) \cdot s_\xi^*(r_{0\xi}, f_{d,\xi}, t) \cdot \boldsymbol{\Phi}_\nu\, \boldsymbol{a}_\nu \boldsymbol{a}_\xi^H\, \boldsymbol{\Phi}_\xi^H \right\} . \tag{5.7}
$$

Therefore, each entry in equation (5.7) (at position α, β in $\boldsymbol{R}$) can be found by:

$$
r_{\alpha\beta} = \frac{1}{NM} \sum_{\nu,\xi}^{Q^2} e^{+j \frac{4\pi}{c_0} f_0(r_{0\nu} - r_{0\xi})} \cdot e^{-j\pi \frac{K}{c_0^2}\left(r_{0\nu}^2 - r_{0\xi}^2\right)} .
$$
$$
\cdot e^{-j \frac{4\pi}{c_0} f_0 [\boldsymbol{d}_\alpha \cdot \boldsymbol{e}_r(\vartheta_\nu, \varphi_\nu) - \boldsymbol{d}_\beta \cdot \boldsymbol{e}_r(\vartheta_\xi, \varphi_\xi)]} .
$$
$$
\cdot \lim_{T_c \to +\infty} \frac{1}{T_c} \int_{T_c} U_\nu U_\xi^* \cdot e^{+j \frac{4\pi}{c_0} K t(r_{0\nu} - r_{0\xi})} \cdot e^{+j\, 2\pi(f_{d,\nu} - f_{d,\xi})\, t} . \tag{5.8}
$$
$$
\cdot e^{-j \frac{4\pi}{c_0} K t [\boldsymbol{d}_\alpha \cdot \boldsymbol{e}_r(\vartheta_\nu, \varphi_\nu) - \boldsymbol{d}_\beta \cdot \boldsymbol{e}_r(\vartheta_\xi, \varphi_\xi)]} .
$$
$$
\cdot e^{+j \frac{4\pi}{c_0} K t \left[\delta_\alpha^2(r_{0,\nu}) - \delta_\beta^2(r_{0,\xi})\right]} \cdot e^{+j \frac{4\pi}{c_0} K (v_{r,\nu} - v_{r,\xi})\, t^2}\, dt .
$$

Generally, exponential functions bear orthogonality [50], [51], [53], [56]:

$$
\lim_{T_c \to +\infty} \frac{1}{T_c} \int_{T_c} e^{+j(\nu - \xi)\omega\, t}\, dt = \delta_{\nu\xi} = \begin{cases} 1 & \text{, for } \nu = \xi \\ 0 & \text{, for } \nu \neq \xi \end{cases} \tag{5.9}
$$

$$
\text{with } \nu,\, \xi \in \mathbb{Z} .
$$

Due to this orthogonality, the integral in equation (5.8), which is a functional in terms of mathematics [43], declines very rapidly for $(r_{0\nu} \neq r_{0\xi})$, $[\boldsymbol{e}_r(\vartheta_\nu, \varphi_\nu) \neq \boldsymbol{e}_r(\vartheta_\xi, \varphi_\xi)]$ or $(f_{d,\nu} \neq f_{d,\xi})$. Especially, several targets in same range cell but at different azimuth bins can be separated by equation (5.8), hence all mixed terms with $\nu \neq \xi$ appear close to zero. Then, with knowledge from section 4.4.3 and $(r_{0\nu} = r_{0\xi})$, $[\boldsymbol{e}_r(\vartheta_\nu, \varphi_\nu) = \boldsymbol{e}_r(\vartheta_\xi, \varphi_\xi)]$ and $(f_{d,\nu} = f_{d,\xi})$, the integral in equation (5.8) can be identified as the squared absolute (normalized) *ambiguity function* broadened by convolution with a *Fresnel integral*, which in turn can be approximated by the pure *ambiguity function* for long chirp durations T_c. Only entries with $\nu = \xi = q$ give:

$$r_{\alpha\beta} = \frac{1}{NM} \sum_{q=1}^{Q} |\varsigma_q|^2 \cdot e^{-j \frac{4\pi}{c_0} f_0 [\boldsymbol{d}_\alpha \cdot \boldsymbol{e}_r(\vartheta_\nu, \varphi_\nu) - \boldsymbol{d}_\beta \cdot \boldsymbol{e}_r(\vartheta_\xi, \varphi_\xi)]}.$$
$$\cdot \left| \chi \left[r = (r_{0\nu} - r_{0\xi}), \, f_d = (f_{d,\nu} - f_{d,\xi}) \right] \right|^2 .$$
(5.10)

Therefore, all mixed terms in equation (5.7), thus $\nu \neq \xi$, are of very low amplitude, wherefore they shall be represented by zero in the following. This is often denoted as *uncorrelated signal scenario* in this case [45], [87], [88]. In case of normalised *ambiguity function*, $|\chi(0,0)| = 1$ is valid. The covariance matrix from equation (5.5) can then be written as:

$$\boldsymbol{R} = \mathcal{E} \left\{ \boldsymbol{s}_p \, \boldsymbol{s}_p^{\mathrm{H}} \right\} = \sum_{q=1}^{Q} |\varsigma_q|^2 \cdot \boldsymbol{a}_q(\Psi_q) \, \boldsymbol{a}_q^{\mathrm{H}}(\Psi_q) =$$
$$= \boldsymbol{A}_q \, \mathrm{diag}\{\ldots, |\varsigma_q|^2, \ldots\} \, \boldsymbol{A}_q^{\mathrm{H}} \quad \text{"uncorrelated" case.}$$
(5.11)

Hereby, the directional vectors $\boldsymbol{a}_q$ serve as columns in the *Vandermonde* matrix $\boldsymbol{A}_q$ [45], [48]. The matrix $\boldsymbol{R}$ is of rank Q [45]. The matrix $\boldsymbol{R}$ is of *Toeplitz* type, thus *persymmetric* or *orthosymmetric* [48], [89]. In addition, $\boldsymbol{R}$ is of rank $Q \leq \Gamma$, thus there are $Q \leq \Gamma$ significant eigenvalues with their eigenvectors spanning the complete signal subspace $\mathcal{S}$ [45]. In case of additive noise contribution, the signal $\boldsymbol{s}_n$ can be described by:

$$\boldsymbol{s}_n(t) = \boldsymbol{s}_p(t) + \boldsymbol{n}(t) \quad . \tag{5.12}$$

In general, the covariance matrix of $\boldsymbol{s}_n(t)$ is the weighted sum of dyadic products of directional vectors and a noise covariance matrix [48].

$$\boldsymbol{R}_n = \mathcal{E} \left\{ \boldsymbol{s}_n \, \boldsymbol{s}_n^{\mathrm{H}} \right\} = \sum_{q=1}^{Q} |\varsigma_q|^2 \cdot \boldsymbol{a}_q(\Psi_q) \, \boldsymbol{a}_q^{\mathrm{H}}(\Psi_q) + \mathcal{E} \left\{ \boldsymbol{n} \, \boldsymbol{n}^{\mathrm{H}} \right\} . \tag{5.13}$$

In case of uncorrelated zero-mean noise contributions between each array element (variances $\sigma_1^2, \ldots, \sigma_\Gamma^2$), $\boldsymbol{R}$ gets diagonally biased or loaded [89], [48]:

$$\boldsymbol{R}_n = \boldsymbol{R} + \mathrm{diag} \left\{ \sigma_1^2, \ldots, \sigma_\Gamma^2 \right\} . \tag{5.14}$$

Since the array consists of discrete elements, the entries of $\boldsymbol{R}$ alter in discrete phase shifts. The diagonal load usually prevents $\boldsymbol{R}$ of implemented systems of becoming *singular*, which features $\boldsymbol{R}_n$ to be of full rank Γ (which is the number of sensors), hence $\boldsymbol{R}$ is invertible [89]. For $\sigma_1^2 = \ldots = \sigma_\Gamma^2 = \sigma_n^2$, $\boldsymbol{R}_n$ is also consequently *persymmetric*. Further, the evolution from $\boldsymbol{R}$ to $\boldsymbol{R}_n$ by equation (5.14) is identical to a so-called *eigenvalue shift* [90].

The representation of covariance values in equation (5.10) not only corresponds to an *uncorrelated signal scenery* but is also supposed to be a natural *Carathéodory representation*, thus all entries can be described by the autocorrelation values parameterised by lags of $(\xi - \eta)$ [45], [87], [88]:

$$r_{\alpha\beta}\left(\alpha - \beta\right) = \sum_{q=1}^{Q} |\varsigma_q|^2 \, e^{-j\,\Psi_q\,(\alpha-\beta)} + \sigma_\alpha^2\,\delta_{\alpha\beta} \, , \quad (\alpha - \beta) \in \mathbb{Z} \, . \quad (5.15)$$

Please note, that $\delta_{\alpha\beta}$ describes a *Kronecker delta function*, thus contributions only occur on the main diagonal where $\alpha = \beta$. The lags are integers and constant along a diagonal of $\boldsymbol{R}$, thus the *Carathéodory representation* leads to the mentioned *Toeplitz* structure. The *Carathéodory theorem* in context of *finite moments* is of special importance for *sparse arrays* such as for the *minimum redundancy arrays* of this thesis [45], [87], [88]. Nevertheless, this is only a feasible approach if equation (5.8) exhibits its fourth line, cancelling all mixed terms belonging to targets at same range but different angles. To achieve this, the phase term needs to run through more than 2π during one chirp:

$$\frac{4\pi}{c_0} \, KT_c \, [\boldsymbol{d}_\alpha \cdot \boldsymbol{e}_r\,(\vartheta_\nu, \varphi_\nu) - \boldsymbol{d}_\beta \cdot \boldsymbol{e}_r\,(\vartheta_\xi, \varphi_\xi)] \geq 2\pi \, ,$$
$$\Rightarrow [\boldsymbol{d}_\alpha \cdot \boldsymbol{e}_r\,(\vartheta_\nu, \varphi_\nu) - \boldsymbol{d}_\beta \cdot \boldsymbol{e}_r\,(\vartheta_\xi, \varphi_\xi)] \geq \frac{c_0}{2\,\Delta F} = \Delta r \, . \tag{5.16}$$

Equation (5.16) shows that this run through 2π is equal to range cell migration, hence the angular deviation causes a range difference larger than one range bin Δr (equation (4.24)).

5.3.2 Covariance matrix of resampled signal

In the later section 11.3, after a resampling process, another type of signal will be provided (equation (11.114)):

$$
s_p(t) = \sum_{q=1}^{Q} U_q \cdot e^{+j \frac{4\pi}{c_0} K t \cdot r_{0q}} \cdot e^{-j \pi \frac{K}{c_0^2} \left(r_{0q}^2 - r_{01}^2 \right)} \cdot
$$
$$
\cdot \operatorname{diag} \left\{ \dots, \delta u_i(t) \, e^{+j 2 \pi \delta \tilde{\varepsilon}_i(t) \, \Delta \tau_{q1}}, \dots \right\} \boldsymbol{a}_q \, .
$$
(5.17)

Hereby, the diagonal matrix in equation (5.17) contains slight phase shifts due to the statistical component of frequency deviation or incorrect curvature correction per channel i, denoted by $\delta \tilde{\varepsilon}_i(t)$ (see section 11.3 for more details). U_q corresponds to the nominal envelope functions, whereas the $\delta u_i(t)$ terms correspond to the statistical amplitude deviations. Nevertheless, the absolute values $\delta u_i(t)$ are the same for all r_{0q}. A single entry in $\boldsymbol{R}$ is then analogously to equation (5.8):

$$
r_{\alpha\beta} = \frac{1}{NM} \sum_{\nu, \xi}^{Q^2} e^{-j \pi \frac{K}{c_0^2} \left(r_{0\nu}^2 - r_{0\xi}^2 \right)} \cdot e^{-j \frac{4\pi}{c_0} f_0 \, [\boldsymbol{d}_\alpha \cdot \boldsymbol{e}_r(\vartheta_\nu, \varphi_\nu) - \boldsymbol{d}_\beta \cdot \boldsymbol{e}_r(\vartheta_\xi, \varphi_\xi)]} \cdot
$$
$$
\cdot \lim_{T_c \to +\infty} \frac{1}{T_c} \int_{T_c} U_\nu U_\xi^* \cdot \delta u_\alpha(t) \, \delta u_\beta^*(t) \cdot e^{+j \frac{4\pi}{c_0} K t (r_{0\nu} - r_{0\xi})} \cdot
$$
$$
\cdot e^{+j \, 2\pi [\delta \tilde{\varepsilon}_\alpha(t) - \delta \tilde{\varepsilon}_\beta(t)] \, (r_{0\nu} - r_{0\xi})} \, dt \, .
$$
(5.18)

Especially for targets in the same range bin, the mixed combinations, hence $\boldsymbol{a}_\nu \neq \boldsymbol{a}_\xi$ for $\nu \neq \xi$, do not vanish anymore. This is often denoted as *coherent signal scenario* in this case [45]. Nevertheless, the phase deviation terms vanish since the integral's value in equation (5.18) usually declines very rapidly for $r_{0\nu} \neq r_{0\xi}$. The amplitude deviations $\delta u_i(t)$ in turn can be represented by the sum of their mean $\delta u_{i,m}$ and a zero-mean component. All mixed terms with means and the zero-mean component decline with the integral. Therefore, the mean values $\delta u_{\alpha,m}$, $\delta u_{\beta,m}$ remain.

$$
r_{\alpha\beta} = \frac{\delta u_{\alpha,m} \cdot \delta u_{\beta,m}^*}{NM} \sum_{\nu, \xi}^{Q^2} \varsigma_\nu \varsigma_\xi^* \cdot e^{-j \frac{4\pi}{c_0} f_0 \, [\boldsymbol{d}_\alpha \cdot \boldsymbol{e}_r(\vartheta_\nu, \varphi_\nu) - \boldsymbol{d}_\beta \cdot \boldsymbol{e}_r(\vartheta_\xi, \varphi_\xi)]} \cdot
$$
$$
\cdot \left| \chi \left(r = r_{0\nu} - r_{0\xi}, 0 \right) \right|^2 \, .
$$
(5.19)

The covariance matrix is then:

$$R = \mathrm{diag}\left\{\ldots, \delta u_{i,m}, \ldots\right\} A_q \, \varsigma \, \varsigma^{\mathrm{H}} \, A_q^{\mathrm{H}} \, \mathrm{diag}\left\{\ldots, \delta u_{i,m}, \ldots\right\} , \quad (5.20)$$

which corresponds to a *coherent signal scenario* [45], [87], [88], however with altered directional vectors. The general ansatz for a *coherent signal scenario* would be [45], [87], [88]:

$$R = A_q \, \varsigma \, \varsigma^{\mathrm{H}} \, A_q^{\mathrm{H}} \quad \text{"coherent" case.} \quad (5.21)$$

In general the *coherent signal scenario* can be recognised with help of equation (5.16) again, however with inverted logic. Therefore, the phase in fourth line of equation (5.8) must not run larger phase differences than 2π. This in turn, would stand for a scenario with range cell migration.

$$\left[d_\alpha \cdot e_r\left(\vartheta_\nu, \varphi_\nu\right) - d_\beta \cdot e_r\left(\vartheta_\xi, \varphi_\xi\right)\right] \geq \frac{c_0}{2\,\Delta F} = \Delta r . \quad (5.22)$$

The reflection coefficients are aligned in the column vector ς. A closer look on equation (5.19), (5.21) however, discloses that the mixed terms $\nu \neq \xi$ do not vanish anymore. This is typical for the *correlated* or *coherent signal scene* [45], [87], [88]. The covariance entries do not exhibit a *natural Carathéodory representation* as in equation (5.15) [45], [87], [88], which will have influence on the signal processing for the following *minimum redundancy arrays* (please find in section 9). The covariance entries are then:

$$r_{\alpha\beta} = \sum_{q=1}^{Q} |\varsigma_q|^2 \, e^{-j\,\Psi_q\,(\alpha-\beta)} + \sum_{\xi \neq \eta}^{Q(Q-1)} \varsigma_\xi \, \varsigma_\eta^* \, e^{-j\,(\Psi_\xi\,\alpha - \Psi_\eta\beta)} + \sigma_\alpha^2 \, \delta_{\alpha\beta} . \quad (5.23)$$

Therefore, the covariance matrix R is not necessarily *Toeplitz* anymore [45]. R is of rank one due to the creation out of a dyadic product of ς [45]. After an eigenvalue decomposition, only one dominant eigenvalue can be expected. Its corresponding eigenvector represents the sum of all significant directional vectors a_q weighted by individual ς_q [45]. Please note, that the properties of a *coherent signal scene* will not only be valid for covariance matrix of resampled signals but also for the dyadic products over vectors of maxima in range compressed MIMO channels. Nevertheless, this section discloses that the classical augmentation techniques as usual for minimum redundancy arrays, fail here. Possible remedies could be *compressed sensing techniques* (see section 9) or methods like cyclic shifting or spatial smoothing (see end of this section).

5.3.3 Eigenanalysis on covariance matrix

No matter if *uncorrelated* or *coherent* signal scenario, $\boldsymbol{R}$ and $\boldsymbol{R}_n$ are *hermitian*, thus $\boldsymbol{R}^{\mathrm{H}} = \boldsymbol{R}$, $\boldsymbol{R}_n^{\mathrm{H}} = \boldsymbol{R}_n$ [89]. This yields the following properties [43], [48], [70]:

$$\boldsymbol{x}^{\mathrm{H}}\, \boldsymbol{R}_{(n)}\, \boldsymbol{x} \in \mathbb{R} \quad \forall\, \boldsymbol{x} \in \mathbb{C}^{\Gamma \times 1}\,, \tag{5.24}$$

$$\mathrm{eig}\left\{ \boldsymbol{R}_{(n)} \right\} \in \mathbb{R}\,, \tag{5.25}$$

$$\boldsymbol{a}_\nu^{\mathrm{H}}\, \boldsymbol{a}_\xi = 0 \quad \nu \neq \xi \quad\text{, for } \boldsymbol{a}_\nu,\ \boldsymbol{a}_\xi \text{ as eigenvectors.} \tag{5.26}$$

Equation (5.24) is principally the output power of the conventional beamformer (see section 5.4.1), equation (5.25) states that all eigenvalues are real. Section 5.4.1 will further show, that $\boldsymbol{R}$ is *semi-positive definite*, $\boldsymbol{R}_n$ *positive-definite* [89]. With the *hermitian* property, this leads to [48]:

$$\boldsymbol{x}^{\mathrm{H}}\, \boldsymbol{R}\, \boldsymbol{x} \geq 0\,, \boldsymbol{x}^{\mathrm{H}}\, \boldsymbol{R}_n\, \boldsymbol{x} > 0 \quad \forall\, \boldsymbol{x} \in \mathbb{C}^{\Gamma \times 1}\,, \tag{5.27}$$

$$\mathrm{eig}\left\{ \frac{\boldsymbol{R} + \boldsymbol{R}^{\mathrm{H}}}{2} \right\} = \mathrm{eig}\left\{ \boldsymbol{R} \right\} \geq 0\,, \tag{5.28}$$

$$\mathrm{eig}\left\{ \frac{\boldsymbol{R}_n + \boldsymbol{R}_n^{\mathrm{H}}}{2} \right\} = \mathrm{eig}\left\{ \boldsymbol{R}_n \right\} > 0\quad. \tag{5.29}$$

Equation (5.13) or (5.14) demonstrate that the directional vectors $\boldsymbol{a}_q$ span the so-called *signal (sub-)space* $\mathcal{S}$ of dimension Γ_s in *uncorrelated* case and of dimension one in *coherent* case [45], whereas the additive noise spans the *noise (sub-)space* $\mathcal{N}$ of dimension Γ_n [45]. Both *sub-spaces* are orthogonal to each other and form a *Hilbert space* of dimension $\Gamma = \Gamma_s + \Gamma_n$ [43], [45]. Now an *eigenvalue decomposition* shall be performed [43], [70], [90].

Uncorrelated case: For the *uncorrelated* signal scenario, the noise-free covariance matrix $\boldsymbol{R}$ can be of maximum rank Γ_s [45]: In case of $Q > \Gamma$ several targets can be found in same direction but different range bins. Therefore, $\boldsymbol{R}$ can exhibit Γ_s real and positive eigenvalues γ_i with an *algebraic* and *geometric multiplicity* of one (compare [43], [45], [70], [90]). In [89] it was stated, that $\boldsymbol{R}_n$ can be considered to be *regular* and *invertible*, thus of full rank Γ. Due to the *eigenvalue shift* [90] of equation (5.14), real and positive eigenvectors can be obtained:

$$\mu_i = \begin{cases} \gamma_i + \sigma_i^2, & i \in [1, \Gamma_s]\ \text{, for } \textit{signal sub-space}\,\mathcal{S}\,, \\ \sigma_i^2, & i \in [\Gamma_s + 1, \Gamma]\ \text{, for } \textit{noise sub-space}\,\mathcal{N}\,. \end{cases} \tag{5.30}$$

In case of $\sigma_i^2 = \sigma^2$ for all $i \in [1, \Gamma]$, the noise sub-space is characterised by one eigenvalue of *algebraic multiplicity* Γ_s. In general, a matrix of full rank Γ with Γ linear independent eigenvectors can be diagonalised [45], [90].

Since $\boldsymbol{R}$ and $\boldsymbol{R}_n$ are *hermitian*, the eigenvectors are always orthogonal to each other [45], [48]. With matrix $\boldsymbol{A}_e$ where the eigenvectors $\boldsymbol{a}_{e,i}$ serve as columns, the eigenvalue decomposition can be written as [43], [45], [48]:

$$\boldsymbol{R}_n = \boldsymbol{R} + \mathrm{diag}\left\{\sigma_1^2, \ldots, \sigma_\Gamma^2\right\} =$$

$$= \sum_{i=1}^{\Gamma} \mu_i \cdot \boldsymbol{a}_e(\Psi_q)\, \boldsymbol{a}_e^{\mathrm{H}}(\Psi_q) = \boldsymbol{A}_e\, \boldsymbol{\Lambda}\, \boldsymbol{A}_e^{\mathrm{H}} \,, \qquad (5.31)$$

$$\text{with } \boldsymbol{\Lambda} = \mathrm{diag}\left\{\mu_1, \ldots, \mu_\Gamma\right\} \,.$$

If the eigenvectors are normalised to length one, they form an orthonormal base, accumulated in matrix $\boldsymbol{A}_e$, where they serve as columns. $\boldsymbol{A}_e$ is automatically *unitary*, thus $\boldsymbol{A}_e^{-1} = \boldsymbol{A}_e^{\mathrm{H}}$ [43]. Now the eigenvectors of S, N are applied explicitly on $\boldsymbol{R}_n$:

$$\boldsymbol{R}_n\, \boldsymbol{a}_{e,i} = \left(\gamma_i + \sigma_i^2\right) \boldsymbol{a}_{e,i} \,, \quad \text{for } \boldsymbol{a}_{e,i} \in \mathcal{S} \,, \qquad (5.32)$$

$$\boldsymbol{R}_n\, \boldsymbol{a}_{e,i} = \sigma_i^2\, \boldsymbol{a}_{e,i} \,, \quad \text{for } \boldsymbol{a}_{e,i} \in \mathcal{N} \,, \qquad (5.33)$$

$$\Rightarrow \boldsymbol{R}\, \boldsymbol{a}_{e,i} = \boldsymbol{0} \,, \text{ for } \boldsymbol{a}_{e,i} \in \mathcal{N} \,. \qquad (5.34)$$

The equations (5.33), (5.34) disclose that the eigenvectors spanning the *noise sub-space* $\mathcal{N}$ are orthogonal to $\boldsymbol{R}$, thus to *signal sub-space* $\mathcal{S}$ [45].

Coherent case: In case of *coherent* signal scenario, however, $\boldsymbol{R}$ can be represented by a *dyadic product* (equation (5.20)), which sets the rank of $\boldsymbol{R}$ automatically to one [45]. Then there is only one dominant eigenvalue left, describing the signal space $\mathcal{S}$ with one eigenvector. $\boldsymbol{R}$ becomes *singular* then. $\boldsymbol{R}$ and $\boldsymbol{R}_n$ are still *hermitian* but not necessarily *Toeplitz* anymore [45].

5.3.4 Covariance matrix in filter design

The covariance matrix is a well-known in stochastic models such as *auto-regressive* (AR), *moving-average* (MA), *auto-regressive moving-average* (ARMA) as e.g. *Yule-Walker equations* [89], [91]. In case of linear prediction/ filtering/ smoothing the solution of *Wiener-Hopf equations* exploits covariance matrices [89], [91].

5.4 Angular processing methods

The listed processing methods are just a spot on a variety of algorithms. Some well-known concepts are explained briefly in appendix of this thesis. For those and others, the reader shall be guided to literature such as [45], [89], [92], [93] and many more.

5.4.1 Conventional beamformer

This method consists of sequential dot-products (inner product) of unknown incoming directional vectors a_q with the corresponding directional vector $a_\nu(\Psi_\nu)$ of a dictionary of test directions. The following quadratic form represents the correlation result, thus quasi the processing output energy as a function of test direction. The output power of the *conventional beamformer* is then [32], [45], [47], [89]:

$$P_{bf}(\Psi_\nu) = a_\nu^{\mathrm{H}} R_n a_\nu . \tag{5.35}$$

In case of a perfect match $a_q = a_\nu$, the output is the reflectivity plus a quasi noise variance $\sigma_{n,bf}^2$ providing maximum signal-to-noise ratio: $P_{bf}(\Psi_\nu) = |\varsigma_q|^2 + \sigma_{n,bf}^2$. By choosing a fine grid of directions ν, the output signal P_{bf} can be interpolated [45], [47]. Beamformer methods adapted to MIMO configurations can also be found in [9], [85].

5.4.2 Augmentation for sparse arrays

As mentioned in section 5.3.1 a *uncorrelated signal scenery* is also supposed to exhibit a natural *Carathéodory representation* (equation (5.15)) of covariance entries [45], [87], [88].

From equation (5.15), it can be derived, that all covariance entries can be described by auto-correlation values [45], [87], [88]. The *Carathéodory theorem* states for the *uncorrelated signal scenario*, that the covariance matrix of a sparse, non-equidistant array signal can be reconstructed if all auto-correlation lags appear at least once [45], [87], [88]. The reconstruction is exact for the absence of noise or for an equal noise variance over all entries [45], [87], [88]. Knowing this, the *Carathéodory theorem* in context of *finite moments* is of special importance for the *minimum redundancy arrays* of this thesis [45], [87], [88]. The more convenient representation of equation (5.15) is the *Toeplitz* structure of covariance matrices [45], [87], [88]:

$$
\boldsymbol{R} = \begin{bmatrix}
r(0) & r(1) & r(2) & \cdots & r(\Gamma_d) \\
r^*(1) & r(0) & r(1) & & \vdots \\
\vdots & & \ddots & & \vdots \\
\vdots & & & \ddots & r(1) \\
r^*(\Gamma_d) & & r^*(1) & & r(0)
\end{bmatrix} . \tag{5.36}
$$

The realisation in detail is demonstrated in section 6.3.3.

5.4.3 Cyclic methods

These methods are especially of interest in *coherent signal* case [45].

Cyclic shifting

The augmentation of a sparse covariance matrix is feasible for the *uncorrelated signal scenario* (section 5.3.1). However, this is not the case for the *correlated* or *coherent scenario* (section 5.3.2). An alternative method filling a *Toeplitz matrix* can be derived from *first order statistics* [45]. Hereby, it does not matter if the signals are of *correlated* or *uncorrelated* type [45]. This is especially feasible for range compressed signals. The signal vector within a range bin shall be $\boldsymbol{s}_{rc}(r_{0q})$, collecting all $s_{rc,i}$. By cyclic shifting, a *Toeplitz matrix* $\boldsymbol{W}(r_{0q})$ at range bin $r = r_{0q}$ can be imposed:

$$
\boldsymbol{W}(r_{0q}) = \begin{bmatrix}
s_{rc,1} & s^*_{rc,2} & s^*_{rc,3} & \cdots & s^*_{rc,\Gamma} \\
s_{rc,2} & s_{rc,1} & s^*_{rc,2} & \cdots & s^*_{rc,\Gamma-1} \\
s_{rc,3} & s_{rc,2} & s_{rc,1} & \cdots & s^*_{rc,\Gamma-2} \\
\vdots & \vdots & \vdots & \vdots & \vdots \\
s_{rc,\Gamma} & s_{rc,\Gamma-1} & s_{rc,\Gamma-2} & \cdots & s_{rc,1}
\end{bmatrix} \tag{5.37}
$$

The fundamental signal $\boldsymbol{s}_{rc}$ would be:

$$
\boldsymbol{s}_{rc}(r_{0q}) = \sum_{q=1}^{Q} \varsigma_q \, \boldsymbol{a}_q \, , \quad \varsigma_q \in \mathbb{R} \, , \, \notin \mathbb{C} \, . \tag{5.38}
$$

Similar to equation (5.11), the matrix from equation (5.37) can be decomposed into [45]:

$$W = A_q \operatorname{diag}\{\ldots, \varsigma_q, \ldots\} A_q^H , \quad \varsigma_q \in \mathbb{R} , \notin \mathbb{C} . \tag{5.39}$$

Nevertheless, this method bears the drawback that the decomposition would only work for real ς_q, not for complex coefficients which would be the general case. However, W is still *hermitian* and exhibits therefore real eigenvalues (compare to section 5.3.3) [48]. In case of complex coefficients ς_q, an *eigenvalue decomposition* would yield:

$$W = \ldots = A_e \operatorname{diag}\{\ldots, \mu_i, \ldots\} A_e^H , \quad \mu_i \in \mathbb{R} , \notin \mathbb{C} . \tag{5.40}$$

Hereby, A_e contains the eigenvectors as columns, μ_i are the real eigenvalues. Nevertheless, one drawback remains anyhow: If the array setup is sparse, complete diagonals of W will stay empty without entries. Therefore, the next method might be more feasible.

Adapted spatial smoothing

This method is a derivate of *spatial smoothing algorithm* [45] which creates a covariance matrix of higher ranks than one in a *coherent signal scenario* [45]. The fundamental idea is to use several array signals shifted by single element positions [45], similar to ESPRIT algorithm [45], [93]. This can either be achieved by selecting partial arrays out of a densely filled array such as a ULA, or by shifting a sparse array laterally as in SAR applications. Hereby, rotational invariance of directional vectors a_q are the fundament of the algorithm [45].

Now, there shall be L sub-arrays, enumerated by l, whose signals (e. g. range compressed, in one range cell at r_{0q}) shall be used for dyadic products [45].

$$
\begin{aligned}
R_l &= s_{rc,l}(r_{0q})\, s_{rc,l}(r_{0q})^H = \\
&= A_q\, C^{l-1} \varsigma \varsigma^H \left(C^{l-1}\right)^H A_q^H + \mathcal{E}\left\{n_l n_l^H\right\} .
\end{aligned}
\tag{5.41}
$$

The diagonal matrix C contains the phase terms necessary for each directional vector shifted [45]:

$$C = \operatorname{diag}\left\{\ldots, \mathrm{e}^{-j\,\Psi_q}, \ldots\right\} , \quad q \in [1; Q] . \tag{5.42}$$

Shifts by l element positions, power C by $(l - 1)$. The average over all dyadic products is then [45]:

$$\boldsymbol{R}_s = \frac{1}{L} \sum_{l=1}^{L} \boldsymbol{R}_l = \boldsymbol{A}_q \, \boldsymbol{R}_\varsigma \, \boldsymbol{A}_q^{\mathrm{H}} + \boldsymbol{N}_L =$$

$$= \boldsymbol{A}_q \left[\frac{1}{L} \sum_{l=1}^{L} \boldsymbol{C}^{l-1} \, \varsigma \varsigma^{\mathrm{H}} \left(\boldsymbol{C}^{l-1} \right)^{\mathrm{H}} \right] \boldsymbol{A}_q^{\mathrm{H}} + \boldsymbol{N}_L \; . \qquad (5.43)$$

The noise contribution $\boldsymbol{N}_L$ depends on the method how the lateral shifts are generated: In case of an array shifted mechanically, the same Tx and Rx combinations form the signal structure, thus the average over all noise contribution is again:

$$\boldsymbol{N}_L = \mathcal{E} \left\{ \boldsymbol{n}_l \boldsymbol{n}_l^{\mathrm{H}} \right\} \; . \qquad (5.44)$$

Whereas if a dense array is used and several sub-arrays are taken out, this will lead to different Tx-Rx combinations in case of a coherent MIMO radar. In this case, the average noise contribution will be in general:

$$\boldsymbol{N}_L = \frac{1}{L} \sum_{l=1}^{L} \mathcal{E} \left\{ \boldsymbol{n}_l \boldsymbol{n}_l^{\mathrm{H}} \right\} \; . \qquad (5.45)$$

A closer look on $\boldsymbol{R}_\varsigma$ and some resorting discloses the following decomposition [45]:

$$\boldsymbol{R}_\varsigma = \left[\frac{1}{L} \sum_{l=1}^{L} \boldsymbol{C}^{l-1} \, \varsigma \varsigma^{\mathrm{H}} \left(\boldsymbol{C}^{l-1} \right)^{\mathrm{H}} \right] =$$

$$= \frac{1}{L} \left[\varsigma, \; \boldsymbol{C}\varsigma, \; \boldsymbol{C}^2 \varsigma, \; \ldots \; \boldsymbol{C}^{L-1} \varsigma, \right] \begin{bmatrix} \varsigma^{\mathrm{H}} \\ (\boldsymbol{C}\varsigma)^{\mathrm{H}} \\ (\boldsymbol{C}^2 \varsigma)^{\mathrm{H}} \\ \vdots \\ (\boldsymbol{C}^{L-1} \varsigma)^{\mathrm{H}} \end{bmatrix} = \boldsymbol{D} \, \boldsymbol{D}^{\mathrm{H}} \; , \qquad (5.46)$$

with the matrix D as [45]:

$$D = \frac{1}{\sqrt{L}} \left[\varsigma, \quad C \varsigma, \quad C^2 \varsigma, \quad \ldots \quad C^{L-1} \varsigma, \right] =$$

$$= \frac{1}{\sqrt{L}} \operatorname{diag}\{\ldots, \varsigma_q, \ldots\} \begin{bmatrix} 1 & z_1 & z_1^2 & \cdots & z_1^{L-1} \\ 1 & z_2 & z_2^2 & \cdots & z_2^{L-1} \\ \vdots & \vdots & \vdots & & \vdots \\ 1 & z_Q & z_Q^2 & \cdots & z_Q^{L-1} \end{bmatrix} = \quad (5.47)$$

$$= \frac{1}{\sqrt{L}} \operatorname{diag}\{\ldots, \varsigma_q, \ldots\} \, A_{q,l}^{\mathrm{T}} \, .$$

The matrix $A_{q,l}^{\mathrm{T}}$ is a *Vandermonde matrix*. Each row in turn is a directional vector with L entries, since z_q can be written as [45]:

$$z_q = \mathrm{e}^{-j\,\Psi_q} \, . \quad (5.48)$$

For small L, $A_{q,l}^{\mathrm{T}} A_{q,l}$ (as kernel of R_ς) is *hermitian* but not *Toeplitz*. Therefore, R_ς is also *hermitian* but not *Toeplitz*:

$$R_\varsigma = D \, D^{\mathrm{H}} =$$

$$= \frac{1}{L} \operatorname{diag}\{\ldots, \varsigma_q, \ldots\} \, A_{q,l}^{\mathrm{T}} \, A_{q,l}^{*} \operatorname{diag}\{\ldots, \varsigma_q, \ldots\}^{\mathrm{H}} \, . \quad (5.49)$$

The same is then valid for $A_q \, R_\varsigma \, A_q^{\mathrm{H}}$, since A_q is again a *Vandermonde matrix*. However, for larger L, the entries in secondary diagonals of $A_{q,l}^{\mathrm{T}} A_{q,l}$ not only become more and more equal, but also close to zero. In this case $A_{q,l}^{\mathrm{T}} A_{q,l}$ alters to a unity matrix (E_Q is a unity matrix of dimension $Q \times Q$). Then R_ς becomes diagonal, $A_q \, R_\varsigma \, A_q^{\mathrm{H}}$ becomes *Toeplitz* respectively.

$$R_\varsigma \rightarrow \frac{1}{L} \operatorname{diag}\{\ldots, \varsigma_q, \ldots\} \, E_Q \operatorname{diag}\{\ldots, \varsigma_q, \ldots\}^{\mathrm{H}} =$$

$$= \frac{1}{L} \operatorname{diag}\{\ldots, |\varsigma_q|^2, \ldots\} \, , \text{ for } L \rightarrow +\infty \, . \quad (5.50)$$

This means that A_q in turn can also be sparsely filled. The sparse $A_q \, R_\varsigma \, A_q^{\mathrm{H}}$ can be filled by copying the few or even single entries per diagonal to all positions on the diagonal. Therefore this method can be applied to a coherent MIMO radar with minimum redundancy.

6 MIMO radar with minimum redundancy

In the following considerations, the array configurations will be regarded in a $\nu\, d_0 = \lambda_0/2$-grid in order to fulfill the *Nyquist criterion* over azimuth, hence avoiding *grating lobes* [32], as proposed in section 3.3.1. Since the array setups will be created by MIMO techniques, they must be treated as monostatic arrays, thus $\nu = 2$ and therefore, the minimum grid spacing is $d_0 = \lambda_0/4$ as fundamental element spacing. Without loss of generality, the array elements shall be distributed along the y-axis of the current coordinate system (whereas the x-axis shall be the looking direction). Their position shall be stored in a row vector $\boldsymbol{y}_{c,nm}$ (c for coefficient, nm to indicate that the elements are created by a Tx-Rx combination in case of a virtual array): Each entry belongs to a position in this d_0-scale along y-axis. At those positions where virtual elements are sited, the entry is 1, otherwise 0. From now on, arrays with Γ positions are examined, whereas only $\Gamma_u = NM \leq \Gamma$ virtual element combinations are generated.

6.1 The principle of minimum redundancy arrays

In previous section 5, the *hermitian* and *Toeplitz* characteristics of the covariance matrices $\boldsymbol{R}$ and $\boldsymbol{R}_n$ in *uncorrelated signal scenario* were examined. In equation (5.13), the covariance matrix was introduced generally.

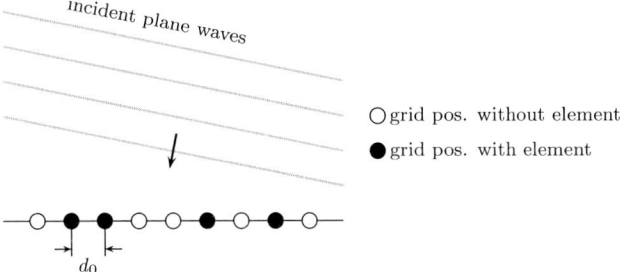

Fig. 6.1 Plane wave scenario as requirement for minimum redundancy array techniques.

Thereby, the directional vectors served as mean values composed to a sum over dyadic products, equation (5.11) (see also [48]).

Basically, the *Vandermonde* characteristics of $\boldsymbol{A}_q$ (matrix with directional vectors $\boldsymbol{a}_q$ as columns) lead to the *Toeplitz* structure of covariance matrices $\boldsymbol{R}$, $\boldsymbol{R}_n$. Since noise effects shall be examined in a later section 8, the noise free covariance matrix $\boldsymbol{R}$ is subject of this section.

This *Toeplitz* structure includes that each secondary diagonal, ξ positions away from the main diagonal, contains the same information $r(\xi)$, $(\Gamma_u - \xi)$-times, however generated by different array element combinations:

$$
\boldsymbol{R} = \begin{bmatrix}
r(0) & r(1) & r(2) & \cdots & r(\Gamma_d) \\
r^*(1) & r(0) & r(1) & & \vdots \\
\vdots & & \ddots & & \vdots \\
\vdots & & & \ddots & r(1) \\
r^*(\Gamma_d) & & & r^*(1) & r(0)
\end{bmatrix} .
\tag{6.1}
$$

On the one hand, this gives a possibility of signal redundancy with respect to high availability of the system output. But on the other hand, hardware expense could also be used to generate signal covariance values of higher lag values. Nevertheless, the plane wave scenario is an absolute requirement for the following minimum redundancy considerations. For fundamental ideas to minimum redundancy arrays, please see also [26], [27], [87], [88], [94], [95], [96].

6.1.1 Principal considerations from radio-astronomy

This is the point where the minimum redundancy (MR) idea comes in [26]. Typical considerations from radio astronomy and their very large telescope arrays were: How can the array signal be reconstructed in case of a single telescope mal-function? Or is there anyway a need to form a full telescope array or is it possible to keep it sparse in order to shrink the hardware expense (compare to [26])?

In apt manner, it can be stated that the full matrix $\boldsymbol{R}$ or $\boldsymbol{R}_n$ respectively, can be filled if at least one element in each diagonal is known. This leads to MR arrays, where the objective function of an optimization is to find a maximum dimension of the array (in order to have the finest angular resolution) with a minimum of redundancy in element spacing combinations. The distribution of element combinations is also called *spatial frequency spectrum* or *spatial frequency sensivity*, which is equivalent to the number of same covariance entries in the corresponding diagonals of $\boldsymbol{R}$. Basically, there are two types of MR arrays which can be distinguished [26].

The first type is the so-called *restricted* MR array [26]: There, within the maximum lateral dimension, each spatial frequency is covered by at least one pair of elements. Therefore, $\boldsymbol{R}$ does not contain any diagonals with only zero entries. The redundancy consists of combinations which occur more often. The array is as large as the maximum element spacing [26].

The second type is denoted as *general* MR array [26]. Contrary to *restricted* MR arrays, missing element combinations are allowed, thus $\boldsymbol{R}$ contains few diagonals with only zero elements, as long as there is a sub-matrix of $\boldsymbol{R}$ of *restricted* MR type [26]. This means for the physical array: There are Tx or Rx elements far outside the actual setup in order to provide additional signal combinations which help to reduce the redundancy within a certain core array [26].

6.1.2 From ULA to sparse array

For a filled ULA of Γ_u elements, there are by combinatorial means

$$\Gamma_u + (\Gamma_u - 1) + \ldots + 1 = \frac{1}{2} \Gamma_u (\Gamma_u + 1) \tag{6.2}$$

different covariance values, including the Γ_u auto-covariance values, to fill one triangular matrix in $\boldsymbol{R}$. If those Γ_u auto-covariance values are subtracted from equation (6.2), this will give in total

$$(\Gamma_u - 1) + (\Gamma_u - 2) + \ldots + 1 = \frac{1}{2} \Gamma_u (\Gamma_u + 1) - \Gamma_u = \frac{1}{2} \Gamma_u (\Gamma_u - 1) \tag{6.3}$$

covariance values on the secondary diagonals. Those in turn form the so-called *spatial frequency spectrum* [26] (compare to Fig. 6.2). Alternatively, this *spectrum* is also denoted as *co-array* [27], [95], which is always symmetric [27], [95]. Γ shall be the maximum number of possible array positions in a completely filled ULA, considering a grid in units of single element spacing d_0. Then Γ_d shall be the maximum (sub-)array dimension which is taken into account, again in units of d_0. For a ULA this is $\Gamma_d = (\Gamma - 1)$. For *restricted* MR arrays, Γ_d can be read from the maximum lateral dimension. Whereas for *general* MR arrays, $\Gamma_d d_0$ is the maximum dimension of the core array respectively, not regarding the outer elements. The maximum feasible dimension of $\boldsymbol{R}$ is $\mathbb{C}^{\Gamma \times \Gamma}$. Then the figure of merit describing the remaining redundancy (with respect to element spacings) in *spatial frequency spectrum* is [26]:

$$\gamma_r = \frac{\Gamma_u \, (\Gamma_u - 1)}{2\,\Gamma_d} \ , \quad \gamma_{mr} = \gamma_r - 1 \ . \tag{6.4}$$

In ideal case, each element spacing occurs exactly once and would directly determine the maximum dimension Γ_d by $1/2\Gamma_u \, (\Gamma_u - 1) = \Gamma_d$. Hence, the minimum value of γ_r is one. The redundancy in general can be described by γ_{mr}. $\boldsymbol{R}$ has got Γ positions on the main diagonal, with $\Gamma - \Gamma_u$ zeros

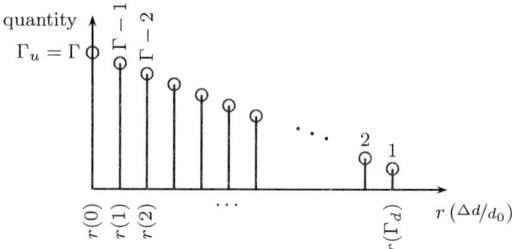

Fig. 6.2 ULA setup in Fig. 3.10: Number of covariance values which is equivalent to the positive side of a *spatial frequency spectrum.*

entries in case of a sparse array with Γ_u elements. The $\Gamma_d = \Gamma - 1$ secondary diagonals of $\boldsymbol{R}$'s upper triangular matrix provide $\Gamma_d + (\Gamma_d - 1) + \cdots + 1 = 1/2\Gamma_d \, (\Gamma_d + 1) = 1/2\Gamma \, (\Gamma - 1)$ positions. A uniform linear array (ULA), like in Fig. 3.10, would bear $\Gamma_u = \Gamma$ and would fill $\boldsymbol{R}$ completely, thus the covariance value $r(1)$ would occur $\Gamma - 1$ times, $r(2)$ $\Gamma - 2$ times, up to $r(\Gamma - 1) = r(\Gamma_d)$ which would be present only once. Fig. 6.2 sketches the number of covariance values for a typical ULA setup.

An equivalent diagram, the *spatial frequency spectrum* [26] or *co-array* [27], [95], can be calculated directly by taking the discrete auto-correlation function over the array weights [27], [95]. In general, this consideration would yield complex auto-correlation functions [95]. However, in the following, the arrays shall bear equal weights of one (no window over the array elements as in [32] or steering as in section 3.3.2). Therefore, the auto-correlation function acf $\{\boldsymbol{y}_c\} = \phi_c$ over the discrete array element positions collected in $\boldsymbol{y}_c$ can serve as *spatial frequency spectrum* or *co-array* [27], [95]. Furthermore, the element spacings correspond to *correlation lags* [27]. Systematically complex element weights would again lead to array steering which has already been described in section 3.3.2.

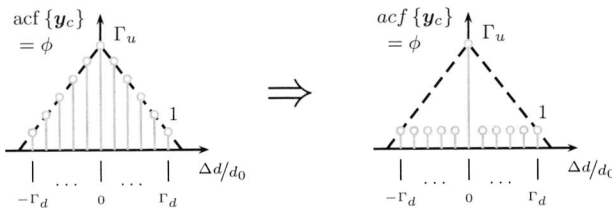

Fig. 6.3 MR idea: Altering the *spatial frequency spectrum* (discrete auto-correlation) from ULA to a sparse array with minimum redundancy.

In the following, the count of redundant covariance values shall be reduced with the objective to increase the lateral array dimension, thus maximize Γ_d with Γ_u elements subject to a discrete convolution due to the coherent MIMO character of the array. Hereby, each virtual element spacing shall occur ideally only once (compare to Fig. 6.3) [26].

The outcome will be sparse arrays, however with a greatest common divisor in element spacing of $d_0 = \lambda_0/4$. With respect to the monostatic character of the coherent MIMO ansatz, the *visible region* is $\Psi \in [-\pi; +\pi]$, thus the angular detection is unique over the complete half space in front of the radar (compare equation (3.35)). Considerations similar to this section had also been published by Pillai [87], [88].

6.1.3 Zero redundancy sub-arrays

There are array configurations with zero redundancy, thus $\gamma_r = 1$ or $\gamma_{mr} = 0$ [25], [26]. However, Bracewell could proof that there are only four linear arrays with zero redundancy possible [25], [26]: The single element ($y_c = [1]$), two elements with spacing d_0 ($y_c = [1,1]$), three elements with spacing $\Delta y = [1, 2]\, d_0$ ($y_c = [1,1,0,1]$) and four elements with spacing $\Delta y = [1, 3, 2]\, d_0$ ($y_c = [1,1,0,0,1,0,1]$) [25], [26] (see Fig. 6.4). All larger linear arrays provide values of $\gamma_{mr} > 0$ [25], [26]. The practically relevant cases (c) (three elements) and (d) (four elements) in Fig. 6.4 will be exploited as basic sub-arrays for MIMO array setups with minimum redundancy (MR) in this thesis.

The discrete auto-correlation of vector y_c gives the *spatial frequency sensivity* or *co-array* (see Fig. 6.4) [25]. The main peaks in Fig. 6.4 (for $\Delta d/d_0 = 0$) stand for the incidence of correlating each element with itself, which occurs then Γ_u times, respectively. The elements left and right correspond to the same element combinations, however, with inverted sense

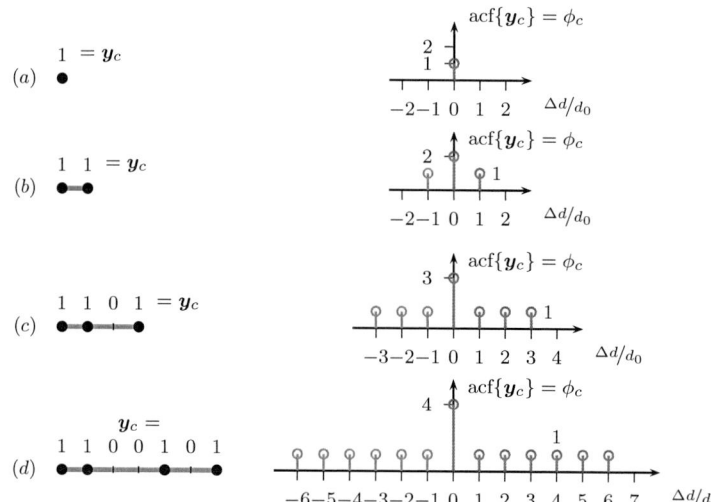

Fig. 6.4 The only four zero redundancy array setups and their discrete auto-correlation $acf\{\boldsymbol{y}_c\} = \phi_c$, giving the corresponding co-arrays [25], [26]. The negative side (gray) represents the complex conjugate correlation values to positive side (orange), thus corresponds to the same element combinations [25], [26], [27], [95].

of direction which leads to complex conjugated values in case of complex (radar) signals. The discrete *dyadic* product (please note $\boldsymbol{y}_c$ is a row):

$$\boldsymbol{\Upsilon} = \boldsymbol{y}_c^{\mathrm{T}}\, \boldsymbol{y}_c \quad , \tag{6.5}$$

gives a *covariance position matrix* $\boldsymbol{\Upsilon}$, thus a matrix, which indicates those entries of the *Toeplitz* type matrices $\boldsymbol{R}$, $\boldsymbol{R}_n$ which can directly be filled. All others must be generated by reconstruction methods later on. The complete ansatz of one main peak with low sidelobe peak levels at the acf shows how closely related the MR idea is to *radar signal design* and *phase coding* like e. g. *Barker codes* etc. [24], [55].

6.1.4 MIMO radar combined with minimum redundancy

In case of a conventional linear array with minimum redundancy as in [26], the procedure is a combinatorial one: The elements at stage (μ) are placed, their possible signal combinations (element lags) are counted and their po-

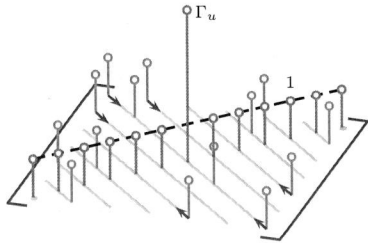

Fig. 6.5 The sum over the diagonals of covariance position matrix Υ gives the co-array (orange).

sitions are eventually replaced. The next element at stage $(\mu + 1)$ is placed for the next signal combination (element lag) which does not occur within the array at stage (μ) [26]. Thereby a certain redundancy must occur in order to satisfy the constraint of a linear array. Bracewell and Moffet imposed the hypothesis that the idea of MR was only applicable in feasible manner to linear arrays [25], [26]. However, the field of *coded aperture imaging* or *uniformly redundant arrays* URA covers a similar idea over two dimensions: In high energy photon detection such like x-ray or γ-ray detectors, aperture blends with transparent and dark fields provide a two dimensional correlation with a single peak and a uniform redundancy distribution [97], [98], [99]. The same idea is covered by *coded apertures* [100], [101], [102].

A completely new constraint occurs if a linear MR array shall be generated by MIMO radar techniques, thus the equivalent virtual array shall bear minimum redundancy: The desired constellation must be generated by the discrete convolution of a Tx and Rx array (see section 3.5.1). Hence, the inverse problem would be, taking a desired array configuration and determining the corresponding Tx and Rx arrays by a de-convolution procedure. However, for many setups as proposed in [26] this de-convolution is not possible due to the fact that they do not contain any periodic basic structure, such like the eight element setup with element spacing $\Delta y = [1, 3, 6, 6, 2, 3, 2]\, d_0$, giving $\Gamma_d = 23$ [26].

A possible remedy is to solve an optimization problem with MR as object and the convolution procedure as subject [28]. However, it could be shown, that exhaustive combinatorial optimization methods are needed to solve this problem [28]. Alternatively, *brute-force* methods were proposed [26]. With respect to *phase coding*, the convolutional character of a MIMO setup leads to *nested phase coding* which will be covered in more detail by a later section [55].

6.2 A design algorithm for a linear MR array

One possible way is to take MR constellations (which are well documented) and to search the minimum number of Tx and Rx elements to realise these setups. Now, the direction of optimization shall be imposed the other way round: A priori it is known that there are N Tx and M Rx elements. The objective function is a maximum dimension along y-axis (compare Fig. 3.10).

6.2.1 Brute force algorithm for restricted linear MR arrays

A brute force algorithm helped to design linear arrays with minimum redundancy of *restricted* type. The start configuration is a fully uniform linear (virtual) array (ULA) which can be generated by a configuration like in Fig. 3.10. It is only allowed to place virtual elements in discrete steps of $d_0 = \lambda_0/4$ along y-axis.

Again, the position is stored digitally in the column vector $\boldsymbol{y}_{c,nm}$ (nm shall indicate a virtual array). At those positions where virtual elements are sited, the entry is 1, otherwise 0. If occasionally two or more virtual elements have the same positions, the entry is equal to the number of virtual elements at this position. Hence, with the pre-defined start configuration, $\boldsymbol{y}_{c,nm}$ is filled with exclusively non-zeros. The combinatorial work can either performed by the *covariance position matrix* $\boldsymbol{\Upsilon}$ equation (6.5) or the *co-array* $\phi_{c,nm}$. Just for recall, the matrix $\boldsymbol{\Upsilon}$ demonstrates, which components of $\boldsymbol{R}$, respectively $\boldsymbol{R}_n$, can be filled in directly. $\boldsymbol{\Upsilon}$ is always symmetric: $\boldsymbol{\Upsilon} = \boldsymbol{\Upsilon}^{\mathrm{T}}$.

Now, the parameters which are acted on, are the $(N + M - 2)$ Tx and Rx element spacings (please note: physical arrays, not virtual ones): One by one, the element spacing is increased by $2 d_0$, because then the virtual elements are manipulated by a shift of d_0 (compare to equations (3.63), (3.68)). After each manipulation, $\boldsymbol{y}_{c,nm}$ and $\boldsymbol{\Upsilon}$ (or $\phi_{c,nm}$ respectively) are re-calculated. Thereby, $\boldsymbol{y}_{c,nm}$ and $\boldsymbol{\Upsilon}$ increase in dimension and contain more and more zeros, since valid places on the y-axis are not occupied. The constraint for the optimization is defined by $\boldsymbol{\Upsilon}$: The array is considered as valid if at least one non-zero element remains in each secondary diagonal, hence the corresponding *co-array* $\phi_{c,nm}$ on $\boldsymbol{y}_{c,nm}$ does not exhibit a zero entry at the corresponding position. Furthermore, all element spacings are affected by the same occurrence in order to separate them by dimensions

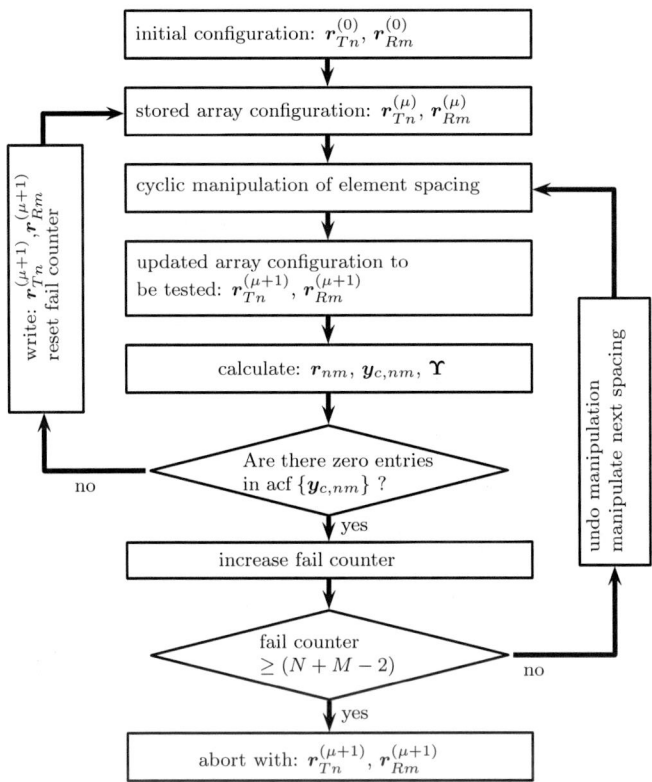

Fig. 6.6 Flow chart of design algorithm for *restricted* MR-MIMO arrays.

which allow a reasonable antenna manufacturing. Fig. 6.6 shows a principle flow chart of the algorithm.

Since $\boldsymbol{\Upsilon}$ is symmetric, it is sufficient to concentrate on the upper or lower triangular matrix of $\boldsymbol{\Upsilon}$. According to that, the *co-array* $\phi_{c,nm}$ is symmetric, too [27], [95]. Therefore, it would be sufficient to concentrate on the positive side of $\phi_{c,nm}$. The main diagonal of $\boldsymbol{\Upsilon}$ automatically contains $\Gamma_u = N M$ non-zeros elements, due to the fact that always $N M$ virtual elements are generated and their variance value (or auto-correlation value) cannot be zero at any time. (In this case, the radar would not deliver measured data

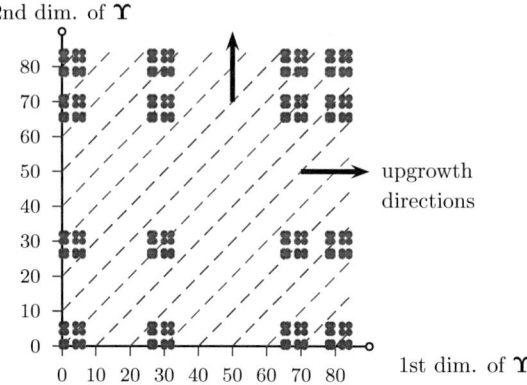

Fig. 6.7 Example for upgrowth of Υ: Indices of covariance position matrix Υ after 216 (112) stages for setup Fig. 6.12. Red dots are non-zero elements.

at all.) Fig. 6.7 shows the matrix Υ (according to their indices) for a configuration of Fig. 6.12. The upgrowth direction for each turn of the generation algorithm is indicated.

If at stage (μ), at least one diagonal of Υ contains only zero elements (or $\phi_{c,nm}$ exhibits at least one zero entry), then the manipulation is considered invalid, the configuration is reset to stage ($\mu - 1$) and another element spacing is manipulated. If the $2\,d_0$ shift on any element spacing leads to this invalidity, then the optimization is aborted and the actual array is considered to be the linear array with maximum extension along y-axis and minimum redundancy which can be reached from the individual start configuration. Hence, the algorithm provides MR arrays of *restricted* type. However, the results highly depend on the start configuration. In case of this example with $N = 4$, $M = 4$, the necessary computational effort for current standard desktop PCs is negligible.

6.2.2 Simulation results

The *brute-force* method described in section 6.2.1 was tested with a configuration of three or four Tx and four Rx elements similar to Fig. 3.10. Therefore the total number of generated signal combinations Γ_u is either $3 \cdot 4 = 12$ (with three Tx elements) or $4^2 = 16$ (with four Tx elements). The ULAs from Fig. 3.10 and 3.11, which both provide a virtual ULA of length $15\,d_0$ shall serve for comparison.

Redundant virtual elements at start configuration: Since redundancy shall be avoided, how do redundant virtual elements in the start configuration affect the final result? One example can be the following setup: $\Delta y_{Tx} = 2\,d_0\,[1, 1, 1]$, $\Delta y_{Rx} = 2\,d_0\,[4, 2, 2]$ as element spacing, which gives a ULA of twelve elements separated by d_0:

$$y_{c,Tx} * y_{c,Rx} =$$
$$= [1, 0, 1, 0, 1, 0, 1] * [1, 0, 0, 0, 0, 0, 0, 0, 1, 0, 0, 0, 1, 0, 0, 0, 1] =$$
$$= [1, 0, 1, 0, 1, 0, 1, 0, 1, 0, 1, 0, 2, 0, 2, 0, 2, 0, 2, 0, 1, 0, 1]$$
$$\Downarrow \text{ spatial sub-sampling with factor } \frac{1}{2}$$
$$y_{c,nm} = [1, 1, 1, 1, 1, 1, 2, 2, 2, 2, 1, 1]\ .$$

$$(6.6)$$

The discrete convolution of the Tx and Rx arrangements gives a vector of non-zeros always spaced by one zero element. The correct virtual array can be derived by taking every second entry, thus a spatial sub-sampling by the factor of $1/2$: The reason for this can be found in $d_i = 1/2\,(d_n + d_m)$ (equation 3.68). If both operations, the discrete convolution and the spatial sub-sampling, are applied at one stage, this shall be denoted by $\circledast$ instead of $*$:

$$y_{c,nm} = y_{c,Tx} \circledast y_{c,Rx} = y_{c,Rx} \circledast y_{c,Tx}\ . \qquad (6.7)$$

An alternative and more compact way of writing can be imposed by the absolute positions along y-axis: $y_{Tx} = [1, 3, 5, 7]\,d_0$, $y_{Rx} = [1, 9, 13, 17]\,d_0$, which gives:

$$y_{nm} = y_{Tx} \oplus y_{Rx} =$$
$$= \frac{1}{2}\,d_0\ \{\,[1, 3, 5, 7, 1, 3, 5, 7, 1, 3, 5, 7, 1, 3, 5, 7] +$$
$$[1, 1, 1, 1, 9, 9, 9, 9, 13, 13, 13, 13, 17, 17, 17, 17]\} =$$
$$= [1, 2, 3, 4, 5, 6, 7, 8, 7, 8, 9, 10, 9, 10, 11, 12]\,d_0\ .$$

$$(6.8)$$

Again the special operator $\oplus$ shall include the cyclic element copying procedure, arranging in ascending order and the division by two.

It can be seen, that in the start configuration four elements are generated twice, which gives $\Gamma_d = 11$ at $\Gamma_u = 16$ and therefore $\gamma_r = 10.91$. If the generation algorithm is fed with this start configuration, the outcome after

112 stages will be: $\boldsymbol{y}_{Tx} = [1,\, 5,\, 13,\, 15]\, d_0$, $\boldsymbol{y}_{Rx} = [1,\, 37,\, 75,\, 81]\, d_0$, hence for the virtual array (see Fig. 6.8):

$$\boldsymbol{y}_{nm} = [1,\, 5,\, 13,\, 15] \oplus [1,\, 37,\, 75,\, 81] =$$
$$= [1,\, 3,\, 7,\, 8,\, 19,\, 21,\, 25,\, 26,\, 38,\, 40,\, 44,\, 45,\, 41,\, 43,\, 47,\, 48]\ . \tag{6.9}$$

Fig. 6.8 shows the result which yields: $\Gamma_d = 47$ at the same Γ_u, thus

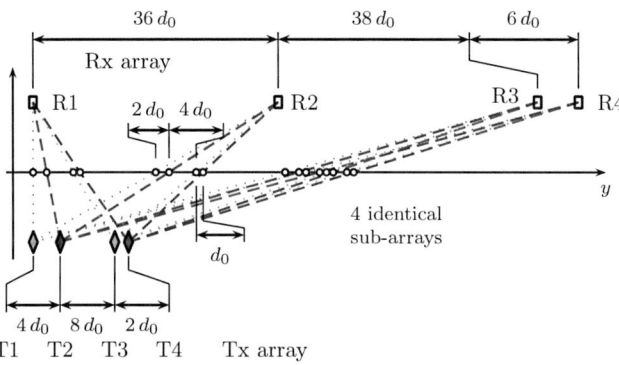

Fig. 6.8 A set of four Tx ($N = 4$) and four Rx elements ($M = 4$), thus $\Gamma_u = 16$, after 112 stages and redundant initial condition results in two nested sub-arrays.

$\gamma_r = 2.55$. It can be observed further, that two sub-arrays become nested in order to fulfill the condition that all element lags must occur at least once (see Fig. 6.9). However, no redundant virtual elements can be found in the final result. Regarding the result in Fig. 6.8, the redundant part within the start configuration cannot be disentangled anymore. This effect is even stronger, the more redundant virtual elements are in the start configuration, demonstrated by the following example: $\boldsymbol{y}_{Tx} = [1,\, 3,\, 5,\, 7]\, d_0$, $\boldsymbol{y}_{Rx} = [1,\, 3,\, 5,\, 7]\, d_0$, which provides a start configuration of $\gamma_r = 20$:

$$\boldsymbol{y}_{nm} = [1,\, 3,\, 5,\, 7] \oplus [1,\, 3,\, 5,\, 7] =$$
$$= [1,\, 2,\, 3,\, 4,\, 2,\, 3,\, 4,\, 5,\, 3,\, 4,\, 5,\, 6,\, 4,\, 5,\, 6,\, 7]\ . \tag{6.10}$$

The brute-force algorithm fed with equation (6.10) gives:

$$\boldsymbol{y}_{nm} = [1,\, 3,\, 33,\, 49] \oplus [1,\, 9,\, 39,\, 43] =$$
$$= [1,\, 2,\, 17,\, 25,\, 5,\, 6,\, 21,\, 29,\, 20,\, 21,\, 36,\, 44,\, 22,\, 23,\, 38,\, 46]\ . \tag{6.11}$$

$\text{acf}\{\boldsymbol{y}_{c,nm}\} = \phi_{c,nm}$

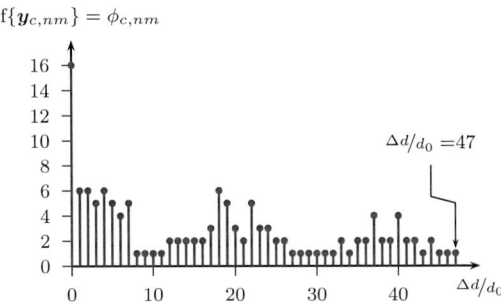

Fig. 6.9 *Spatial frequency spectrum* (only positive side) for resulting array in Fig. 6.8 with redundant initial condition, $\gamma_r = 2.55$.

In this case, all four sub-arrays become nested and redundant virtual elements remain, in particular at $21\,d_0$. The figure of merit is then $\gamma_r = 2.67$.

The incomplete disentangling can also be observed, if additionally lacks occur in the start configuration: $\boldsymbol{y}_{Tx} = [1, 9, 11, 13]\,d_0$, $\boldsymbol{y}_{Rx} = [1, 3, 11, 19]\,d_0$, thus $\boldsymbol{y}_{nm} = [1, 5, 6, 7, 2, 6, 7, 8, 6, 10, 11, 12, 10, 14, 15, 16]\,d_0$. Then it takes 121 stages for the algorithm to come to:

$$\boldsymbol{y}_{nm} = [1, 25, 29, 33] \oplus [1, 3, 49, 61] =$$
$$= [1, 13, 15, 17, 2, 14, 16, 18, 25, 37, 39, 41, 32, 43, 45, 47] \,. \tag{6.12}$$

A non-redundant ULA as start configuration: Regarding the preceding results, it is of advantage if the start configuration is without any redundant virtual elements or lacks which could be achieved by a ULA. In addition, after a closer look on the initial setups behind Fig. 6.8 and 6.10, one array (Tx or Rx) should be consequently larger than the other one, in order to put virtual sub-arrays next to each other instead of nesting them. The antenna configuration from Fig. 3.10 would provide such a proper initial ULA:

$$\boldsymbol{y}_{nm} = [1, 3, 5, 7] \oplus [1, 9, 17, 25] =$$
$$= [1, 2, 3, 4, 5, 6, 7, 8, 9, 10, 11, 12, 13, 14, 15, 16] \tag{6.13}$$

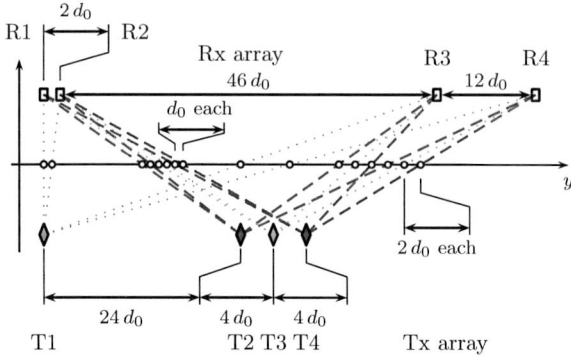

Fig. 6.10 A set of four Tx ($N = 4$) and four Rx elements ($M = 4$), thus $\Gamma_u = 16$, after 121 stages with redundant initial condition thinned out with lacks, results in four nested sub-arrays.

The redundancy is the same as for a standard ULA $\gamma_r = 8$. But after 216 turns, the outcome is then:

$$
\begin{aligned}
\boldsymbol{y}_{nm} &= [1,\, 3,\, 9,\, 13] \oplus [1,\, 53,\, 131,\, 157] = \\
&= [1,\, 2,\, 5,\, 7,\, 27,\, 28,\, 31,\, 33,\, 66,\, 67,\, 70,\, 72,\, 79,\, 80,\, 83,\, 85]\ ,
\end{aligned}
\tag{6.14}
$$

which is a result of nearly the doubled length of the previous virtual arrays. The figure of merit is now $\gamma_r = {}^{120}\!/_{84} = 1.43$. Basically, the Tx elements stay close together and serve as elementary block, thus as basis,

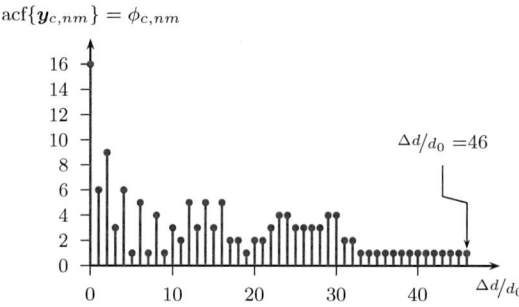

Fig. 6.11 *Spatial frequency spectrum* (only positive side) for setup in Fig. 6.10 after 121 stages.

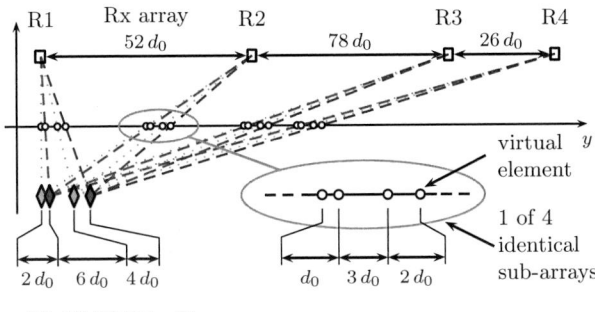

Fig. 6.12 A set of four Tx ($N = 4$) and four Rx elements ($M = 4$), thus $\Gamma_u = 16$, with *restricted minimum redundancy* after 216 (112) stages with non-redundant ULA initial condition (not to scale).

for the virtual array: The virtual elements in each of the four sub-arrays have the zero-redundancy element spacing $[1, 3, 2]\, d_0$, thus the Tx elements $[1, 3, 2]\, 2\, d_0$. The Rx elements are also separated by multiples of the zero redundancy elements spacings: $13\, [2, 3, 1]\, 2\, d_0$. In section 6.3 it will be demonstrated why the factor 13 is the only possible one for this configuration. Alternatively, the Rx elements can also be placed by the Tx spacing sequence, again with the factor 13, thus $13\, [1, 3, 2]\, 2\, d_0$, which gives an al-

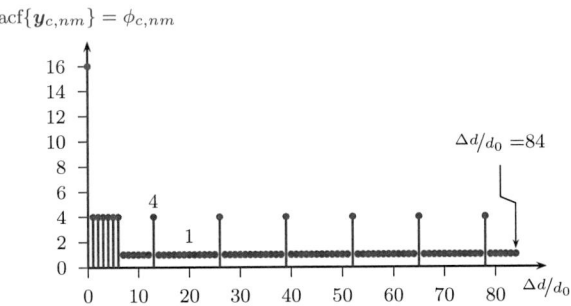

Fig. 6.13 *Spatial frequency spectrum* (only positive side) for setup in Fig. 6.12 (*restricted minimum redundancy*) after 216 (112) stages.

ternative virtual array resulting in the same spatial frequency distribution like in Fig. 6.13:

$$y_{nm} = [1, 3, 9, 13] \oplus [1, 27, 105, 157] =$$
$$= [1, 2, 5, 7, 14, 15, 18, 20, 53, 54, 57, 59, 79, 80, 83, 85] \ , \quad (6.15)$$

Principally, only the two inner virtual sub-arrays have switched their relative positions to the outer sub-arrays. In both cases $\gamma_r = 1.43$, which is

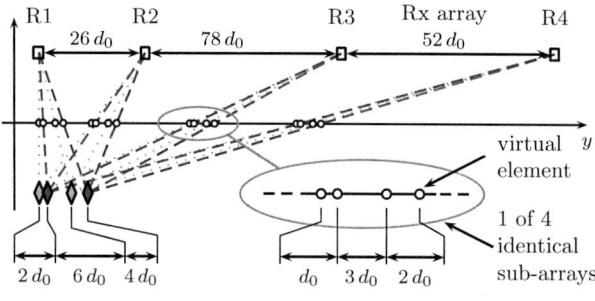

Fig. 6.14 Alternative set of four Tx ($N = 4$) and four Rx elements ($M = 4$), thus $\Gamma_u = 16$, with *restricted minimum redundancy* after 216 (112) stages with non-redundant ULA initial condition (not to scale).

very close to Moffet's theoretical forecast where $\gamma_{r,min} \in [1.217; 1.332]$ for large Γ_u [26].

The maximum dimension of the virtual array is therefore achieved by separating the Rx elements as far as possible. Of course, the assignment of Tx and Rx can also be used vice-versa. Hence, it can be stated that one array (Tx or Rx) should serve as basis, the other one for expanding means.

Zero redundancy sub-array in start configuration: If initial configuration already contains a Tx array of zero redundancy with restriction to only alter the Rx element spacings, then the number of turns reduces to 112 with the same results of Fig. 6.12, 6.13, 6.14.

Lack of elements in start configuration: If redundant virtual elements are avoided in initial setups, but lacks remain, the result are again the same as in Fig. 6.12, 6.13, 6.14. One example could be $\boldsymbol{y}_{Tx} = [1, 9, 11, 13]\, d_0$, $\boldsymbol{y}_{Rx} = [1, 3, 11, 19]\, d_0$, giving a start configuration of:

$$\begin{aligned}\boldsymbol{y}_{nm} = \boldsymbol{y}_{Tx} \oplus \boldsymbol{y}_{Rx} = \\ = [1, 2, 5, 6, 7, 8, 10, 11, 12, 14, 15, 16]\, d_0 \,.\end{aligned} \tag{6.16}$$

6.2.3 Relaxation for general arrays

The algorithm in Fig. 6.6 can be relaxed in order to generate *general* MR arrays. If the first secondary diagonal only contains zero elements at stage $(\mu + 1)$, then a distinction of cases is necessary: If the zero diagonal is close to the edges of $\boldsymbol{\Upsilon}^{(\mu+1)}$, but the dimensions of a core, thus a sub-matrix of $\boldsymbol{\Upsilon}^{(\mu+1)}$, are still greater or equal to the former matrix $\boldsymbol{\Upsilon}^{(\mu)}$, then the manipulation is accepted. Additionally, it is required to alter more than one of the $(N + M - 2)$ Tx and Rx element spacings at one stage. The hope is, that a core-matrix is still growing to some extent, whereas the *spatial frequency spectrum* can become sparse for large element lags. However, tests with the configuration from Fig. 6.12 have shown, that immediately zero diagonals close to the main diagonal are created due to the convolution procedure, thus the idea of a growing core has to be refused for the setup idea of $\Delta\boldsymbol{y}_{Tx} = [1, 3, 2]\, 2\, d_0$, thus for the basic idea of putting each virtual sub-array next to each other. Therefore, a setup as in Fig. 6.12 can be considered as optimum subject to $N = 4$ and $M = 4$, with $\gamma_r = 1.43$. However, in section 6.3.2 another type of nested virtual arrays will be shown, which provide low redundancy with arrays of *general* type.

6.3 Analytical approach

A closer look on the combinatorial relations of the final results in Fig. 6.12 and 6.14 shall serve as fundament for a deterministic ansatz concerning the *spatial frequency sensivities* in Fig. 6.13 and their significance as possible global minimum for the case $N = 4$, $M = 4$. The derived principles are then tested with other constellations such as $N = 3$, $M = 4$. Further, those principles can then be generalized for all virtual arrays with minimum redundancy.

In both cases Fig. 6.12 and 6.14, the Tx element spacing was $[1, 3, 2]\, 2\, d_0$ and the Rx element spacing $13\,[1, 3, 2]\, 2\, d_0$ or $13\,[2, 3, 1]\, 2\, d_0$ which gave four virtual sub-arrays of zero redundancy, separated by $13\,[1, 3, 2]\, d_0$ or $13\,[2, 3, 1]\, d_0$ respectively, which is again a scaled version of the four element zero redundancy array setup (Fig. 6.4). Thus, both results show a structure of nested zero redundancy cases. (Of course, the assignment of Tx and Rx

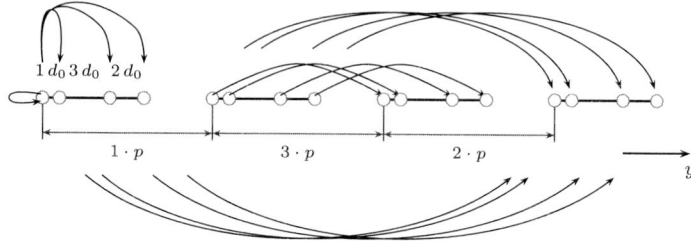

Fig. 6.15 A MR-MIMO setup with $N = 4$ and $M = 4$ and its fundamental combinatorial relations. Please note: The relations between the sub-arrays at the very edge and their direct neighbours in the middle were left away for reduction of picture complexity.

elements could also swap places.) In order to create a recipe for generating the afore mentioned structures without any search algorithm, the possible and necessary combinatorial relations between the virtual sub-arrays shall be examined in more detail. Hereby, Fig. 6.15 sketches the main cases which shall be projected onto Fig. 6.19: The $M = 4$ virtual sub-arrays themselves are identical. Therefore, their inner redundancy is the same and must occur M-times. Hence, the values from $d/d_0 = 0$ to 6 in Fig. 6.19 correspond to the co-array of zero redundancy arrays (Fig. 6.4), scaled by M.

The co-array of one virtual sub-array was derived from the discrete auto-correlation function in section 6.1.3. For the case $N = 4$, in total $1/2 N (N - 1) = 6$ element combinations are possible. Further, there are always $N = 4$ possibilities of correlating each element with itself. The negative parts of the acfs in Fig. 6.4 had the logical significance of the same element combinations like in positive case, just with swapped logical direction which in turn ends in the complex conjugated signal in case of complex radar signals. Therefore, according to Fig. 6.4, a four element zero redundancy sub-array delivers $p = N (N - 1) + 1 = 13$ signals. Their pattern can be found aligned and coloured in Fig. 6.19. Hereby, the first three, $13 = p$, $26 = 2p$ and $39 = 3p$ correspond to the relations between each virtual sub-array and its direct neighbour (Fig. 6.15). The two cases $52 = 4p = (3 + 1) p$ and $65 = 5p = (3 + 2) p$ belong to the two possibilities of combining elements of two virtual sub-arrays with one sub-array in between. The very last one, $78 = 6p = (3 + 2 + 1) p$ is the covariance between the two outer sub-arrays over the two remaining in between. The factor 6 does not occur accidentally here. Actually, there are again

$1/2\,M\,(M-1) = 6$ possibilities of combining virtual sub-arrays with each other. This can also be shown analytically: The vector with digital coefficients 0 or 1, $\boldsymbol{y}_{c,nm}$ can be described by the convolution:

$$\boldsymbol{y}_{c,nm}\,(\xi) = \boldsymbol{y}_{c,Tx}\,(2\xi)\,\circledast\,\boldsymbol{y}_{c,Rx}\,(2\xi) =$$
$$= [1,1,0,0,1,0,1] *$$
$$\left[1,\ \underbrace{0,\ldots 0}_{12\text{ or }25\text{ zeros}}\ ,1,\underbrace{0,\ldots 0,1}_{38\text{ zeros}}\ ,\underbrace{0,\ldots 0}_{25\text{ or }12\text{ zeros}}\ ,1\right].$$ \hfill (6.17)

Hereby, the vectors $\boldsymbol{y}_{c,Tx}$ and $\boldsymbol{y}_{c,Rx}$ correspond to the discrete distribution of Tx and Rx elements (please the down-sampling by the factor of two). The parameter $\xi = {}^{d}/d_0$ shall characterize the coordinate direction by integer steps. The discrete auto-correlation $\phi_{c,nm}\,(\xi)$ of $\boldsymbol{y}_{c,nm}$ is then (by exploitation of commutative and associative laws of convolution operation [57]):

$$\phi_{c,nm}\,(\xi) = \boldsymbol{y}_{c,nm}\,(\xi) * \boldsymbol{y}_{c,nm}\,(-\xi) =$$
$$= \{\boldsymbol{y}_{c,Tx}\,(2\xi)\,\circledast\,\boldsymbol{y}_{c,Rx}\,(2\xi)\} * \{\boldsymbol{y}_{c,Tx}\,(-2\xi)\,\circledast\,\boldsymbol{y}_{c,Rx}\,(-2\xi)\} =$$
$$= \{\boldsymbol{y}_{c,Tx}\,(2\xi) * \boldsymbol{y}_{c,Tx}\,(-2\xi)\}\,\circledast\,\{\boldsymbol{y}_{c,Rx}\,(2\xi) * \boldsymbol{y}_{c,Rx}\,(-2\xi)\} =$$
$$= \phi_{c,nn}\,(2\xi)\,\circledast\,\phi_{c,mm}\,(2\xi)\,.$$ \hfill (6.18)

This corresponds to the discrete version of *Wiener-Lee relation* [57] known from communication theory from linear and time-invariant (LTI) filter outputs. It describes the auto-correlation of a filter's output signal by the convolution of two auto-correlation functions, the one of the filter's input signal (here $\phi_{c,nn}$) and the one over the filter impulse response (here $\phi_{c,mm}$) [51], [57]. Please note, that the roles of $\phi_{c,nn}$ and $\phi_{c,mm}$ can also be switched [51], [57]. Hence, the Tx constellation can be regarded as input vector, the Rx constellation as filter impulse response, or vice versa [51], [57]. Consequently according to *Wiener-Lee relation* [57], the spectral

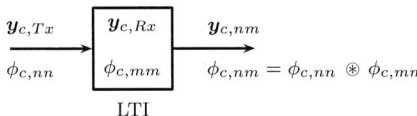

Fig. 6.16 Discrete auto-correlation function $\phi_{c,nm}$ as filter output according to *Wiener-Lee relation* [57].

representation of equation (6.18) is a product of the corresponding *power density functions* [51], [57], or in case of *spatial sampling*, the product of squared absolute array factors [31], [32], [37]. This has already been demonstrated in sections 3.2, 3.3, 3.4, and among others, explicitly in equation (3.57):

$$|AF_{TRx}|^2 = |AF_{Tx}|^2 \cdot |AF_{Rx}|^2 \ . \tag{6.19}$$

6.3.1 MR-MIMO setup as nested minimum redundancy approach

The analytical approach demonstrated that the optimal results in Fig. 6.12 and 6.14 can be obtained by nesting a zero redundancy array (scaled by the factor 2) by a braced version of a zero redundancy array. Hereby, the discrete acf of the short array was aligned by convolution by means of the large array's acf. The same procedure is valid for any combination of two arrays, of any redundancy. In case of $N, M \in \{1, 2, 3, 4\}$, the Tx and Rx

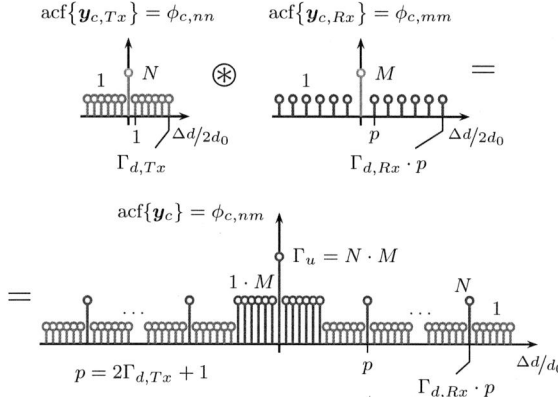

Fig. 6.17 Co-array for setup in Fig. 6.12 and 6.14 represented by the convolution of Tx and Rx co-array.

array bear maximum dimension, hence zero redundancy [26]. In general, it can be stated that [26]:

$$\Gamma_{d,Tx} \le \frac{1}{2} N (N - 1) \ , \tag{6.20}$$

$$\Gamma_{d,Rx} \le \frac{1}{2} M (M - 1) \ , \tag{6.21}$$

$$p = 2\Gamma_{d,Tx} + 1 \le N (N - 1) + 1 \ , \tag{6.22}$$

$$\Gamma_d = \Gamma_{d,Rx} \cdot p + \Gamma_{d,Tx} \ . \tag{6.23}$$

For the optimal case, the large array's acf is braced as widely such that the short array's acf can be aligned again and again. A similar behaviour can be found with nested phase codes in radar signal theory [55]. There, the total acf of e. g. a nested Barker code is again a convolution of the two acfs of the fundamental codes.

Now, in order to proof the latest insights, the properties of an optimal system with $N = 3$ transmitters and $M = 4$ receivers shall be predicted analytically and then compared to the algorithms output. The Tx element spacings shall be again of zero redundancy $[1, 2] \, 2d_0$, the bracing factor $p = N (N - 1) + 1 = 7$. The Rx element spacing can be chosen to $p [1, 3, 2] \, 2d_0$ or $p [2, 3, 1] \, 2d_0$, thus $[1, 3, 2] \, 14d_0$ or $[2, 3, 1] \, 14d_0$. A possible realisation could be $\mathbf{y}_{Tx} = [1, 3, 7] \, d_0$, $\mathbf{y}_{Rx} = [1, 15, 57, 85] \, d_0$ or $\mathbf{y}_{Rx} = [1, 29, 71, 85] \, d_0$, which delivers a virtual array $\mathbf{y}_{nm} = [1, 2, 4, 8, 9, 11, 29, 30, 32, 43, 44, 46] \, d_0$ or $\mathbf{y}_{nm} = [1, 2, 4, 15, 16, 18, 36, 37, 39, 43, 44, 46] \, d_0$ respectively. Both deliver a *spatial frequency spectrum* like in Fig. 6.18, resulting in $\gamma_r = 1.47$. The final question of this section is now: Are the presented configurations delivering results like in Fig. 6.19 or Fig. 6.18 optimal, thus do they represent the minimum achievable redundancy? In general, the ideal maximum length of a *restricted* MR-MIMO array $\Gamma_{d,ideal} = 1/2 \, NM (NM - 1)$ can be expressed by the sum of maximum array length Γ_d and the number of redundant signal combinations Γ_r:

$$\begin{aligned}
\Gamma_{d,ideal} &= \Gamma_d + \Gamma_r \ , \\
\Rightarrow \Gamma_r &= \Gamma_{d,ideal} - \Gamma_d = \\
&= \frac{1}{2} \Gamma_u (\Gamma_u - 1) - \Gamma_{d,Rx} \cdot p - \Gamma_{d,Tx} = \\
&= \frac{1}{2} NM (NM - 1) - \Gamma_{d,Rx} \cdot (2 \, \Gamma_{d,Tx} + 1) - \Gamma_{d,Tx} \ .
\end{aligned} \tag{6.24}$$

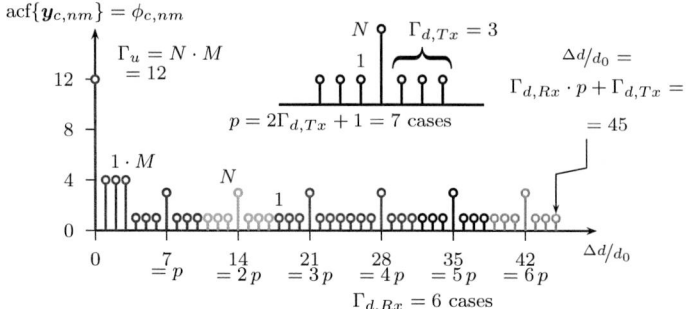

Fig. 6.18 *Spatial frequency spectrum* (only positive side) for optimal setup with $N = 3$, $M = 4$: Maximal array lengths $\Gamma_{d,Tx}$, $\Gamma_{d,Rx}$ provide a maximum possible length of virtual array with $\gamma_r = 1.47$.

Equation (6.24) discloses that the terms $\Gamma_{d,Tx}$, $\Gamma_{d,Rx}$ ought to be maximized for minimizing Γ_r. Those expressions can be taken from equations (6.20), (6.21), (6.22). In case of zero redundancy Tx and Rx arrays, hence $N, M \in \{1, 2, 3, 4\}$, it can be concluded that:

$$\Gamma_{r,min} = \frac{1}{2} \left[N^2 (M - 1) + M^2 (N - 1) + M + N - 2\,MN \right] \quad , \text{ (6.25)}$$

With a step back to combinatorics, this can be proofed: Overall there are $\Gamma_u = MN$ virtual elements, hence without the MN auto-covariances, there are $1/2\,MN\,(MN - 1)$ distinct element combinations. A closer look on Fig. 6.19 or Fig. 6.18 also discloses the remaining redundancy: In classical MR array case [26], there would only be a peak of MN amplitude at $d/d_0 = 0$, whereas all other peaks should occur with amplitude one. However, it was demonstrated that due to the convolution character of a virtual array, there is a must for $1/2\,N\,(N - 1)$ *spatial frequencies* (element lags) with amplitude M, thus $(M - 1)$ higher as needed. Additionally, there are always $1/2\,M\,(M - 1)$ elements with amplitude N, hence $(N - 1)$ times higher as

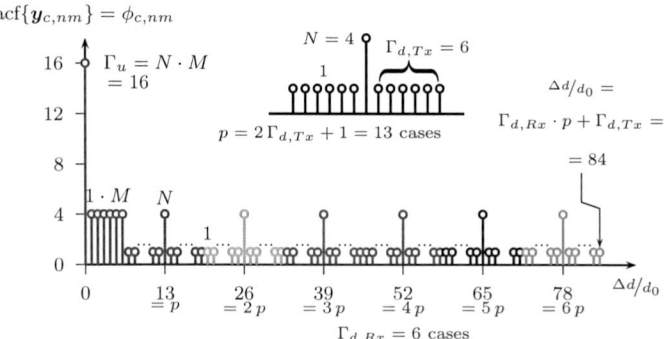

Fig. 6.19 *Spatial frequency spectrum* (only positive side) for optimal setup in Fig. 6.12 and 6.14 with $N = 4$, $M = 4$: Maximal array lengths $\Gamma_{d,Tx}$, $\Gamma_{d,Rx}$ provide a maximum possible length of virtual array with $\gamma_r = 1.43$.

required. Therefore, the inherent redundancy which has at least to occur for MIMO virtual arrays is:

$$\Gamma_r = \frac{1}{2} N \, (N - 1) \, (M - 1) + \frac{1}{2} M \, (M - 1) \, (N - 1) = \ldots =$$
$$= \frac{1}{2} \left[N^2 \, (M - 1) + M^2 \, (N - 1) + M + N - 2 \, MN \right] = \qquad (6.26)$$
$$= \Gamma_{r,min} \quad \text{q.e.d.}$$

signal combinations. Please note, that all arrays provided by this method were of *restricted* type [26]. The above consideration with inspiration from Moffet who had demonstrated that all arrays with N, $M > 4$ exhibit higher redundancy, the array configuration of 6.19 can be considered as global optimum with $\gamma_r = 1.43$ for linear *restricted* MR-MIMO radar systems. Therefore, it can be proclaimed for the number of redundant signal combinations:

$$\Gamma_r \geq \frac{1}{2} \left[N^2 \, (M - 1) + M^2 \, (N - 1) + M + N - 2 \, MN \right] , \qquad (6.27)$$

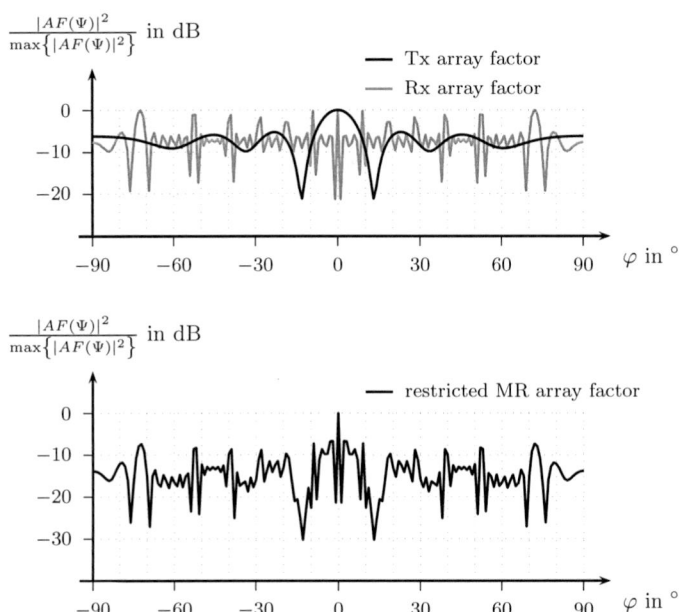

Fig. 6.20 Array factors of nested restricted MR arrays according to setups like in Fig. 6.12 or 6.14, $N = 4$, $M = 4$ (without *augmentation*): Array factors of sparse Tx and Rx arrays (top), array factor of complete virtual array with *restricted minimum redundancy* (bottom). For *augmentation* see section 6.3.3.

$$
\begin{aligned}
\gamma_r &\geq \frac{\Gamma_{d,ideal}}{\Gamma_{d,max}} = \frac{\Gamma_u \left(\Gamma_u - 1\right)}{2 \max \left\{\Gamma_d\right\}} = \frac{\Gamma_u \left(\Gamma_u - 1\right)}{2 \left(\Gamma_{d,ideal} - \Gamma_{r,min}\right)} = \cdots = \\
&= \frac{NM \left(NM - 1\right)}{\left(M^2 - M - 1\right) \cdot N \left(N - 1\right) + M \left(M - 1\right)} = \\
&= \frac{NM \left(NM - 1\right)}{\left(N^2 - N + 1\right) \cdot M \left(M - 1\right) + N \left(N - 1\right)} \geq 1.43 \, .
\end{aligned}
\tag{6.28}
$$

Equation (6.28) demonstrates further that Tx and Rx elements have equal effects on redundancy. Furthermore, it could be demonstrated that the overall behaviour is similar to nested codes in radar codes [55] or channel coding [103].

6.3.2 MR-MIMO setup as braced and nested zero redundancy approach

In the preceding section, the discrete acf of the Tx array was copied next to each other by a convolution operation with the Rx configuration. This distribution was similar to Tx array, however braced by the factor $p = 1/2\,N\,(N-1)$. Both together provided the discrete acf of the overall virtual array. Hence, one array was kept at zero redundancy, the other one was taken to put the virtual sub-arrays optimally. However, a closer look on Fig. 6.12 or 6.14 discloses a technical problem: One couple of transmitters is separated by $2\,d_0$, thus by $\lambda_0/2$ in this case. For large wavelengths, the different Tx elements can be fed by e. g. coaxial cables. However, the higher the frequency is chosen, the more losses occur in cables, such as ohmic losses $\sim \sqrt{f}$, dielectric losses $\sim f$ [58], [104]. This can be encountered by e. g. special cables with air as dielectric (and just a few dielectric rings for centring the inner conductor) to some extent.

Nevertheless, these methods are limited and in frequencies higher than 60 GHz, another type of waveguide becomes more and more attractive: The rectangular hollow waveguide (compare to section B.2, or e. g. [37]). The implemented systems (see section 10) for experimental results operate in so-called E-band (60 GHz to 90 GHz), explicitly around 75 GHz which gives a wavelength of $\lambda_0 = c_0/f_0 = 4\,\text{mm}$. In this spectral region, a very feasible type of antenna is the (pyramidal) horn antenna, which in fact is a tapered hollow waveguide.

However, such a rectangular waveguide must bear an inner dimension of more than half of the free-space wavelength ($a \geq \lambda_0/2$) in order to allow on field mode propagating. At the same time, two Tx elements shall be spaced by $\lambda_0/2$ which shows the inherent mechanical conflict for such a waveguide. In addition, such waveguides are standardised in order to cover certain frequency bands. In this special case, an E-band waveguide is of dimension $a \times b = 3.10\,\text{mm} \times 1.55\,\text{mm}$ (WR-12 flange, $a = 120\,\text{mil}$). The next higher one for W-band would be $a \times b = 2.54\,\text{mm} \times 1.27\,\text{mm}$ (WR-10 flange, $a = 100\,\text{mil}$). But in both cases the separation of $\lambda_0/2 = 2\,\text{mm}$ cannot be fulfilled.

Further, the horn antenna is a so-called *aperture antenna* [31], [32]. Acceptable gain and radiation pattern can only be achieved by aperture dimension larger than λ_0, which is not possible for one lateral direction, such as the azimuth direction, in this case. In order to put it in a nutshell, for the wanted frequency band, array setups like in Fig. 6.12 or 6.14 cannot be realized properly. Therefore, this section shall show a remedy to this by designing virtual arrays with slightly higher redundancy, but relaxed mechanical boundary conditions.

Now, the basic idea is to take a minimum redundancy (or even zero redundancy) Tx configuration and brace it by an integer factor $\varrho \leq (M-1)$. In case of a zero redundancy array of $N = 4$, the new Tx element spacing

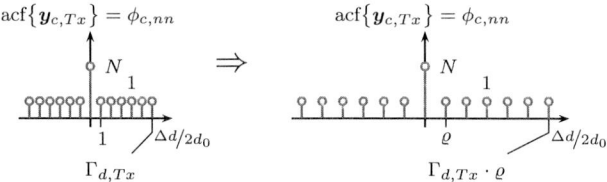

Fig. 6.21 The Tx array is braced by a factor ϱ. In this special case, the original Tx array is of zero redundancy, hence $\Gamma_{d,Tx} = 1/2\, N\, (N-1)$.

is then ϱ [1, 3, 2] $2d_0$. This leads to a systematic spatial sub-sampling in the virtual sub-arrays. The discrete acf of the Tx array exhibits lacks on the $d_0 = \lambda_0/4$ grid. Now, the spatial sub-sampling, which would again lead to grating lobes in the field of view, must be filled by reduced Rx element spacing and the relations between the virtual sub-arrays. The Rx element spacing is now a little more complex. The very first $(\varrho-1)$ smallest spacings in Rx array must be chosen out of:

$$\Delta y_{Rx} \in \{\Gamma_{d,Tx} \cdot \varrho + 1, \ldots, \Gamma_{d,Tx} \cdot \varrho + (\varrho - 1)\}\, 2d_0 \; . \tag{6.29}$$

Then, if all separations of equation (6.29) were aligned, the maximum Rx array dimension would be $\Gamma_{d,Rx}\, 2d_0$ with:

$$\Gamma_{d,Rx}^{(\varrho-1)} = \sum_{i}^{\varrho-1} \Delta y_{Rx,i} = [2\,\Gamma_{d,Tx} + 1] \cdot \frac{\varrho\,(\varrho - 1)}{2} \; . \tag{6.30}$$

Now this additional spacing is added to the set of element spacing in equation (6.29) but must be neighboured by the two smallest separations in equation (6.29). This secures a constant re-fill of the discrete acf $\phi_{c,nm}$. Furthermore this doubles the total length of Rx array. Fig. 6.23 demonstrates that the outcome is a minimum redundancy array of general type, thus the larger array (here Rx) dominates the total length of virtual ar-

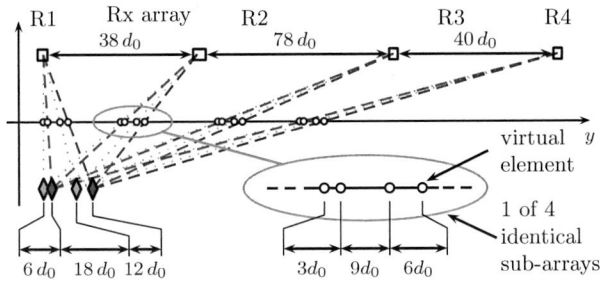

Fig. 6.22 Virtual array of *general minimum redundancy*: Generation by nesting of braced arrays. $N = M = 4$, $\varrho = 3$.

$\mathrm{acf}\{\boldsymbol{y}_c\} = \phi_{c,nm}$

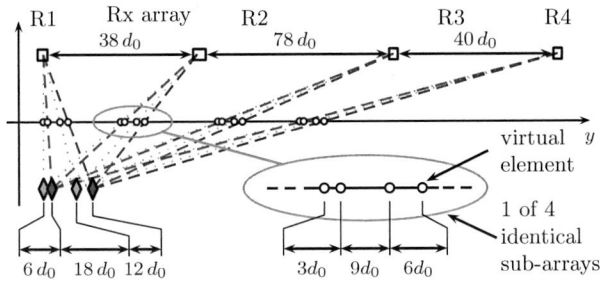

Fig. 6.23 *Spatial frequency spectrum* (only positive side) for setup in Fig. 6.22: Nesting of braced acfs ($\rho = 3$) giving a *general minimum redundancy* array of $\gamma_r = 1.54$.

ray. The maximum length of Rx array is $\Gamma_{d,Rx}\, 2d_0$, thus the total length of virtual array $\Gamma_d\, d_0$ is then respectively:

$$\Gamma_{d,Rx}^{(\varrho)} = 2\,\Gamma_{d,Rx}^{(\varrho-1)} = (2\,\Gamma_{d,Tx} + 1) \cdot \varrho\,(\varrho - 1) = \Gamma_d\,. \qquad (6.31)$$

Now, there are ϱ Rx element spacings, which finish the refill of $\phi_{c,nm}$. For the overall discrete acf, this leads to a nesting of braced acfs, which gives a filled spatial frequency distribution again. Furthermore, at the very end for high spatial frequencies, a sub-sampled part remains, thus the outcome are *general* MR-MIMO arrays, which usually bear a higher redundancy than *restricted* ones [26]. Anyhow, a setup like in Fig. 6.22 provides a figure

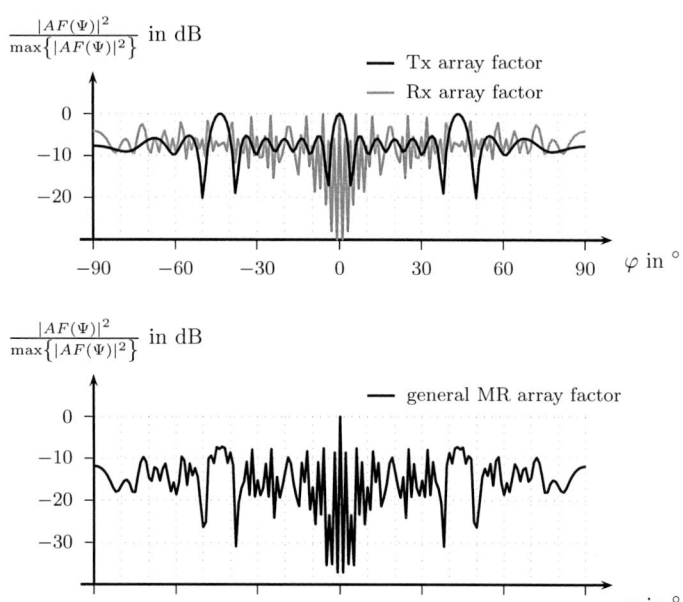

$\dfrac{|AF(\Psi)|^2}{\max\{|AF(\Psi)|^2\}}$ in dB

Fig. 6.24 Array factors of nested braced MR arrays according to setups like in Fig. 6.22 , $N = 4$, $M = 4$, $\rho = 3$ (without *augmentation*): Array factors of sparse Tx and Rx arrays (top), array factor of complete virtual array with *general minimum redundancy* (bottom). For *augmentation* see section 6.3.3.

of merit of $\gamma_r = {}^{(16 \cdot 15)}\!/_{(2 \cdot 78)} = 1.54$, which is slightly larger than for the *restriced* array. Again, the zero redundancy arrays of $N = M = 4$ provide the global minimum of γ_r. Fig. 6.24 shows the pure array factors of such a constellation without *augmentation techniques* [45], [87], [88], which will be explained in the following section. If the process of elongation was to be continued, another Rx element spacing could be introduced by:

$$\Gamma^{(\varrho+1)}_{d,Rx} = 2\,\Gamma^{(\varrho)}_{d,Rx} + 1 = 2\,(2\,\Gamma_{d,Tx} + 1) \cdot \varrho\,(\varrho - 1) + 1\,. \qquad (6.32)$$

However, combinatorics become more and more complicated. More promising is the idea of cascading such virtual arrays of minimum redundancy. The idea is equivalent to equation (6.32). The virtual array is shifted lat-

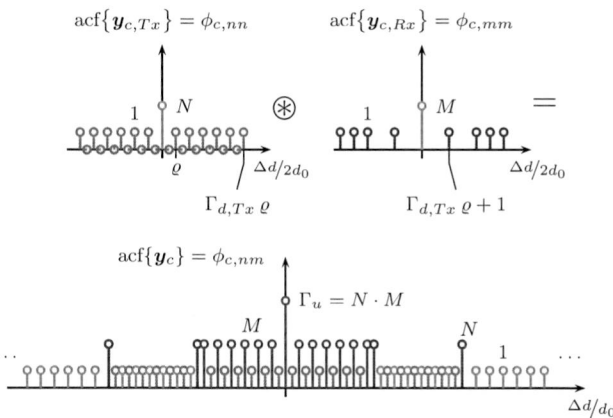

Fig. 6.25 Co-array for setup in Fig. 6.12 and 6.14 represented by the convolution of Tx and Rx co-array.

erally with Γ_d as fundamental parameter under MR considerations. This is demonstrated in chapter 7.

6.3.3 Augmentation of sparse covariance matrix

Up to now, only the count of certain signal combinations has been discussed. However, how are the values for filling the signal covariance matrix, gained exactly? A possible way in case of *uncorrelated signal scenario* (there the covariance matrix is *Toeplitz* [45], [87], [88]) shall be presented here [45], [87], [88]: The data acquisition and pre-processing stage provides the signal vector (matrix due to parameter t) $s_n(t)$ (equation (5.3)) which consists of $\Gamma_u = (NM)$ signals over time (rows). Those can be braced or extended to Γ signals by multiplication with a bracing matrix $\boldsymbol{B}_c$.

$$s_{sp}(t) = \boldsymbol{B}_c\, s_n(t)\,. \tag{6.33}$$

The rows filled in are zero vectors. However, the signals are now aligned properly in the sparse $\lambda_0/4$-grid, defined by y_c. The sparsely filled covariance matrix $\boldsymbol{R}_{sp}$ can then be expressed by:

$$\boldsymbol{R}_{sp} = \mathcal{E}\left\{s_{sp}(t)s_{sp}^{\mathrm{H}}(t)\right\} = \boldsymbol{B}_c\,\mathcal{E}\left\{s_n(t)s_n^{\mathrm{H}}(t)\right\}\boldsymbol{B}_c^{\mathrm{H}}\,. \tag{6.34}$$

The sparsely filled matrix $\boldsymbol{R}_{sp}$ is completed to a densely filled matrix $\boldsymbol{R}$ ($\boldsymbol{R}_n$ in noisy case) by taking the non-zero elements of each diagonal, forming the mean value and copying it to all diagonal positions. This mean value is in detail:

$$r^*_{mr}(\xi) = r_{mr}(-\xi) = \frac{1}{\phi_{c,nm}(\xi)} \sum_{\nu}^{\phi_{c,nm}(\xi)} r_{\nu}(\xi) \text{ , with } \xi \in \mathbb{Z}_0 . \quad (6.35)$$

There are as many summands as the *spatial frequency sensivity* $\phi_{c,nm}(\Delta d/d_0 = \xi)$ indicates for the lag ξ (please note: $\phi_{c,nm}(-\xi) = \phi_{c,nm}(\xi)$ [27], [95]). The only difference to a completely filled ULA configuration is, that the counts on axis of ordinate in *spatial frequency spectrum* $\phi_{c,nm}$ are much lower for MR arrays compared to uniformly filled, dense arrays. Therefore, systematic errors will have more impact on before-mentioned mean values and therefore on total quality of $\boldsymbol{R}$.

A closer look is needed to explain the bracing matrix $\boldsymbol{B}_c$. Its index c shall demonstrate that it only contains zero and one entries and can be derived from equation (6.17). $\boldsymbol{B}_c$ depends on the Tx and Rx configurations and can be decomposed into a *Kronecker product* [48]:

$$\boldsymbol{B}_{c,Rx} \otimes \boldsymbol{B}_{c,Tx} = \boldsymbol{B}_c \in \{0, 1\}^{[(2\,\Gamma_{d,Tx}+1)\cdot(2\,\Gamma_{d,Rx}+1)]\times\Gamma_u} . \quad (6.36)$$

For the example of array configurations from equation (6.15) ($N = M = 4$), $\boldsymbol{B}_{c,Rx}$ and $\boldsymbol{B}_{c,Tx}$ can be expressed by:

$$\boldsymbol{B}_{c,Tx} = \boldsymbol{B}_{c,Rx} = \begin{bmatrix} 1 & 0 & 0 & 0 \\ 0 & 1 & 0 & 0 \\ 0 & 0 & 0 & 0 \\ 0 & 0 & 0 & 0 \\ 0 & 0 & 1 & 0 \\ 0 & 0 & 0 & 0 \\ 0 & 0 & 0 & 1 \\ 0 & 0 & 0 & 0 \\ \vdots & \vdots & \vdots & \vdots \\ 0 & 0 & 0 & 0 \end{bmatrix} , \quad (6.37)$$

$$\boldsymbol{B}_{c,Rx} \in \{0, 1\}^{(2\,\Gamma_{d,Rx}+1)\times M} \text{ , } \boldsymbol{B}_{c,Tx} \in \{0, 1\}^{(2\,\Gamma_{d,Tx}+1)\times N} .$$

The *Kronecker product* serves as matrix convolution operation [43], [48]. Please note, that $\Gamma_{d,Tx}$ and $\Gamma_{d,Rx}$ correspond to the maximum lags in transmitter or receiver *co-array*. By a closer inspection, $\boldsymbol{y}_{c,Tx}$ can be found in columns of $\boldsymbol{B}_{c,Tx}$, $\boldsymbol{y}_{c,Rx}$ in $\boldsymbol{B}_{c,Rx}$ and $\boldsymbol{y}_{c,nm}$ in $\boldsymbol{B}_c$ respectively.

Equation (6.35) can be expressed completely by matrices. $\boldsymbol{B}_c$ braces the signal vector to a sparse one. The needed gap between element responses (parameter ξ) is provided by a shift matrix $\boldsymbol{B}_s$. For covariance entries of element gap ξ, $\boldsymbol{B}_s$ is defined by:

$$\boldsymbol{B}_s(\xi) = \begin{bmatrix} \boldsymbol{0}_{\xi \times (\Gamma - \xi)} & \boldsymbol{0}_{\xi \times \xi} \\ \boldsymbol{E}_{(\Gamma - \xi)} & \boldsymbol{0}_{(\Gamma - \xi) \times \xi} \end{bmatrix} \in \{0,1\}^{\Gamma \times \Gamma} \tag{6.38}$$

Again, the number of possible array positions in fundamental d_0-grid is denoted by Γ, In this case: $\Gamma = (2\,\Gamma_{d,Tx} + 1) \cdot (2\,\Gamma_{d,Rx} + 1)$. $\boldsymbol{E}_{(\Gamma - \xi)}$ is a unity matrix (eye) of dimension $(\Gamma - \xi) \times (\Gamma - \xi)$. The entries for ξ-th diagonal in the covariance matrix, $r_{mr}(\xi)$, are then (compare also to equation 6.35):

$$\begin{aligned} r_{mr}(\xi) &= \frac{1}{\phi_{c,nm}(\xi)}\, \mathcal{E}\left\{ \boldsymbol{s}_{sp}^{\mathrm{H}}\, \boldsymbol{s}_{sp,shifted}(\xi) \right\} = \\ &= \frac{1}{\phi_{c,nm}(\xi)}\, \mathcal{E}\left\{ \boldsymbol{s}_n^{\mathrm{H}}\, \boldsymbol{B}_c^{\mathrm{H}}\, \boldsymbol{B}_s(\xi)\, \boldsymbol{B}_c\, \boldsymbol{s}_n \right\} . \end{aligned} \tag{6.39}$$

According to the considerations presented in section 5.3.1 including the noise model, which results into bias by noise variances on the main diagonal elements only, equation (6.39) can be rewritten to:

$$r_{mr}(\xi) = \frac{1}{\phi_{c,nm}(\xi)} \sum_{q=1}^{Q} |\varsigma_q|^2 \left[\boldsymbol{a}_q^{\mathrm{H}}\, \boldsymbol{B}_c^{\mathrm{H}}\, \boldsymbol{B}_s(\xi)\, \boldsymbol{B}_c\, \boldsymbol{a}_q \right] .\, . \tag{6.40}$$

The calculation of equation (6.39) is only needed to be performed for the upper triangle of $\boldsymbol{R}_n$, since the lower triangle is the complex conjugate:

$$\begin{aligned} r_{mr}(-\xi) &= \frac{1}{\phi_{c,nm}(-\xi)}\, \mathcal{E}\left\{ \boldsymbol{s}_n^{\mathrm{H}}\, \boldsymbol{B}_c^{\mathrm{H}}\, \boldsymbol{B}_s(-\xi)\, \boldsymbol{B}_c\, \boldsymbol{s}_n \right\} = \\ &= \frac{1}{\phi_{c,nm}(\xi)}\, \mathcal{E}\left\{ \boldsymbol{s}_n^{\mathrm{H}}\, \boldsymbol{B}_c^{\mathrm{H}}\, \boldsymbol{B}_s^{\mathrm{H}}(\xi)\, \boldsymbol{B}_c\, \boldsymbol{s}_n \right\} = \\ &= \frac{1}{\phi_{c,nm}(\xi)} \sum_{q=1}^{Q} |\varsigma_q^2| \left[\boldsymbol{a}_q^{\mathrm{H}}\, \boldsymbol{B}_c^{\mathrm{H}}\, \boldsymbol{B}_s^{\mathrm{H}}(\xi)\, \boldsymbol{B}_c\, \boldsymbol{a}_q \right] = \\ &= r_{mr}^*(\xi) . \end{aligned} \tag{6.41}$$

Please note, that the *hermitian* of $\boldsymbol{B}_c$ or $\boldsymbol{B}_s$ is equal to the corresponding transposed version. Furthermore, it is valid: $\boldsymbol{B}_c(-\xi) = \boldsymbol{B}_c^{\mathrm{T}}(\xi) = \boldsymbol{B}_c^{\mathrm{H}}(\xi)$. Just by substituting $\boldsymbol{s}_n(t)$ by a vector of only ones in equation (6.39), also the *spatial frequency spectrum* $\phi_{c,nm}(\xi)$ could be determined.

As next step, the values of $r_{mr}(\xi)$ must be placed properly into covariance matrix $\boldsymbol{R}_{n,mr}$. This can be achieved by a matrix of zeros containing only ones on the ξ-th diagonal. By closer look, it can be seen that this matrix is identical to the transposed (*hermitian*) version of the shifting matrix $\boldsymbol{B}_s(\xi)$. For the main diagonal of $\boldsymbol{R}_{n,mr}$, the shift matrix becomes a unity matrix. Please note, that for the simplest noise model in section 5.3.1, the main diagonal was biased with noise variances (equation 5.14). These noise variances are also averaged to σ_{mr}^2 and added to $r_{mr}(0)$. With help of equation (6.41), the complete matrix $\boldsymbol{R}_{n,mr}$ can then be written as:

$$
\boldsymbol{R}_n = \boldsymbol{E}_\Gamma \left[r_{mr}(0) + \sigma_{mr}^2 \right] + \sum_{\xi=1}^{\Gamma-1} \left[\boldsymbol{B}_s(\xi) \, r_{mr}^*(\xi) + \boldsymbol{B}_s^{\mathrm{H}}(\xi) \, r_{mr}(\xi) \right] = \ldots
$$

$$
= \sigma_{mr}^2 \, \boldsymbol{E}_\Gamma + \sum_{\xi=-\Gamma+1}^{\Gamma-1} \frac{1}{\phi_{c,nm}(\xi)} \, \boldsymbol{B}_s^{\mathrm{H}}(\xi) \left\{ \sum_{q=1}^{Q} |\varsigma_q|^2 \left[\boldsymbol{a}_q^{\mathrm{H}} \, \boldsymbol{B}_c^{\mathrm{H}} \, \boldsymbol{B}_s(\xi) \, \boldsymbol{B}_c \, \boldsymbol{a}_q \right] \right\} .
$$

$$(6.42)$$

In general, beyond the introduced method, the filling of correlation lags and completing a *Toeplitz* matrix was also denoted as *matrix augmentation* by Pillai [45], [87], [88]. Also similar principles of copying matrix entries by matrices such as $\boldsymbol{B}_c$, can already be found for the general case in publications of Pillai [87], [88]. However, the proposed approach is feasible for the *uncorrelated signal scenario* [45], [87], [88] and has been developed especially for arrays with minimum redundancy [26]. For the *coherent scenario* a cascading solution shall be presented in the following section. This cascaded solution is a combination of spatial smoothing algorithm [45] with methods of minimum redundancy [26].

7 Cascaded minimum redundancy arrays

In section 6.3, the fundamental character of MIMO radars with minimum redundancy were examined. Thereby section 6.3 presented the largest lateral dimensions with zero redundancy structures (N, $M \in \{3, 4\}$). For even larger dimensions, more elements must be added in more and more complex steps. Alternatively, the idea of cascading several MR-MIMO arrays was introduced at the end of section 6.3.2. In both cases, *restricted* and *general* MR array, the analytical approach in section 6.3 has proofed that the overall discrete auto-correlation or co-array can be represented by the convolution of discrete Tx and Rx array auto-correlation functions [55], [95]:

$$\phi_{c,nm}(\xi) = \cdots = \phi_{c,nn}(2\xi) \circledast \phi_{c,mm}(2\xi) \ . \tag{7.1}$$

In case of cascading, thus shifting the MR-MIMO array, another convolution with a discrete co-array formed by the discrete auto-correlation of shift positions (counted in d_0-grid) completes equation (7.1).

7.1 The linearly cascaded minimum redundancy array

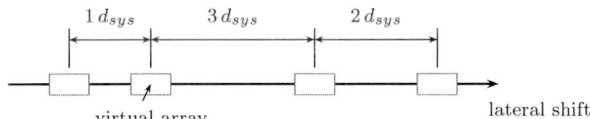

Fig. 7.1 Principle of cascading virtual arrays laterally under MR aspects.

7.1.1 Cascading MR-MIMO arrays by nesting

In same chronology like in section 6.3, cascading arrays by nesting one after another shall be examined. In Fig. 6.17, the principle had already been presented: A complete co-array is convoluted by another one with larger separations such that copies of the first are aligned without overlap (hence unnecessary redundancy) in *spatial frequency spectrum*. This method is feasible again for the general *uncorrelated signal scenario* [45], [87], [88]. According to the principle in section 6.3.1, the fundamental distance of the cascading co-array d_{sys} is:

$$p_{sys} = (2\,\Gamma_d + 1) \,, \tag{7.2}$$

$$d_{sys} = p_{sys}\, d_0 = (2\,\Gamma_d + 1)\, d_0 \,. \tag{7.3}$$

In this case, it does not matter if the fundamental array is of *restricted* or *general* type. The only difference is again the definition of Γ_d, which must be chosen accordingly to Fig. 6.19, 6.23. Fig. 7.1 demonstrates the fundamental principle. Whereas Fig. 7.2 shows the most interesting versions of $L = 3$ or $L = 4$ lateral positions, since those can be selected in zero redundancy manner.

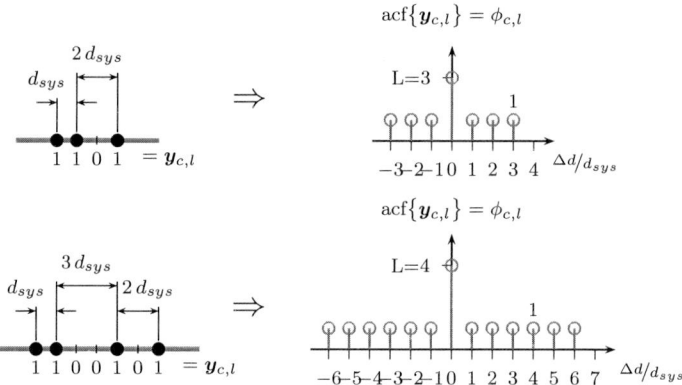

Fig. 7.2 Lateral shift positions for cascading, thus nesting of MR-MIMO arrays. Again as *restricted* MR array, with zero redundancy. For sake of clearness sketched in d_{sys}-grid.

For the *spatial frequency spectrum* of the overall synthetic aperture, this yields:

$$\phi_{c,sys}\left(\xi\right) = \cdots = \left[\phi_{c,nn}\left(2\xi\right) \circledast \phi_{c,mm}\left(2\xi\right)\right] * \phi_{c,sys}\left(\xi\right) \,, \tag{7.4}$$

with $\phi_{c,l}\left(\xi\right)$ as co-array of lateral shift positions in d_0-grid. The lateral dimension of the shifted, and therefore synthetic, aperture is $\Gamma_{sys}\, d_{sys}$:

$$\Gamma_{sys} \leq \frac{1}{2}L\left(L - 1\right) \,. \tag{7.5}$$

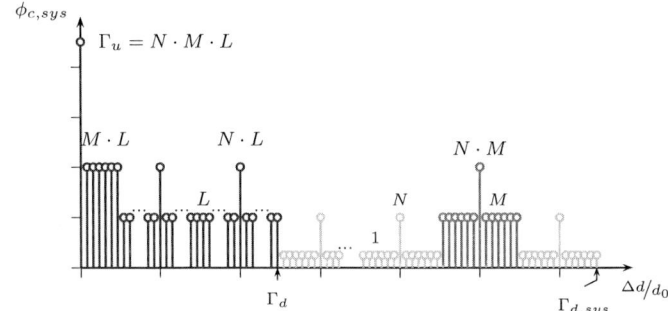

Fig. 7.3 *Spatial frequency spectrum* (only positive side) after cascading a restricted MR virtual array of $N = M = 4$ by $L = 4$ lateral positions.

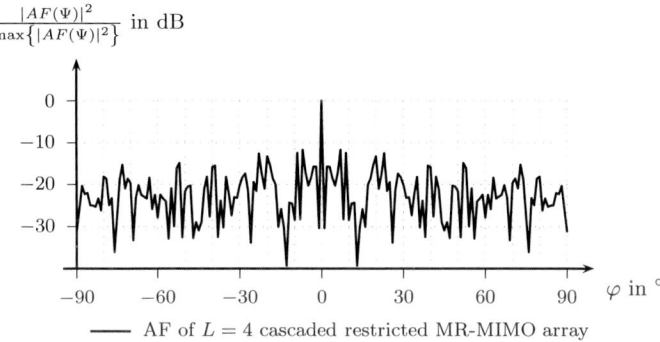

— AF of $L = 4$ cascaded restricted MR-MIMO array

Fig. 7.4 Array factor after cascading an optimal restricted MR virtual array of $N = M = 4$ by $L = 4$ lateral positions. $4 \cdot 16 = 64$ virtual elements cover an aperture length of $1098 \, d_0$.

Hereby, L is the total number of positions. Then, according to

$$\Gamma_d = p \cdot \Gamma_{d,Rx} + \Gamma_{d,Tx} = \Gamma_{d,Rx} \left(2 \, \Gamma_{d,Tx} + 1 \right) + \Gamma_{d,Tx} \, , \qquad (7.6)$$

the same consideration can be applied on $\Gamma_{d,sys}$:

$$\Gamma_{d,sys} = p_{sys} \cdot \Gamma_{sys} + \Gamma_d = \Gamma_{sys} \left(2 \, \Gamma_d + 1 \right) + \Gamma_d \, . \qquad (7.7)$$

In case of $N = M = L = 4$, thus nested zero redundancy arrays in all levels, this yields $\Gamma_{d,sys} = 1098$. However in this special case, only $16 \cdot 4 = 64$

virtual elements are involved. However, with a fundamental distance of $d_0 = 0.95 \, \mathrm{mm}$, a synthetic aperture of $1.043 \, \mathrm{m}$ can be generated. The figure of merit gives the following minimal redundancy:

$$\gamma = \frac{64 \cdot 63}{2 \cdot 1098} = 1.84 \, . \tag{7.8}$$

Nevertheless, the different MR approaches can also be mixed: A *general* MR array can be cascaded under *restricted* MR considerations. The *general* MR array of section 6.3.2 with $N = M = 4$ exhibits $\Gamma_d = 78$ (core without zero values in co-array). Shifted by four lateral positions with $\Gamma_{sys} = 157$ as fundamental multiple of d_0 ends in the array factor of Fig. 7.5. For later

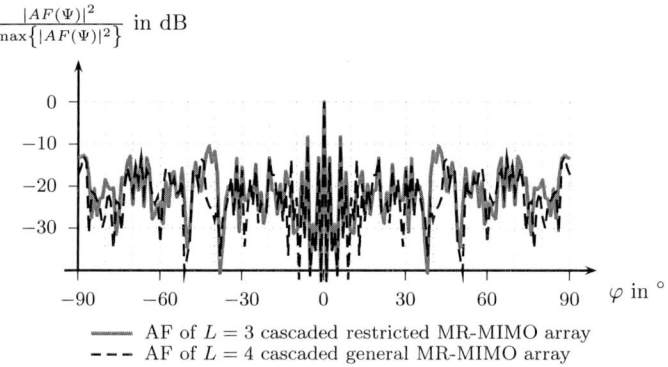

$\dfrac{|AF(\Psi)|^2}{\max\{|AF(\Psi)|^2\}}$ in dB

——— AF of $L = 3$ cascaded restricted MR-MIMO array
– – – AF of $L = 4$ cascaded general MR-MIMO array

Fig. 7.5 Array factor after cascading an optimal *general* MR virtual array of $N = M = 4$ by $L = 3$, $L = 4$ lateral positions. $4 \cdot 16 = 64$ virtual elements cover an aperture length of $549 \, d_0$ or $1020 \, d_0$ respectively.

experiments, the case $N = M = 4$, $L = 3$ is of practical relevance. There, a virtual array of *general* type $\Gamma_d = 78$, hence $\Gamma_{sys} = 157$ is shifted to three positions by a linear motion slide, which gives $\Gamma_{d,sys} = 3 \cdot 157 + 78 = 549$. With $d_0 = 0.95 \, \mathrm{mm}$, the total lateral dimension is then $0.522 \, \mathrm{m}$.

The principle of this section shows that the same mathematical methods as introduced in section 6.3 can be interleaved in a tree structure, thus multi-levels. Theoretically, the cascaded block of virtual arrays could again be cascaded in blocks under MR considerations. However, possible errors will affect the results more and more (see *Gauss error propagation* [43]). Besides that, for larger lateral dimensions, the need of absolutely plane wave fronts increases additionally. Therefore, the cascading into several levels is limited in praxis.

7.1.2 Cascading braced MR-MIMO arrays by nesting

In section 6.3, braced arrays were convoluted. Thereby, the bracing factor ϱ was limited to $\varrho \leq (M-1)$ in order to fill all remaining gaps in *spatial frequency sensivity*. It is also possible to choose $\varrho > (M-1)$. Then the Rx array configuration must contain again:

$$\Delta y_{Rx} \in \{\Gamma_{d,Tx} \cdot \varrho + 1, \ldots \Gamma_{d,Tx} \cdot \varrho + (M-2), \} \, 2d_0 \,. \tag{7.9}$$

At first instance, the single virtual array is then sub-sampled. This can be encountered by shifting the array again laterally. The additional spacings from $\Gamma_{d,Tx} \cdot \varrho + (M-1)$ to $\Gamma_{d,Tx} \cdot \varrho + \varrho$ are the separations of lateral shift positions. Then again, the total sum over both parts is the largest Rx spacing in Rx configuration: $\Gamma_{d,Tx} \cdot \varrho + (\varrho + 1) \, \varrho/2$. Nevertheless, this exhibits the fundamental problem of this method: Array configuration design and lateral positions are nested. Therefore the design is much more complicated than in section 7.1.1.

Nevertheless, a derivative of this idea helps especially for the *coherent signal scenario* [45], [87], [88]. The *spatial smoothing algorithm* was introduced in section 5.4.3 for especially this case. Now, the virtual array from Fig. 6.22 shall be examined. This general MR array could now be shifted laterally by many steps of $\lambda_0/4$. Tests for this thesis regarding *spatial smoothing algorithm* have shown that at least 60 shifts would be necessary for this array (Fig. 6.22) in order to obtain acceptable results with approximately 20 dB processing gain in angular processing. However, all design efforts for sparse arrays with minimum redundancy would be gone in this case. Furthermore many element positions and therefore their signal would be generated many times, although once would have been sufficient. Now the same considerations as for minimum redundancy can be imposed: Are there sparse shifting schemes in order to generate each virtual element only once or a few times along the lateral shift? The answer is yes and the solution is the same as proclaimed in section 6.3.

If the indication vector $\boldsymbol{y}_{c,nm}$ of the sparse general type array from Fig. 6.22 is taken and its co-array is examined, it can be seen that each element position is addressed once or just a few times, subject to a maximum lateral expansion. The discrete auto-correlation $\phi_{c,nm}$ could therefore also be interpreted as position vector $\boldsymbol{y}_{c,sys}$ for a virtual array. This can be written as:

$$\boldsymbol{y}_{c,sys}(\xi) \overset{!}{=} \phi_{c,nm} = \boldsymbol{y}_{c,nm}(\xi) * \boldsymbol{y}_{c,nm}(-\xi) = \\ = \boldsymbol{y}_{c,nm}(\xi) * \boldsymbol{y}_{c,l}(\xi) \,. \tag{7.10}$$

The correlation can be expressed by a convolution [43]. The sparse shifting sequence $y_{c,l}$ is nothing else but the vector $y_{c,nm}$ in inverted order. Thus, in case of e. g. the sparse array of *general* type MR (Fig. 6.22) consisting of $\Gamma_u = NM$ virtual elements (16 in this case), the shifting by Γ_u sparse lateral positions delivers a synthetic array shown in Fig. 7.6, whose fundamental shape is already known from the co-array $\phi_{c,nm}$ belonging the virtual array of Fig. 6.22. Therefore the mathematics behind $\phi_{c,nm}$ can be applied on $y_{c,sys}$ equivalently. In section 6.3.2, it was demonstrated that the length of

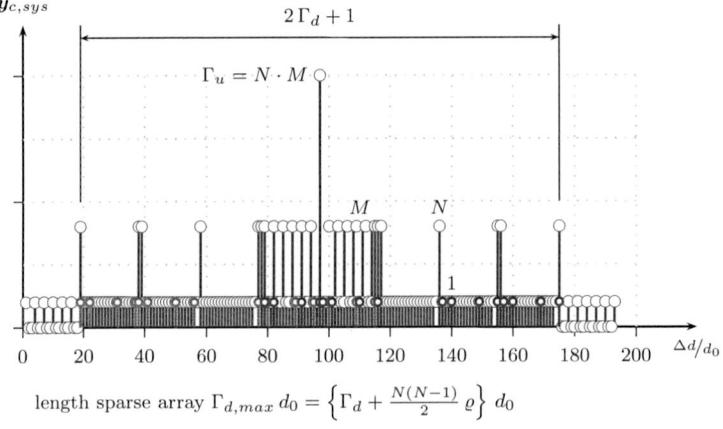

length sparse array $\Gamma_{d,max}\, d_0 = \left\{ \Gamma_d + \frac{N(N-1)}{2}\, \varrho \right\} d_0$

Fig. 7.6 Frequency of array element assignments for a general MR virtual array with $\Gamma_u = NM$ elements shifted laterally by Γ_u sparse positions. Selection of elements for the very left case in orange, for the very right one in blue.

inner core (w. r. t. *spatial frequency spectrum*) of such a *general* type array is:

$$\Gamma_d = [N\,(N-1)+1]\cdot \varrho \cdot (\varrho - 1)\ ,\ \varrho \leq (M-1)\ . \tag{7.11}$$

Please note that the virtual array itself is even longer:

$$\Gamma_{d,max} = \Gamma_d + \frac{N\,(N-1)}{2}\, \varrho =$$

$$= [N\,(N-1)+1]\cdot \varrho \cdot (\varrho - 1) + \frac{N\,(N-1)}{2}\, \varrho = \tag{7.12}$$

$$= N\,(N-1)\cdot \varrho \cdot (\varrho - 1/2) + \varrho \cdot (\varrho - 1)\ .$$

The obtained distribution $y_{c,sys}$ now describes a densely filled, one dimensional, synthetic virtual array whose inner core is of total length $2\Gamma_d + 1$. For the chosen example of $N = M = 4$ Fig. 6.22), the sparse array elements on the one hand are positioned at:

$$y_{nm} = \{1, 4, 13, 19, 20, 23, 32, 38, 59, 62, 71, 77, 79, 82, 91, 97\}\, d_0\ , \quad (7.13)$$

whereas the lateral shifts can be found at

$$y_l = \{1, 7, 16, 19, 21, 27, 36, 39, 60, 66, 75, 78, 79, 85, 94, 97\}\, d_0\ . \quad (7.14)$$

Now a conventional array processing method could be applied to this inner core, or the proposed *adapted spatial smoothing algorithm*. For the latter, the sparse distribution $y_{c,nm}$ must be mapped onto $y_{c,sys}$ (by steps of d_0), which can be done

$$\begin{aligned}
L_s &= 2\Gamma_d + 1 - \Gamma_d - \frac{N\,(N-1)}{2}\,\varrho = \\
&= \cdots = N\,(N-1)\cdot\varrho\cdot(\varrho - {}^3/_2) + \varrho\cdot(\varrho - 1)\ \text{times.}
\end{aligned} \quad (7.15)$$

The principle is sketched in Fig. 7.6: orange for start, blue for stop position. Then the matrix D from equation (5.47) can be imposed directly by using the L_s sparse signal vectors (gained by the mentioned mapping/shifting procedure) as columns for $D \in \mathbb{C}^{NM \times L_s}$. Then, the matrix R_ς from equation (5.46) can be determined by $R_\varsigma = D\,D^H$. R_ς in turn, is sparse. However, as proposed in section 5.4.3, it will be close to *Toeplitz* and can therefore be filled by copying the diagonal entries similar to section 6.3.3. Results will be presented in section 12.

7.2 The two dimensional cascaded minimum redundancy case

The step from one dimension to two dimensions is consequent: Discrete auto-correlation functions had been convoluted with each other. They had depended on one parameter due to the same lateral and one-dimensional shift. Nevertheless, the (linear) virtual array could also be shifted in two lateral dimensions. Hereby, the shift along the linear virtual array dimension is chosen as proposed in sections 7.1.1, 7.1.2. However, the second dimension (not necessarily orthogonal to first shift direction) can be selected in *restricted* MR manner, in order to obtain lower minimum redundancy as for the *general* type. In terms of mathematics and according to equation (7.4), this can be described by:

$$s_{c,sys} = y_{c,nm}\, e_1\, *\, u_{c,l1}\, e_1\, *\, v_{c,l2}\, e_2\ ,$$

$$\Rightarrow \phi_{c,sys}\,(\xi) = s_{c,sys}(\xi)\, *\, s_{c,sys}(-\xi) = \ldots$$

$$= \phi_{c,nm}(\xi)\, e_1\, *\, [u_{c,l1}(\xi)\, *\, u_{c,l1}(-\xi)]\, e_1\, *\, [v_{c,l2}(\xi)\, *\, v_{c,l2}(-\xi)]\, e_2 =$$

$$= \phi_{c,nm}(\xi)\, e_1\, *\, \phi_{c,l1}(\xi)\, e_1\, *\, \phi_{c,l2}(\xi)\, e_2\ ,$$

$$(7.16)$$

with $\phi_{c,l1}\,(\xi)$ and $\phi_{c,l2}\,(\xi)$ as discrete auto-correlation functions of lateral shift positions $u_{c,l1}$, $v_{c,l2}$ in d_0-grid along the fundamental vectors e_1, e_2 spanning the plane synthetic aperture. There are several possibilities for planar shifts according to achieve e. g. circular, rectangular or hexagonal aperture configurations [95]. The hexagonal one provides the same performance with respect to visible region and grating lobes like other configurations, however, with lowest amount of element positions [95]. This can be compared to spatial sampling theorem in two dimensions [34]. Nevertheless, since the experimental results were gained with a linear motion slide, the purely rectangular configuration had been chosen, thus $e_1 \perp e_2$.

Another effect is borne by equation (7.16): Up to now, Tx and Rx configuration had been fixed together due to mechanical reasons like e. g. the milling of horn antennas out of metal blocks (compare to section 10) or stiff fixations due to vibrations coming from harsh environment such as a helicopter turbine etc. (compare again to section 10). Furthermore, the antenna cross-talk or coupling effects stay more or less the same for the complete shift procedure. However, equation (7.16) gives another possibility. The Tx and Rx arrays could be shifted independently in two different directions e_1, e_2, resulting in the same overall discrete auto-correlation function $\phi_{c,nm}$, which might be of interest for bistatic systems of variable bistatic angles. However, the requirements for virtual array approximation are still mandatory such as in [21] or the phase front curvature which must not differ more than $\pi/8$ from plane projection [31].

The shape of two dimensional spatial frequencies shall be flat close to one with a single peak around $\Delta x \approx \Delta y \approx 0$. This closes the circle to the field of uniformly redundant arrays (URA) or coded apertures.

7.3 Description of cascading

In terms of mathematics, the cascading is related to the bracing procedure in section 6.3.3. In one dimensional case (section 7.1), a bracing matrix analog to equation (6.37) can be imposed:

$$
\boldsymbol{B}_{c,l1} = \begin{bmatrix} 1 & 0 & 0 & 0 \\ 0 & 1 & 0 & 0 \\ 0 & 0 & 0 & 0 \\ 0 & 0 & 0 & 0 \\ 0 & 0 & 1 & 0 \\ 0 & 0 & 0 & 0 \\ 0 & 0 & 0 & 1 \\ 0 & 0 & 0 & 0 \\ \vdots & \vdots & \vdots & \vdots \\ 0 & 0 & 0 & 0 \end{bmatrix}, \ \in \{0, \, 1\}^{[L_1 \, (L_1 - 1) + 1] \times L_1} \ . \tag{7.17}
$$

The signal vectors $\boldsymbol{s}_{n,l1}(t)$ measured at L_1 different lateral positions (numbered by l_1) are collected as rows in $\boldsymbol{s}_{n,L1}(t)$. This vector (matrix due to parameter t) is braced to $\boldsymbol{s}_{L1}(t)$ with mostly rows of zero (analogously to equation (6.36)):

$$
\boldsymbol{s}_{L1}(t) = (\boldsymbol{B}_{c,l1} \otimes \boldsymbol{B}_c) \, \boldsymbol{s}_{n,l1}(t) = (\boldsymbol{B}_{c,l1} \otimes \boldsymbol{B}_c) \begin{bmatrix} \vdots \\ \boldsymbol{s}_{n,l1}(t) \\ \vdots \end{bmatrix} ,
$$
$$
= \boldsymbol{B}_{c,L1} \, \boldsymbol{s}_{n,L1}(t) \, , \quad \boldsymbol{s}_{n,L1}(t) \in \mathbb{C}^{(NML_1) \times 1} \ . \tag{7.18}
$$

Hereby, the matrix $\boldsymbol{B}_c$ is the bracing for one single measurement of the virtual array. Whereas $\boldsymbol{B}_{c,l1}$ describes the lateral shift positions (see section 7.1). A *Kronecker product* of both gives a bracing matrix $\boldsymbol{B}_{c,L1}$ of the complete cascaded array. For angular signal processing, it can be written:

$$
\mathcal{E} \left\{ \boldsymbol{s}_{L1}(t) \, \boldsymbol{s}_{L1}^{\mathrm{H}}(t) \right\} =
$$
$$
= \boldsymbol{B}_{c,L1} \, \mathcal{E} \left\{ \boldsymbol{s}_{n,L1}(t) \, \boldsymbol{s}_{n,L1}^{\mathrm{H}}(t) \right\} \boldsymbol{B}_{c,L1}^{\mathrm{T}} = \tag{7.19}
$$
$$
= \boldsymbol{B}_{c,L1} \, \boldsymbol{R}_{n,L1} \, \boldsymbol{B}_{c,L1}^{\mathrm{T}} \ .
$$

The index $L1$ has already suggested the lateral shift along the main axis of array. In two dimensional cascading, a second shift direction, denoted by $L2$, is added. Now, the signals acquired along $L1$ are collected in columns

$s_{n,L1,l2}(t)$, as for the one dimensional case, and are then used to form the matrix $S_{n,L1,L2}$:

$$S_{L1,L2}(t) = B_{c,L1}\, S_{n,L1,L2}\, B_{c,L2}^{\mathrm{T}}\,,$$

$$S_{n,L1,L2} = \begin{bmatrix} & \vdots & \\ \cdots & s_{n,l1,l2}(t) & \cdots \\ & \vdots & \end{bmatrix} \in \mathbb{C}^{(NML_1)\times L_2}\,. \tag{7.20}$$

The bracing matrix $B_{c,L2}$ along second lateral direction is constructed equivalently to $B_{c,L1}$:

$$\begin{aligned} B_{c,L2} &= (B_{c,l2} \otimes B_{c,2})\,, \\ B_{c,L2}^{\mathrm{T}} &= (B_{c,l2}^{\mathrm{T}} \otimes B_{c,2}^{\mathrm{T}})\,, \text{ see [48].} \end{aligned} \tag{7.21}$$

Hereby, $B_{c,2}$ could be e. g. $[1,1,0,1,0,0,0]$, the zero redundancy constellation in d_0-steps, transformed to a matrix expression (like in equation (7.17)). This gives a zero redundancy configuration with three steps along second lateral dimension. It can be cascaded with minimum redundancy again, analogously to $B_{c,l1}$. However, in this case, there are in total $L2$ steps along second lateral dimension. In terms of signal processing, this yields:

$$\begin{aligned} \mathcal{E}\left\{ S_{L1,L2}(t)\, S_{L1,L2}^{\mathrm{H}}(t) \right\} &= \\ = B_{c,L1}\, \mathcal{E}\left\{ S_{n,L1,L2}(t)\, B_{c,L2}^{\mathrm{T}}\, B_{c,L2}\, S_{n,L1,L2}^{\mathrm{H}}(t) \right\} B_{c,L1}^{\mathrm{T}} &= \\ = B_{c,L1}\, \mathcal{E}\left\{ S_{n,L1,L2}(t)\, E_{L2\times L2}\, S_{n,L1,L2}^{\mathrm{H}}(t) \right\} B_{c,L1}^{\mathrm{T}} &= \\ = B_{c,L1}\, \mathcal{E}\left\{ \sum_{l2=1}^{L2} s_{n,L1,l2}(t)\, s_{n,L1,l2}^{\mathrm{H}}(t) \right\} B_{c,L1}^{\mathrm{T}} &= \\ = B_{c,L1}\, \left\{ \sum_{l2=1}^{L2} R_{n,L1,l2} \right\} B_{c,L1}^{\mathrm{T}} &= \\ = B_{c,L1}\, R_{n,L1,L2}\, B_{c,L1}^{\mathrm{T}}\,. \end{aligned} \tag{7.22}$$

Equation (7.22) demonstrates that the overall two dimensional covariance matrix is a sum over all covariance matrices formed by the one dimensional signal vectors along $L1$ direction. The augmentation of the overall covariance matrix is then again provided by the mean over all non-zero elements in the corresponding diagonals (see equation (6.35)).

Please note, that the vectors corresponding to $L1$ and $L2$ are not necessarily orthogonal to each other in actual environment. They can also span an e. g. hexagonal synthetic aperture [34], [95]. Nevertheless, both vectors must not be collinear to each other.

7.4 Relation to Coded aperture imaging

The presented principles are not only closely related to coding theory [55], [97], [103] but also to general considerations on two dimensional arrays exhibiting a uniform redundancy distribution (or low sidelobe level) called *uniform redundancy array* (URA) [97], [98], [99] or *coded aperture* [100], [101], [102]. The principle idea of coded apertures/ coded masks was the attempt of saving effort (reducing the amount of expensive detectors) in x-ray or γ-ray imaging [101], [105], [106], [107], [108]. Some also depicted this topic as *multiple pinhole masks* [109]. This topic was as well present in x-ray computer tomography [110], [111] or x-ray or γ-ray telescope design [112], [113]. Later the *uniform redundancy array* (URA) were linked to minimization [114] which closes the circle to the topic of compressive sensing again.

8 Noise and error propagation

In the preceding sections the signal structure was introduced. However, specific noise effects or error propagation need to be examined in more detail. First, the effects of non-white noise on virtual elements' signals shall be demonstrated, then the propagation of error terms is examined.

8.1 Noise contribution to virtual elements

In equation (5.12), it was assumed that each of the individual $\Gamma_u = NM$ MIMO channels is biased by independent noise contributions:

$$\boldsymbol{s}_n(t) = \boldsymbol{s}_p(t) + \boldsymbol{n}(t) \, , \, \boldsymbol{n}(t) \in \mathbb{C}^{(NM) \times 1} \, . \tag{8.1}$$

However, this would be valid for a filled monostatic array of $\Gamma_u = NM$ transceiver (TRx) elements. Contrary to that, the convolution principle behind coherent MIMO radar, as introduced in sections 3.4 and 9, discloses that not only the $(N \cdot M)$ signals but also the $(N \cdot M)$ noise contributions are driven by N transmitters plus M receivers $(N + M)$. Therefore, the noise contributions are no longer independent from each other, hence not only the main diagonal of covariance matrix is biased but also some secondary diagonals. In case of additive noise, the noise contribution per MIMO channel can be written as:

$$\boldsymbol{n}(t) = \boldsymbol{A}_{sys} \, \boldsymbol{n}_{TRx}(t) \, ,$$

$$\boldsymbol{n}_{TRx}(t)^{\mathrm{T}} = \begin{bmatrix} \ldots & n_{Rm}(t) & \ldots & n_{Tn}(t) & \ldots \end{bmatrix}^{\mathrm{T}} \, ,$$

$$\text{with} \quad \boldsymbol{A}_{sys} = \begin{bmatrix} \boldsymbol{1}_{N \times 1} \, \boldsymbol{p}_1^{\mathrm{T}}, & \boldsymbol{E}_N \\ \vdots & \vdots \\ \boldsymbol{1}_{N \times 1} \, \boldsymbol{p}_m^{\mathrm{T}}, & \boldsymbol{E}_N \\ \vdots & \vdots \\ \boldsymbol{1}_{N \times 1} \, \boldsymbol{p}_M^{\mathrm{T}}, & \boldsymbol{E}_N \end{bmatrix} \in \{0, \, 1\}^{(NM) \times (N+M)} \, . \tag{8.2}$$

The vector $\boldsymbol{1}_{N \times 1}$ is a column containing N ones, whereas $\boldsymbol{p}_m$ is a column of $(M - 1)$ zeros, however a single one entry at mth position is picked out. $\boldsymbol{E_N}$ is a $N \times N$ unity matrix (eye).

Hereby, the receiver noise $n_{Rx}(t)$ is mainly thermal noise (kTB), thus can be modelled as white noise within the bandwidth B [32], [104]. Whereas, the possible transmitter noise $n_{Tx}(t)$ can be more complex: Spurials, harmonics due to output stage nonlinearities, amplitude noise of output power amplifier, phase noise of signal generation, AM to FM conversion by filters etc., thus in general this noise contribution must be taken into account as non-white noise [32], [67], [104]. As next step, the non-white transmitter noise is supposed to be uncorrelated to the transmitted, reflected and received signals:

$$\mathcal{E}\left\{s_p(t)\,n^{\mathrm{H}}(t)\right\} \approx \mathbf{0} \ . \tag{8.3}$$

This must be considered as approximation since e. g. in case of transmitted harmonics, there are of course deterministic relations between the wanted signal and its harmonic derivatives. Therefore, in this model, it is assumed that the transmitter noise is radiated, reflected and received independently from the desired signal. However, if the non-white noise components are created by signal generation and distribution, the Tx noise components can be considered correlated to each other in general. Nevertheless, the additive white receiver noise is uncorrelated to the additive non-white transmitter noise:

$$\mathcal{E}\left\{n_{Rx}(t)\,n_{Tx}^{\mathrm{H}}(t)\right\} = \mathbf{0} \ . \tag{8.4}$$

For the single noise contributions, it can be stated [48]:

$$\mathcal{E}\left\{n_{Rx}\,n_{Rx}^{\mathrm{H}}\right\} = \mathrm{diag}\left\{\dots,\,\sigma_{Rx}^2,\,\dots\right\} = \mathbf{N}_{Rx} \ , \tag{8.5}$$

$$\begin{aligned}\mathcal{E}\left\{n_{Tx}\,n_{Tx}^{\mathrm{H}}\right\} &= \mathbf{m}_{Tx}\,\mathbf{m}_{Tx}^{\mathrm{H}} + \mathcal{E}\left\{(n_{Tx} - \mathbf{m}_{Tx})\,(n_{Tx} - \mathbf{m}_{Tx})^{\mathrm{H}}\right\} = \\ &= \boldsymbol{\sigma}_{Tx}\,\boldsymbol{\sigma}_{Tx}^{\mathrm{H}} = \mathbf{N}_{Tx} \ .\end{aligned} \tag{8.6}$$

However, the transmitter noise has been radiated by an antenna. Those in turn cannot radiate constant terms (sources with $f = 0$) [31], [32]. Therefore, the overall transmitter noise contribution n_{Tx} in signal processing can be considered to be zero mean ($\mathbf{m}_{Tx} = \mathbf{0}$). Nevertheless, the covariance matrix $\mathbf{N}_{Tx}$ in equation (8.6) is supposed to be completely filled. Now, both noise contributions are aligned into one vector:

$$n_{TRx}(t) = \begin{bmatrix} n_{Rx}(t) \\ n_{Tx}(t) \end{bmatrix} \ . \tag{8.7}$$

The bias noise matrix for eigenvalue shift as introduced in equation (5.13), (5.14), with help of equation (8.2), must then be written as:

$$\mathcal{E}\left\{n(t)\,n^{H}(t)\right\} = A_{sys}\,\mathcal{E}\left\{n_{TRx}(t)\,n_{TRx}(t)^{H}\right\}\,A_{sys}^{H}\,, \tag{8.8}$$

$$\mathcal{E}\left\{n_{TRx}(t)\,n_{TRx}^{H}(t)\right\} = \begin{bmatrix} \text{diag}\left\{\dots,\sigma_{Rx}^{2},\dots\right\}, & \mathbf{0}_{M\times N} \\ \mathbf{0}_{N\times M}, & \mathcal{E}\left\{n_{Tx}(t)\,n_{Tx}^{H}(t)\right\} \end{bmatrix} =$$

$$= \begin{bmatrix} \text{diag}\left\{\dots,\sigma_{Rx}^{2},\dots\right\}, & \mathbf{0}_{M\times N} \\ \mathbf{0}_{N\times M}, & \sigma_{Tx}\,\sigma_{Tx}^{H} \end{bmatrix}. \tag{8.9}$$

By evaluation of equation (8.8), the additive noise matrix from equation (8.8) can be decomposed into [43]:

$$\mathcal{E}\left\{n(t)\,n^{H}(t)\right\} = \begin{bmatrix} \ddots & 0 & 0 \\ 0 & \sigma_{Rm}^{2}\cdot\mathbf{1}_{N\times N} & 0 \\ 0 & 0 & \ddots \end{bmatrix} + \mathbf{1}_{M\times M}\otimes\left[\sigma_{Tx}\,\sigma_{Tx}^{H}\right]\,. \tag{8.10}$$

Hereby, $\otimes$ represents a *Kronecker product* (see [48]), which gives an $(NM)\times(NM)$ matrix filled with copies of $N_{Tx} = \sigma_{Tx}\,\sigma_{Tx}^{H}$. The first part, however, is a $(NM)\times(NM)$ *Jordan matrix* filled with $\sigma_{Rm}^{2}\cdot\mathbf{1}_{N\times N}$ as *Jordan boxes* on the main diagonal [43]. Please note, $\mathbf{1}_{i\times j}$ are matrices of dimension $i\times j$ filled with ones. If another assumption was now applied, such as the transmitter noise was not correlated pairwise, the fundamental transmitter covariance matrix N_{Tx} would become a diagonal matrix. The overall noise covariance matrix in equation (8.10) would not only get biased on the main diagonal but also on every secondary diagonal with distance of multiples of $\Delta d/d_{0} = N$. The conventional monostatic array with $\Gamma_{u} = NM$ transceiver

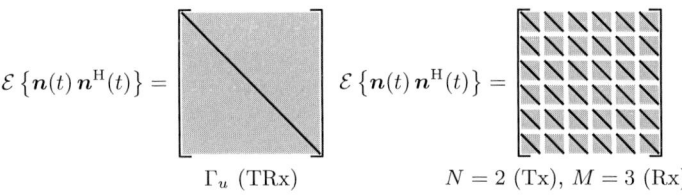

$$\mathcal{E}\left\{n(t)\,n^{H}(t)\right\} = \qquad\qquad \mathcal{E}\left\{n(t)\,n^{H}(t)\right\} =$$

$$\Gamma_{u}\ (\text{TRx}) \qquad\qquad N = 2\ (\text{Tx}),\ M = 3\ (\text{Rx})$$

Fig. 8.1 Noise contribution: Comparison of completely filled monostatic array (left) with a virtual MIMO array of $N = 2$, $M = 3$ (right).

elements (TRx) could also exhibit transmitter noise. This would result in a noise covariance matrix of dimension $\Gamma_u \times \Gamma_u$. In case of uncorrelated Tx noise, only its main diagonal becomes biased. Similar considerations can be found for the thermal receiver noise. Then the *Jordan boxes* in equation (8.10) degrade to dimension 1×1 which results to a diagonal load with noise variances. Equation (8.1) is fully fulfilled then.

8.1.1 Systematic errors

Unlike statistical errors, such as noise contributions, wrong antenna element placement or imperfect (range dependent) curvature filtering (see section 5.2) causes systematic errors. The effect of those errors can be explained briefly with help of equation (4.71):

$$r_{nqm} \approx 2 \left[r_{0q} + \frac{1}{4r_{0q}} \left(d_n^2 + d_m^2 \right) - \frac{1}{2} \left(d_n + d_m \right) \cdot e_r \left(\vartheta_q, \varphi_q \right) \right] =$$
$$= 2 \left[r_{0q} + \delta_i^2 \left(r_{0q} \right) - d_i \cdot e_r \left(\vartheta_q, \varphi_q \right) \right] . \tag{8.11}$$

In case of small shifts applied to Tx and Rx positions, hence $d_n + \Delta d_n$ and $d_m + \Delta d_m$ in equation (8.11), a corrupted bistatic path length $r_{nqm} + \Delta r_{nqm}$ can be obtained. The shift Δr_{nqm} is then:

$$\Delta r_{nqm} = \frac{1}{2r_{0q}} \left(d_n \cdot 2\Delta d_n + d_m \cdot 2\Delta d_m + \Delta d_n^2 + \Delta d_m^2 \right)$$
$$- \left(\Delta d_n + \Delta d_m \right) \cdot e_r(\vartheta_q, \varphi_q) \approx \tag{8.12}$$
$$\approx \frac{1}{2r_{0q}} \left(d_n \cdot 2\Delta d_n + d_m \cdot 2\Delta d_m \right) - \left(\Delta d_n + \Delta d_m \right) \cdot e_r(\vartheta_q, \varphi_q)$$

Hereby, the terms Δd_n^2, Δd_m^2 could be neglected. The very first term in equation (8.14) describes a systematic error in phase front curvature which is independent of direction of arrival. Please note, that the approximation in equation (8.14) is equivalent to considerations leading *total differential* [43] or *Gaussian error propagation law* [43].

At first glance, the deviations with factor two suggest the MIMO radar to be very sensitive to such errors, or at least more than a classical completely filled monostatic array. However, if we impose equation (8.11) for a classical monostatic array, thus $d_n = d_m$, this yields:

$$r_{nqn} \approx 2 \left[r_{0q} + \frac{1}{4r_{0q}} 2d_n^2 - \frac{1}{2} 2d_n \cdot e_r \left(\vartheta_q, \varphi_q \right) \right] . \tag{8.13}$$

With a shift $\Delta \boldsymbol{d}_n$, this results in shift of Δr_{nqn}:

$$\Delta r_{nqn} \approx \frac{1}{2r_{0q}} 2\boldsymbol{d}_n \cdot 2\Delta \boldsymbol{d}_n - 2\Delta \boldsymbol{d}_n \cdot \boldsymbol{e}_r \left(\vartheta_q, \, \varphi_q \right) \ . \tag{8.14}$$

Therefore, the monostatic array is even more sensitive to a $\Delta \boldsymbol{d}_n$ than the proposed MIMO ansatz (by the factor two). The reason for this is simple: If a miss-placement of a Tx or Rx element took place, the error would only effect the transmit or receive path in MIMO case whereas it affects both for the standard monostatic case. The coherent MIMO approach is therefore less sensitive to element miss-placement than any standard monostatic array configuration. Furthermore, there are $N + M$ possible element shifts for the MIMO radar, whereas there are $\Gamma = N M$ possible shifts for a comparable monostatic configuration.

In the following, the effect of this error to the covariance matrix is examined. The resulting error phase for our MIMO radar is then:

$$\Delta \phi_{nqm} = \frac{4\pi}{c_0} \left(Kt + f_0 \right) \Delta r_{nqm} \approx \frac{4\pi}{c_0} f_0 \, \Delta r_{nqm} \ . \tag{8.15}$$

Since the range cell migration was assumed to be small anyhow, tiny shifts in antenna positions are even more negligible. Then the plane wave representation in equation (5.3) must be amended by diagonal matrices $\Delta \boldsymbol{\Lambda}$, $\Delta \boldsymbol{\Phi}_q$:

$$\tilde{\boldsymbol{s}}_p(t) = \sum_{q=1}^{Q} s_q(r_{0q}, f_{d,q}, t) \cdot \boldsymbol{\Phi}_q \left(\vartheta_q, \varphi_q, t \right) \Delta \boldsymbol{\Lambda}(\Delta \boldsymbol{d}_n, \Delta \boldsymbol{d}_m, r_{0q})$$
$$\cdot \Delta \boldsymbol{\Phi}_q \left(\vartheta_q, \varphi_q, t \right) \boldsymbol{a}_q \left(\vartheta_q, \, \varphi_q \right) \ . \tag{8.16}$$

$$\Delta \boldsymbol{\Lambda} = \mathrm{diag} \left\{ \ldots, \mathrm{e}^{+j \frac{4\pi}{c_0} f_0 \frac{1}{2r_{0q}} [2 \, \boldsymbol{d}_n \cdot \Delta \boldsymbol{d}_n + 2 \, \boldsymbol{d}_m \cdot \Delta \boldsymbol{d}_m]}, \ldots \right\} \ , \tag{8.17}$$

$$\Delta \boldsymbol{\Phi}_q = \mathrm{diag} \left\{ \ldots, \mathrm{e}^{-j \frac{4\pi}{c_0} f_0 \, 2\Delta \boldsymbol{d}_i \cdot \boldsymbol{e}_r (\vartheta_q, \varphi_q)}, \ldots \right\} ;,$$

$$\text{with } \Delta \boldsymbol{d}_i = \frac{1}{2} \left(\Delta \boldsymbol{d}_n + \Delta \boldsymbol{d}_m \right) \ . \tag{8.18}$$

Both matrices can be put together as $\Delta \boldsymbol{A}_q = \Delta \boldsymbol{\Lambda} \, \Delta \boldsymbol{\Phi}_q$. If equation (5.11) is imposed again, this yields in general:

$$\tilde{\boldsymbol{R}} = \sum_{q=1}^{Q} |\varsigma_q|^2 \cdot \Delta \boldsymbol{A}_q \, \boldsymbol{a}_q \, \boldsymbol{a}_q^{\mathrm{H}} \, \Delta \boldsymbol{A}_q^{\mathrm{H}} \; , \; \text{uncorrelated case,} \tag{8.19}$$

$$\tilde{\boldsymbol{R}} = \sum_{q=1}^{Q} |\varsigma_q|^2 \cdot \Delta \boldsymbol{A}_q \, \boldsymbol{a}_q \, \boldsymbol{a}_q^{\mathrm{H}} \, \Delta \boldsymbol{A}_q^{\mathrm{H}} + \tag{8.20}$$

$$+ \sum_{\eta \neq \xi}^{Q(Q-1)} \varsigma_\eta \, \varsigma_\xi^* \cdot \Delta \boldsymbol{A}_\eta \, \boldsymbol{a}_\eta \, \boldsymbol{a}_\xi^{\mathrm{H}} \, \Delta \boldsymbol{A}_\xi^{\mathrm{H}} \; , \; \text{coherent case.} \tag{8.21}$$

The equations (8.17), (8.18) demonstrate that only N possible shifts of transmitters and M possible shifts of receiver positions, cause NM combinations for the diagonal matrices. The reason for this is again the convolutional character of coherent MIMO radar. In case of non-plane wavefronts due to e. g. multipath propagation, multiple reflections etc. also each receiver signal becomes biased with a complex factor, which again affects all NM signal combinations and results finally in a diagonal matrix like $\Delta \boldsymbol{A}_q$. Therefore, those errors can be collected together in a matrix $\Delta \boldsymbol{A}_q$. An imperfect correction of quadratic phase front terms (section 5.2) would also lead to a $\Delta \boldsymbol{A}_q$ bias matrix.

The terms independent of angle of arrival will be compensated by a calibration procedure presented in section 11. The terms depending on the direction of arrival must be discriminated into two cases: Those caused by miss placement of antenna elements to each other were minimized mechanically due to milling horn antennas out of blocks of brass (section 10). The others due to arbitrarily shaped wave fronts by multiple reflections, cannot be compensated since they highly depend on the scenario, the site of radar sensor and target as well as the main reflectors like e. g. walls etc. But in this case, the intended principle of minimum redundancy would not work anyway (compare to section 6.1, Fig. 6.1).

8.2 Propagation of errors

8.2.1 Error propagation in virtual arrays

Overall, the systematic errors add with the noise contributions. Therefore if equation (5.13) is imposed again, it can be obtained for the *uncorrelated signal scenario*:

$$\tilde{R}_n = \sum_{q=1}^{Q} |\varsigma_q|^2 \cdot \Delta A_q \, a_q \, a_q^{\mathrm{H}} \, \Delta A_q^{\mathrm{H}} + \mathcal{E}\left\{ n \, n^{\mathrm{H}} \right\} . \tag{8.22}$$

Whereas, for the *coherent scenario*, the equation must be amended to

$$\tilde{R}_n = \sum_{q=1}^{Q} |\varsigma_q|^2 \cdot \Delta A_q \, a_q \, a_q^{\mathrm{H}} \, \Delta A_q^{\mathrm{H}} + \sum_{\eta \neq \xi}^{Q(Q-1)} \varsigma_\eta \, \varsigma_\xi^* \cdot \Delta A_\eta \, a_\eta \, a_\xi^{\mathrm{H}} \, \Delta A_\xi^{\mathrm{H}} +$$
$$+ \mathcal{E}\left\{ n \, n^{\mathrm{H}} \right\} .$$
$$\tag{8.23}$$

The convolutional character of virtual errors becomes especially obvious at noise and error contribution.

8.2.2 Error propagation in virtual arrays with minimum redundancy

Unlike completely filled arrays such as *uniform linear arrays* (ULA), missing virtual elements' signals were derived from the existing ones. Therefore, not only thermal noise but also possible non-white noise contributions (equation (8.9)) or systematic errors (equation (8.19)) are loaded on the virtual elements' signals but they are also propagated systematically during determination of missing elements by use of the others. The noise and error model has been introduced in this section, equation 6.42 can be altered accordingly to:

$$\tilde{R}_n = \sum_{\xi=-\Gamma+1}^{\Gamma-1} \frac{1}{\phi_{c,nm}(\xi)} \, B_s^{\mathrm{H}}(\xi) \, r_{mr}(\xi) ,$$
$$\text{with } r_{mr}(\xi) = \sum_{q=1}^{Q} |\varsigma_q|^2 \, a_q^{\mathrm{H}} \, \Delta A_q^{\mathrm{H}} \, B_c^{\mathrm{H}} \, B_s(\xi) \, B_c \, \Delta A_q \, a_q + \tag{8.24}$$
$$+ \mathcal{E}\left\{ n_{TRx}^{\mathrm{H}} \, A_{sys}^{\mathrm{H}} \, B_c^{\mathrm{H}} \, B_s(\xi) \, B_c \, A_{sys} \, n_{TRx} \right\} .$$

8.2.3 Errors in cascaded virtual arrays

Basically, equation (8.24) is valid for cascaded arrays, too. However, the bracing matrices correspond to the composed matrices presented in section 7.3, as well as the directional vectors $\boldsymbol{a}_q$, as well as the error matrices $\Delta\boldsymbol{A}_q$ contain NML_1 values due to lateral shift. The noise vector $\boldsymbol{n}_{TRx}$ contains $L_1 \cdot (N + M)$ values. $\boldsymbol{A}_{sys}$ must be adapted accordingly.

Equation (8.24) bears the fundamental limitation of (cascaded) minimum redundancy arrays. Noise terms get correlated into element responses and decrease the quality of signal. Independent array elements exhibit uncorrelated noise and are therefore superior to MR array cascaded in a few steps.

9 MIMO radar with compressed sensing

9.1 A short introduction to compressed sensing

A possible way for signal reconstruction in especially coherent signal scenarios can be provided by means of compressed sensing. A fundamental
idea behind signal analysis is the decomposition of a measured signal into
functions or *wavelets* whose properties are well-known a-priori [43], [56].
The signal (sequence, vector) to be analyzed, $s(t)$, is compared to a class of
fundamental functions $\psi(\omega, t)$ by forming an *inner product*, whereby $\psi(t)$
serves as *integration kernel* (in complex conjugated from) [43]. The outcome
is usually a sequence of *functionals* denoted as spectrum and represents the
needed weighting of the fundamental functions $\psi(t)$ to compose $s(t)$ again
[43]. Typical examples are *Fourier* or *Laplace* transformation [43], [56].

Hereby, the sequence (vector) $s(t)$ shall be part of a *Hilbert space* $\mathbb{H}$ over
the *field* $\mathbb{C}^\Gamma$ (appendix A.1, [43]). The fundamental functions $\psi(\omega, t)$ shall
span a vector space $\mathbb{V}$ over $\mathbb{C}^{\Gamma_u}$, thus of dimension $\Gamma_u \leq \Gamma$, and therefore
$\mathbb{V} \subseteq \mathbb{H}$. If Γ_u *linearly independent* functions can be discovered among
all fundamental functions, they can serve as *basis* for $\mathbb{V}$. Then, for all
vectors $s \in \mathbb{V}$, there is a unique representation by a linear combination of
basis functions [43]. For alternative vectors $s \in \mathbb{H}$, only the *projection* of
s onto $\mathbb{V}$ can be represented uniquely by a linear combination of the *basis*
vectors. The remaining difference vector is denoted as *residual res* $\in \mathbb{W}$,
with $\mathbb{V} \cup \mathbb{W} \equiv \mathbb{H}$, but $\mathbb{V} \cap \mathbb{W} = \emptyset$. The mapping by analysis is then lossy
and the synthesis becomes ambiguous. Later this case will be denoted as
incomplete dictionary, where *dictionary* stands for the *basis* of $\mathbb{V}$ [115]. In
the desirable case $\mathbb{V} \equiv \mathbb{H}$, thus $\Gamma_u = \Gamma$, the *basis* of $\mathbb{V}$ is also a basis
for $\mathbb{H}$ and all $s \in \mathbb{H}$ can be represented uniquely by a linear combination
[43], which provides us with a unique *analysis* and *synthesis* operation.
Later on, this case will be denoted as *complete dictionary* [115], [116]. In
case of a continuous distribution of basis functions (like e. g. for *Fourier*
transformation), the general *analysis* integral is then:

$$
S(\omega) = \langle s(t),\, \psi(\omega, t) \rangle = \int\limits_{-\infty}^{+\infty} s(t) \cdot \psi^*(\omega, t)\, dt \quad , \text{analysis integral.} \quad (9.1)
$$

The *synthesis* is then vice versa:

$$s(t) = \langle S(\omega),\, \psi^*(\omega,t) \rangle = \int\limits_{-\infty}^{+\infty} S(\omega)\cdot\psi(\omega,t)\, d\omega + res \quad \text{, synthesis integral.}$$

(9.2)

In case of $\mathbb{V} \equiv \mathbb{H}$, the residual is zero. If the system, or its base respectively, is orthogonal, then it can be stated [43], appendix A.1.4:

$$\langle \psi_\nu,\, \psi_\eta \rangle = \begin{cases} \|\psi_\nu\|_2^2, & \text{for } \nu = \eta\,, \\ 0, & \text{for } \nu \neq \eta\,. \end{cases}$$

(9.3)

If $\|\psi_\nu\|_2^2 = 1$ (*Euclidean norm*), then the *basis* is even *orthonormal* [43]. For *compressed sensing theory*, Mallat and Donoho imposed synonymous vocabulary here: Each *basis* function is indicated as *atom* accumulated in a so-called *dictionary*, which is equivalent to the *basis* of a vector space [115], [116], [117]. The coefficients or functionals found by *analysis* can be tagged as *weights* and the *synthesis* equation (9.2) is declared as *atomic decomposition* [115], [116].

The next stept, is the discretization in time, frequency or space. The synthesis integral from equation (9.2) alters then to a discrete sum:

$$s\,[\xi] = \sum_{\nu=0}^{\Gamma-1} S\,[\nu] \cdot \psi_\nu\,[\xi] + res\,.$$

(9.4)

This can also be represented by a matrix-vector operation [115], describing a signal vector s such as the range compressed signal in one range cell $s_{rc}(r_{0q})$: The basis functions Ψ_ν, thus atoms, are the directional vectors a_ν, collected in directional matrix A_ν and the spectral weights $S[\nu]$ are their corresponding complex reflection factors ς. Equation (9.4) altered to matrix-vector notation yields then [115]:

$$s_{rc}\,(r_{0q}) = A_\nu\,\varsigma + res \text{ , discrete synthesis.}$$

(9.5)

The atoms serve as columns for $A_\nu \in \mathbb{C}^{\Gamma \times Q}$, whereas s_{rc} and ς are column vectors. In case of $\Gamma = Q$ ($\mathbb{V} \equiv \mathbb{H}$), *res* is again zero and $A_{\nu,u}$ shall be briefly depicted by A_ν as well as s_u by s. A case of special interest for *compressed sensing* is $\mathbb{V} \equiv \mathbb{H}$ but with $\Gamma_o > \Gamma$ *atoms*, with a so-called *overcomplete dictionary* (overdetermined) [115]. Here, there are still

Γ *linearly independent atoms* forming a *basis*, however there are additionally $\Gamma_o - \Gamma$ *linearly dependent atoms*. At first glance, there is no advantage in these additional atoms, the more so as they cause now ambiguities in *synthesis*. But exactly these ambiguities offer another degree of freedom with respect to find the shortest representation of s, thus with the lowest amount of *atoms* and their *weights*. This is also denoted as the L_0 quasi norm [118]. Nevertheless, the aims of *compressed sensing* are various. For each case the dictionaries need adaption and of course not all assignments can always be fulfilled at once [115], [119]:

- **Signal separation** and **De-noising**: A standard problem in signal processing is separating one or more signals from each other and their noise contribution in case of insufficient *signal-to-noise ratio*(SNR). This is also of special interest regarding the *stability* of signals and the systems which they are used in (see e. g. sensor noise problem, *robust control theory*).

- **Signal reconstruction**: Signals shall be reconstructed in case of lost samples or un-equidistant sub-sampling.

- **Sparse representation**: The shortest possible representation of a signal helps to save memory expense in case of data acquisition or reduces the channel capacity requirements in case of transmission. Furthermore, a short representation is a feasible premise for fast processing.

However, *compressed sensing* is very often designated to be capable to create *super-resolution* [115], [119]. The reader shall be advised caution at this stumbling block. Resolution capabilities in the wide sense of separating objects into different cells, bins, voxels etc., which is the assigned task of technical systems like e. g. radar, sonar, x-ray or magnetic resonance tomography etc., is always inherently linked to signal bandwidth. Good examples are: A *continuous-wave* (CW) radar used for Doppler measurements in e. g. traffic control. It has not got any range resolution capabilities due to the lack of signal bandwidth [24], [46], [47], [67]. Further, a *side-looking synthetic aperture radar* (SAR) or *side-looking sonar* gain their azimuth resolution by sampling a large *Doppler* (or *slow-time*) *bandwidth* by covering a long pass-by trajectory [61], [71]. Now, if the examined signals are *deterministic* and they are known exactly in a certain interval, they can be extrapolated arbitrarily in time, frequency or spatial domain [43]. However, the intrinsic character of a measurement is not only a statistical nature, but more important, that it takes place discretely in time or space. This means that only a finite amount of information is provided. But knowing a function mathematically exactly is equal to an infinite number of measurements. Hence, the underlying functions are formally estimated or approximated, highly depends on the observation duration, bandwidth and number of samples to what extend the signal can be extrapolated beyond the measured boundaries. Therefore the term *super-resolution* must always be seen with that amendment and is very often just a refinement of already collected signal information or a peak plot of discrete coefficients for basis functions [120].

Principally, the objective is always to determine the full spectral representation and therefore the full signal vector $s_{rc}(r_{0q})$ in this range bin on fundament of a sparse representation $s_{rc,u}(r_{0q})$ [115], [118], [121]:

$$s_{rc}(r_{0q}) = A_\nu\,\varsigma \;, \text{ with } s_{rc,u}(r_{0q}) = A_{\nu,u}\,\varsigma = B_c^{\mathrm{T}}\,A_\nu\,\varsigma \;. \tag{9.6}$$

Since one fundamental idea behind compressed sensing is the representation of a signal with a minimum number of weights belonging to (complex) atoms, the mathematical expression can be denoted as minimal L_0-norm, thus a vector or weights with mostly zero elements and only a few nonzero ones [115], [118], [119], [121], [122], [123], [124], [125], [126], [127]. Nevertheless, finding the L_0 is very often not straight forward and connected to greedy algorithms [115], [118], [119], [121], [122], [123], [124], [125], [126], [127].

$$\|\varsigma\|_0 \quad \text{subject to} \quad A_{\nu,u}\varsigma = s_{rc,u} \;. \tag{9.7}$$

Therefore Chen, Saunders, Donoho etc. suggested to search the minimum L_1 norm instead [115], [118], [119], [121], [122], [123], [124], [125], [126], [127].

$$\|\varsigma\|_1 \quad \text{subject to} \quad A_{\nu,u}\varsigma = s_{rc,u} \;. \tag{9.8}$$

This can be considered as convex optimization problem, which can be solved by means of e. g. linear programming, such as simplex methods, interior point methods etc. [115], [118], [119], [121], [122], [123], [124], [125], [126], [127]. In noisy case the problem formulation can be altered but must be solved by quadratic programming [115], [118], [119], [121], [122], [123], [124], [125], [126], [127].

$$\|\varsigma\|_1 \quad \text{subject to} \quad \|A_{\nu,u}\varsigma - s_{rc,u}\|_2^2 \le \sigma^2 \;. \tag{9.9}$$

Fig. 9.1 demonstrate the different projections. Directional vectors span-

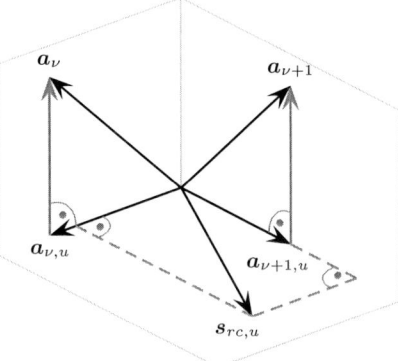

Fig. 9.1 Orthogonal projection of sparse signal $s_{rc,u}$ onto sparse directional vectors $a_{\nu,u}$. Those in turn are an orthogonal projection of a_ν. This projection secures the shortest Euclidean distance between a_ν and $a_{\nu,u}$ [43].

ning the full *Hilbert space* and their projections onto the sub-space, spanned by the sparse array signals. This projection secures the shortest Euclidean distance between a_ν and $a_{\nu,u}$ [43].

9.2 Methods for compressed sensing

For this thesis, the assignments of signal reconstruction and de-noising especially for angular processing will be the most important ones. According to equation (4.86), the representative signal under test shall be the signal within a range bin after compensation of curvature:

$$
s_{rc}(f) = \sum_{q=1}^{Q} \tilde{\sigma}_q \cdot e^{-j \, \pi \, \frac{K}{c_0^2} \, r_{0q}^2} \cdot psf\left(f - f_{r,q} - f_{d,q}\right) \cdot
$$
$$
\cdot e^{+j \, 2k_0 r_{0q}} \cdot a_q\left(\vartheta_q, \varphi_q\right) \ .
$$
(9.10)

The reader shall be guided to a few common state of the art methods for compressed sensing. Please note, that in the following the general atoms ψ_ν shall be denoted by a_ν and represent directional vectors of array signals. The directional vectors coming from the object q are about to be determined and shall be denoted by a_q. The measurement is performed by arrays with non-equidistant and highly sparse virtual arrays like in chapter 9 (therefore index $_u$).

9.2.1 Method of frames

A method, well-known from *linear regression calculus* [43], is the *method of frames* (MOF). It is controversal if MOF belongs to compressed sensing, however, it usually serves as introductional method to compressed sensing [118]. Therefore, it shall be mentioned for the sake of completeness.

The *method of frames* (MOF) tackles the mathematical problem by use of *Euclidian norm* or L_2-norm [118], appendix A.1.1:

$$\min \|\varsigma\|_2 \quad \text{subject to} \quad \boldsymbol{A}_{\nu,u}\,\varsigma = \boldsymbol{s}_{rc,u} \ . \tag{9.11}$$

The ansatz for solving the problem is the same as for *least squares* or *pseudo-inverse matrix* [43], [118]:

$$\varsigma = \left(\boldsymbol{A}_{\nu,u}^{\mathrm{H}}\boldsymbol{A}_{\nu,u}\right)^{-1}\boldsymbol{A}_{\nu,u}^{\mathrm{H}}\,\boldsymbol{s}_{rc,u} \quad . \tag{9.12}$$

For the sake of completeness, equation (9.11) must be differentiated twice with respect to $\boldsymbol{S}$ a second time in order to show if the solution provided by equation (9.12) is a minimum or a maximum. The second derivative is:

$$2\,\boldsymbol{A}_{\nu,u}^{\mathrm{H}}\boldsymbol{A}_{\nu,u} \overset{?}{\geq} 0 \quad . \tag{9.13}$$

If $\boldsymbol{A}_{\nu,u}^{\mathrm{H}}\boldsymbol{A}_{\nu,u}$ is *semi-positive definite*, hence $\boldsymbol{A}_{\nu,u}^{\mathrm{H}}\boldsymbol{A}_{\nu,u} \geq 0$, the problem is *convex* and the solution is a local minimum (or possibly a *saddle point*). However, if $\boldsymbol{A}_{\nu,u}^{\mathrm{H}}\boldsymbol{A}_{\nu,u}$ is *positive definite*, thus $\boldsymbol{A}_{\nu,u}^{\mathrm{H}}\boldsymbol{A}_{\nu,u} > 0$, the problem is *strictly convex*, thus there is only one unique solution which is automatically the global minimum [43].

The method of frames provides a straight forward solution without any iterations, however, it usually does not preserve the sparsity of ς [115]. The geometric idea behind MOF and equation (9.11) is minimizing the radius of an *Euclidean sphere* (or *ball*) touching the set of hyperlanes $\boldsymbol{A}_{\nu,u}\varsigma$ around the centre point $\boldsymbol{s}_{rc,u}$ (see [128]).

9.2.2 Best ortho basis

This is a method which came out of focus during the last years. The fundamental idea is to select an orthogonal basis for representing the signal vector $\boldsymbol{s}_{rc}$ with respect to minimize an entropy cost function [115], [129]. Since *best ortho basis* BOB is not covered in detail in this thesis, the reader shall be referred to [115], [129], [130].

9.2.3 Matching pursuit

A well-known procedure is the so-called *matching pursuit* (MP), which is an iterative decomposition on the fundament of L_∞-norm [115], [119], [116], appendix A.1.1. The problem formulation is similar to equation (9.5) [115]. According to the introduced conventions, the problem can be formulated to [115], [119]:

$$\boldsymbol{s}_{rc,u} = \boldsymbol{A}_{\nu,u}\,\varsigma^{(\mu)} + \boldsymbol{res}^{(\mu)} \quad, \tag{9.14}$$

$$\boldsymbol{s}_{rc}^{(\mu)} = \boldsymbol{A}_{\nu}\,\varsigma^{(\mu)} \quad. \tag{9.15}$$

The wanted solution for $\boldsymbol{s}_{rc}$ shall be element in $\mathbb{V}$, which starts at dimension

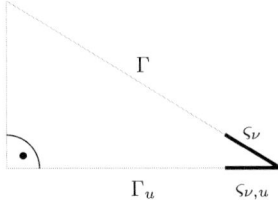

Fig. 9.2 Central dilation in matching pursuit: The projection of residual onto basis vectors spanning sparse signal space must be multiplied by $\frac{\Gamma}{\Gamma_u}$ to obtain the correct weights for atoms of full *Hilbert space*.

zero and is successively increased, whereas the residual $\boldsymbol{res}$ is element of the disjunctive vector space $\mathbb{W}$ whose dimension will be decreased gradually from Γ_u to zero [115], [116], [119]. (Please note that Γ_u stands for the number or sparse array elements.) The graduation is achieved by forming the inner products of the residual $\boldsymbol{res}$ with the dictionary $\boldsymbol{A}_{\nu,u}$. The atom $\boldsymbol{a}_{\nu,u}$ with highest L_∞-norm (highest absolute correlation value) is chosen and its complex correlation value $\varsigma_{\nu,u}$ serves as weight for the corresponding atom at iteration (μ) [115], [116], [119]. Hereby, the observed vector $\boldsymbol{s}_{rc,u}$ serves as starting residual $\boldsymbol{res}$:

$$\left\| \boldsymbol{A}_{\nu,u}^{\mathrm{H}}\,\boldsymbol{res}^{(\mu)} \right\|_\infty = \|\varsigma_{\nu,u}^{(\mu)}\|_2 \;\to\; \varsigma_\nu^{(\mu)} = \frac{\Gamma}{\Gamma_u}\,\varsigma_{\nu,u}^{(\mu)} \;, \tag{9.16}$$

with $\boldsymbol{res}^{(0)} = \boldsymbol{s}_{rc,u}$.

The correction factor Γ/Γ_u must be applied due to the different lengths of atoms and their projections ($\Gamma_u \ll \Gamma$), (see Fig. 9.1 and Fig. 9.2). Then, the coefficient $\varsigma_{\nu,u}^{(\mu)}$ updates the estimated signal $s_{rc}^{(\mu)}$, as well as the residual $res^{(\mu)}$ [115], [119]:

$$s_{rc}^{(\mu+1)} = s_{rc}^{(\mu)} + a_\nu \, \varsigma_\nu^{(\mu)} \, , \qquad (9.17)$$

$$res^{(\mu+1)} = res^{(\mu)} - a_{\nu,u} \, \varsigma_{\nu,u}^{(\mu)} = res^{(\mu)} - B_c^T \, a_\nu \, \varsigma_\nu^{(\mu)} \, . \qquad (9.18)$$

The estimated signal is subtracted from the observation vector and the loop starts again with equation (9.16). After Γ runs, the procedure can be aborted, thus for each of the Γ atoms a coefficient was computed [115]. Please note, the bracing matrix B_c introduced for MR-MIMO serves as projection matrix with its transposed version. The total number of atoms determines the needed number of correlation turns. For the computational effort, this means $\Gamma^2 \, \Gamma_u$ multiplications (see equation (9.16)).

Donoho promised perfect results in case of an orthogonal dictionary but an unclear scenario in case of a non-orthogonal basis [115]. However, own experiments have shown that also non-orthogonal dictionaries can be used successfully for radar array processing. Always the best-correlating atom is chosen for updating $s^{(\mu)}$ and afterwards this signal part is subtracted from the residual which means, there are still the alternative candidates with lower correlation value in the dictionary, however the residual does not contain the corresponding signal parts anymore. So with an orthogonal dictionary, *matching pursuit* finds the only possible and unique representation of s in $\mathbb{H}$. With a non-orthogonal dictionary, the representation is indeed not unique, however, *matching pursuit* finds the best L_0 quasi-norm of S, hence the minimum number of non-zeros, for S.

For the sake of completeness, the *orthogonal matching pursuit* (OMP) shall be explained briefly. After Γ turns of *matching pursuit*, all non-zero weights are put into a least-squares problem again [115]:

$$\min_{\varsigma_\nu} \left\| s_{rc,u} - \sum_\nu B_c^T \, a_\nu \, \varsigma_\nu \right\|_2^2 \, , \, \forall \, \varsigma_\nu \neq 0 \, . \qquad (9.19)$$

Donoho had realized that errors occurring during the first steps of algorithm exhibit the problem of subtracting a wrong atom which in turn affects all later correlation results [115], [119]. For MR-MIMO array processing, this could be observed for especially the case of braced and nested MR configurations, due to the inherent sub-sampling. The suppressed grating lobes can sum additively for targets close together. Then the correlation values referring to the atom at grating lobe can be higher than the weights for the actual atoms. A wrong decision is applied, the wrong atom is subtracted and the algorithm tries to balance this by assigning a non-zero value to all remaining weights to come.

9.2.4 Reduced matching pursuit

Again *matching pursuit* provides a straight forward solution. However, the computational effort of $\mathcal{O}(\Gamma^2\,\Gamma_u)$ multiplications is very exhaustive. Nevertheless, there is margin for improvements, accumulated in *reduced matching pursuit* (RMP).

A possible reduction for MP is to have a closer look on the remaining residual $\boldsymbol{res}^{(\mu+1)}$. The corresponding signal, belonging to the coefficient $\varsigma_\nu^{(\mu)}$ has been removed at stage (μ). Thus, it is unnecessary to correlate the residual again with the atom $\boldsymbol{a}_{\nu,u}$ at stage $(\mu+1)$, since the residual $\boldsymbol{res}^{(\mu+1)}$ is orthogonal to $\boldsymbol{a}_\nu^{(\mu)}$. So after each stage the correlation matrix $\boldsymbol{A}_{\nu,u}^{\mathrm{H}}$ can be reduced by one atom, thus by one row in equation (9.16). After Γ iterations, the same result like for conventional matching pursuit can be achieved. However, the computational effort reduces to:

$$\Gamma_u \cdot [\Gamma + (\Gamma - 1) + \ldots + 1] =$$
$$= \Gamma_u \cdot \sum_i^\Gamma i = \Gamma_u \cdot \frac{\Gamma \cdot (\Gamma + 1)}{2} \quad \text{multiplications.} \tag{9.20}$$

By comparison with the conventional MP, this is a reduction of multiplications to:

$$\frac{\Gamma + 1}{2\,\Gamma} \approx \frac{1}{2} \quad \text{for large } \Gamma. \tag{9.21}$$

Additionally, a further reduction of effort is possible: At each stage a signal part (atom with weight) is subtracted from the residual, i. e. also the corresponding partial signal energy is subtracted from the total signal energy. Therefore the energy of the signal residual $\boldsymbol{res}$ itself can serve as residual $E^{(\mu)}$, too. If this is carried out in signal domain, not in spectral domain, the procedure is independent of the orthogonality of dictionary:

$$E^{(\mu+1)} = E^{(\mu)} - \frac{1}{2} \cdot \|\boldsymbol{a}_\nu\varsigma_\nu\|_2^2 \ , \text{ with } E^{(0)} = \frac{1}{2}\frac{\Gamma}{\Gamma_u} \cdot \boldsymbol{s}_{rc,u}^{\mathrm{H}}\,\boldsymbol{s}_{rc,u} \ . \tag{9.22}$$

Hereby, the initial energy term $E^{(0)}$ must be scaled, due to the fact that only Γ_u elements provide a signal, whereas it is pretended that there are Γ elements in total (Fig. 9.2).

In the special case of an orthogonal (orthonormal) dictionary, the subtracted energy can directly be calculated from the actual weight ς_ν:

$$E^{(\mu+1)} = E^{(\mu)} - \frac{1}{2} \cdot \|\varsigma_\nu\|_2^2 \quad . \tag{9.23}$$

If the scenery is dominated by a few strong point scatterers and most energy is concentrated in a few coefficients, the energy residual saturates very quickly after a few iterations, which reduces the needed turns to a couple, much lower than Γ. Therefore RMP will provide a very efficient and quick tool for compressed sensing.

9.2.5 Simplex method

Equation (9.7) describes the fundamental idea of compressed sensing with L_0-quasi-norm [118]:

$$\min \|\varsigma\|_0 \quad \text{subject to} \quad \boldsymbol{A}_{\nu,u}\,\varsigma = \boldsymbol{s}_{rc,u} \quad . \tag{9.24}$$

The expression $\boldsymbol{A}_\nu\,\varsigma = \boldsymbol{s}_{rc}$ describes a polyhedron in $\mathbb{C}^\Gamma$ [128], a so-called *simplex* [115] or $\boldsymbol{A}_{\nu,u}\,\varsigma = \boldsymbol{s}_u$ the projection of this polyhedron onto the subspace $\mathbb{C}^{\Gamma_u}$. The requirement to satisfy the L_0 condition, links this problem to the field of *simplex methods*, thus linear programming [43], [115].

9.2.6 Basis pursuit

This method is very often implemented with overcomplete dictionaries [115], [119]. However, searching the dictionary with respect to minimal L_0 usually requires a high combinatorial effort, thus Chen, Donoho and Saunders suggested an alternative method using L_1-norm which alters the non-linear problem of equation (9.7) to a *convex optimization* problem with a much larger variety of algorithms [115], [118], [119]:

$$\min \|\varsigma\|_1 \quad \text{subject to} \quad \boldsymbol{A}_{\nu,u}\,\varsigma = \boldsymbol{s}_{rc,u} \quad . \tag{9.25}$$

The intended L_0-norm, however, is not a *convex function* [115], [118], [119], [128]. Again equation (9.25) can be solved by linear programming methods, such as *simplex method* [43]. Donoho proposed an algorithm, where the necessary atoms for a representation of $\boldsymbol{s}_{rc}$ are successively discarded or substituted by other atoms with respect to reduce the number of necessary atoms at constant or even lower error to s [115]. The geometric idea behind this is, that the algorithm tests the boundary of the simplex by pointing at different verteces (extremal points) until no improvement is possbible any more [115]. This can require a greedy algorithm [115], [118]. For noisy signals, equation (9.9) is more feasible [115], [118]:

$$\min \|\varsigma\|_1 \quad \text{subject to} \quad \|A_{\nu,u}\,\varsigma - s_{rc,u}\|_2^2 \leq \sigma^2 . \tag{9.26}$$

Nevertheless, this a quadratic program due to the quadratic subject [43], [128]. The objectives can be linked to subjects by e. g. Lagrange method, Log-Barrier method etc. [43], [131]. It can be solved by e. g. Newton method [43], [131].

9.3 Sparse setups and compressed sensing

The basic intention is now to create a sparse virtual array with non-equidistant element spacings even larger than λ_0, but in such manner, that the greatest common divisor of all element spacings is $d_0 = \lambda_0/4$. The most promising ansatz might be to choose the element positions randomly [132]. However, in section 3.5, it was demonstrated that this is not completely possible for a virtual array.

Nevertheless, the idea behind random positions is, that no systematic, thus periodic, under-sampling in steps of larger than $\lambda_0/4$ occurs [132]. This feature can be provided alternatively by exploiting prime numbers, which must be applied at two stages: First the Tx array (or Rx array) elements are separated by prime number multiples of d_0 (multiplied by two due to the discrete convolution requirements), like in case of $N = 4$ e. g. $\Delta y_{Tx} = 2d_0$ [3, 5, 7]. This gives M identical virtual sub-arrays with prime number multiples of d_0 as spacing. Now, the separation of these has to be a prime number multiple of d_0, too. Therefore, the Rx array is chosen with spacings of prime number multiples of d_0, which should be disjunctive to the set of prime numbers selected for Tx array, added to the total length of each virtual sub-array (and again multiplied by two for convolution). With the aforementioned example in $N = 4$ and a sub-array length of $15d_0$, this could yield e. g. $\Delta y_{Rx} = 2d_0$ [11 + 15, 13 + 15, 17 + 15] = $2d_0$ [26, 28, 32] in case of $M = 4$. The outcome is a virtual array of maximum lateral dimension $101d_0$ (compared to $15d_0$ in Fig. 3.10): $y_{nm} = [1, 4, 9, 16, 27, 30, 35, 42, 55, 58, 63, 70, 87, 90, 95, 102]\,d_0$ (Fig. 9.3). A general description in closed mathematical form can be imposed according to the approach presented in section 6.3 and 7.

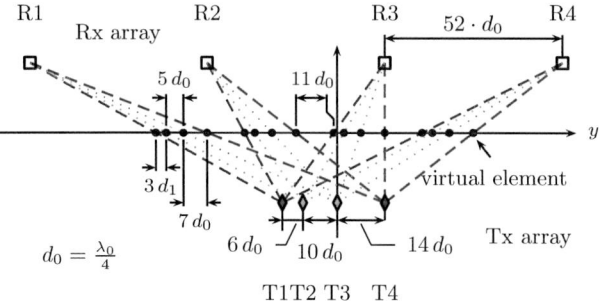

Fig. 9.3 16 virtual elements distributed as sparse array: Each 4 element sub-array has got the element spacings $3\,d_0$, $5\,d_0$, $7\,d_0$, the sub-arrays are all spaced with $11\,d_0$ (not to scale).

Simulative tests have shown that best results can be achieved if the fraction Γ/Γ_u is an integer. Then the cross-correlation values in equation (9.16) are low and the remaining artefacts in azimuth are reduced significantly. With some efforts, this can also be derived from the orthogonality of two exponential functions. Therefore, it can be of advantage to extrapolate the virtual array beyond its outer limits. The example from Fig. 9.3 with lateral dimension $101d_0$ thus $\Gamma = 102$ gives better results with $\Gamma = 7 \cdot 16 = 112$.

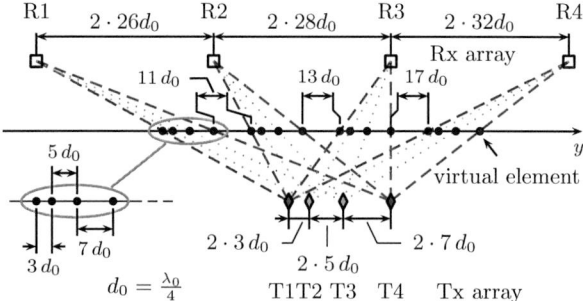

Fig. 9.4 16 virtual elements distributed as sparse array: Each 4 element sub-array has got the element spacings $3\,d_0$, $5\,d_0$, $7\,d_0$, the sub-arrays are spaced with $11\,d_0$, $13\,d_0$, $17\,d_0$ (not to scale).

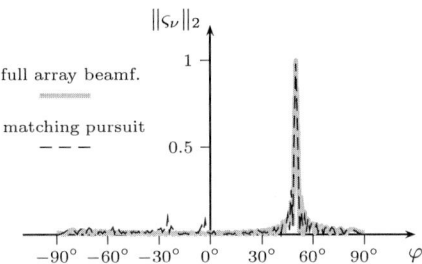

Fig. 9.5 Simulated results, slice at r_q=const., one target at 50 °: comparison of *matching pursuit* (dashed, black) with 16 sparse elements out of 93 possible element positions to a conventional beamformer on the corresponding 93 element ULA (lightgray).

At first glance, the principle of separating the virtual sub-arrays could be expanded arbitrarily. However, with increasing Γ and only $\Gamma u \ll \Gamma$ sampling points, the signals cannot be assigned properly to a unique atom. Artefacts occur again and the procedure gets very sensitive to uncorrelated noise (Therefore the use of s_r, s_e). The reason for this is, that the ∞-norm output rapidly decreases and no concrete maximum can be found anymore.

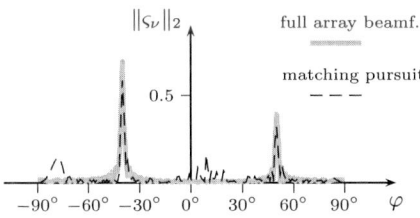

Fig. 9.6 Simulated results, slice at r_q=const., two targets at -40 ° and 50 °: comparison of *matching pursuit* (dashed, black) with 16 sparse elements out of 93 possible element positions to a conventional beamformer on the corresponding 93 element ULA (lightgray).

Correlated noise due to the convolution of virtual array principle however, results in systematic errors, but still gives the old MP results.

9.4 Simulation results

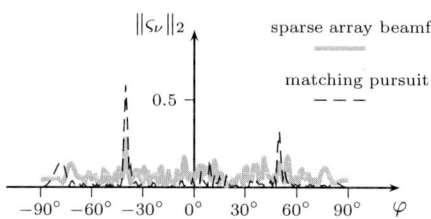

Fig. 9.7 Simulated results, slice at r_q=const., two targets at -40 ° and 50 °: comparison of *matching pursuit* (dashed, black) with 16 sparse elements out of 93 possible element positions to a conventional beamformer on the corresponding 93 element ULA (lightgray).

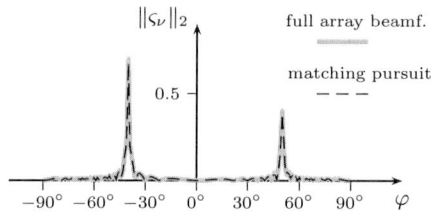

Fig. 9.8 Simulated results, slice at r_q=const., two targets at -40 ° and 50 °: comparison of *matching pursuit* (dashed, black) with 16 sparse elements out of 112 possible element positions to a conventional beamformer on the corresponding 112 element ULA (lightgray).

Fig. 9.8 shows the comparison of MP to the full conventional beamformer with two targets. This cannot be solved easily with a conventional interferometry solution as proposed in section 9.3. Further, it can be seen, that the 16 non-uniform MP solution perfectly matches a beamformer result of a

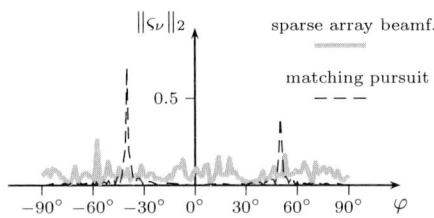

Fig. 9.9 Simulated results, slice at r_q=const., two targets at -40 ° and 50 °: comparison of *matching pursuit* (dashed, black) with 16 sparse elements out of 112 possible element positions to a conventional beamformer on the sparse 16 element array (lightgray).

full 112 element array. Fig. 9.9 shows the MP result with the same targets as in Fig. 9.8 but compared to a conventional beamformer over the sparse array, thus 16 non-uniform elements.

9.5 CS as remedy for coherent signal scenario

Compressed sensing (CS) techniques are used in this thesis as a possible remedy for the *coherent signal scenario*, where the mixed dyadic products do not vanish anymore (see section 5.3.2) and the *augmentation* (section 6.3.3) of sparse *Toeplitz matrices* fails.

In section 12, some more results of CS techniques shall be presented. Nevertheless, CS applied to MR virtual arrays must be examined in more detail. The CS are applied directly to the sparse array signals, without any augmentation, hence the most likely weighted basis functions (atoms) are mapped onto those sparse vectors. The maximum likelihood can also be expressed by the sparse array factor of the MR setups, such as in Fig. 6.24 for *restricted* case or in Fig. 6.24 for *general* case (both repeated in Fig. 9.10). In both cases, the MR setup guarantees a single peak with a factor of approximately 10 dB higher than any other peak. This means that the basis function searched for, correlates ten times better than any other basis function. This indeed is directly the decision fundament for *matching pursuit*, but also indirectly for *basis pursuit* since it describes that there is only one closest cross-polytope with lowest energy. Nevertheless, especially for the *general* MR type, aliasing effects in multi-target scenarios can occur. This can be explained again by a closer look on Fig. 9.10: For restricted

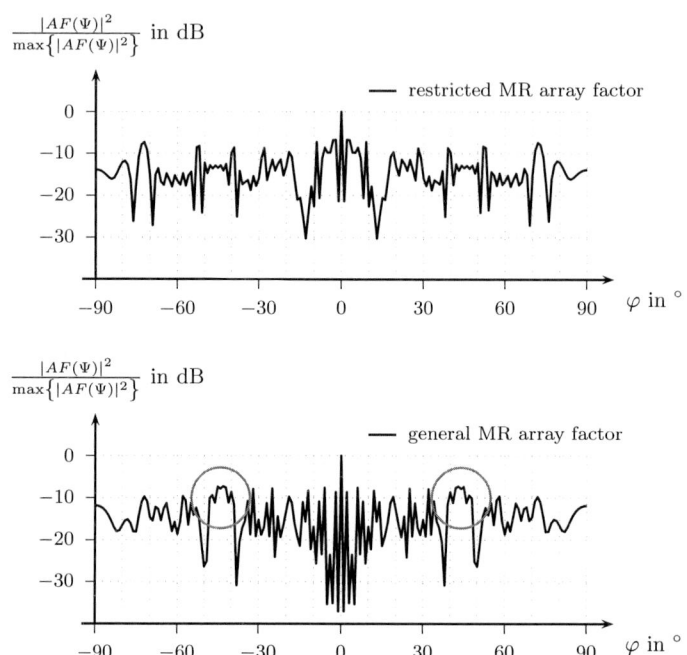

Fig. 9.10 Array factors of nested restricted MR arrays without *augmentation* (top) and nested braced MR arrays arrays (bottom): Tx, Rx and the complete general MR array: $N = 4$, $M = 4$, belonging to setup in Fig. 6.12 and 6.22 respectively. Gray circles: possible coherent superposition of grating lobes.

MR first, all side-peaks are effectively very narrow, thus two target close to each other (with respect to their azimuth angles) usually do not contribute to a coherent superposition of side-peaks or grating lobes. However, this becomes more likely for the *general* MR case (Fig. 9.10 bottom). There, the grating lobe region (here around $\pm 44\,^\circ$, marked with gray circles) is larger and the state of their phases make it possible that coherent grating lobe superposition of two targets close to each other can occur. In this case the overall array factor can bear higher grating lobes than the two target peaks. Therefore, also CS methods decide for the wrong weighted basis function. This effect will be demonstrated in section 12. There, the advantage of *adapted spatial smoothing algorithm* (section 5.4.3) comes again to operation.

Finally for more detailed insights on compressed sensing techniques, the reader shall be guided to [115], [116], [118], [119], [121], [122], [123], [124], [125], [126], [132], [133].

10 Implemented systems

Two systems are presented in the following section. They are mounted on different platforms: The SUM radar sensor and the helicopter landing aid radar, in the following denoted as HeLAR. Both systems were implemented with similar hardware in their frontends. The experimental hardware was manufactured for E-(60 GHz-90 GHz) or W-band (75 GHz-110 GHz). Transmission lines for these bands were rectangular hollow waveguides (see appendix B.2), [30], [104]. Therefore a feasible form of antenna is the pyramidal horn antenna, since this is basically an expanded hollow waveguide. Both systems use pyramidal horn antennas as transmit and receive antennas. Hereby, the dimensions of aperture A, B, the dimension of hollow waveguide a, b (a =3.1 mm, b =1.55 mm in E-band) as well as the taper length L determine the radiation characteristics [31], [32]. It is feasible to assume that those waveguides as well as the antennas are excited by standard TE_{10} mode [31], [32]. According to the coordinate system in Fig. 10.1, the directivity $D(\vartheta, \varphi)$ can be written in spherical coordinates [32]:

$$D(\vartheta,\varphi) = \left(\frac{1 + \cos\vartheta}{2}\right)^2 \cdot \frac{4\pi}{\lambda^2} \cdot AB \cdot \frac{1}{8} \left|F_1(\nu_x, \sigma_a)\right|^2 \left|F_0(\nu_y, \sigma_b)\right|^2 \quad . \quad (10.1)$$

device	manufacturer	f in GHz	annotations
VCO	G. Microw.	12 - 18	$P_{out} \geq$10 dBm
SP4T switch	Miteq	6-18	absorbtive, isolation 60 dB
multiplier, x6	RPG	60-90	factor 6, P_{out} =6 dBm
PA, WR-12	Miteq	72-76	G =20 dB, P_{out} =15 dBm
LNA	RPG	75-110	F =4.5 dB, G =35 dB, P_{1dB} =0 dBm
harm. mixer	RPG	70-90	P_{LO} =14 dBm, L =25 dB

Tab. 10.1 Short list of E- and K-band hardware for both systems.

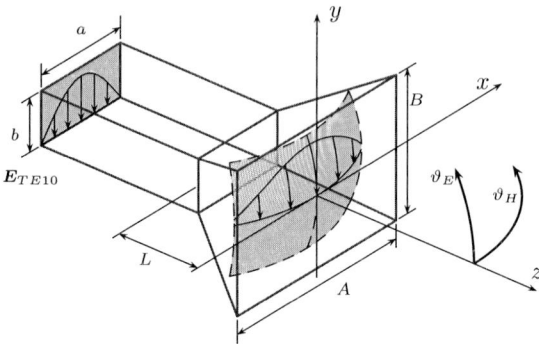

Fig. 10.1 Horn antenna derived from rectangular hollow waveguide and typical coordinate system [32].

Please note that $F_1(\nu_x, \sigma_a)$ and $F_0(\nu_y, \sigma_b)$ are *Fresnel integrals* incorporating geometric parameters of the pyramidal horn.

$$F_0(\nu_y, \sigma_b) = \int\limits_{-1}^{+1} e^{-j\frac{\pi}{2}\sigma_b^2 \cdot \chi^2} \cdot e^{+j\pi\nu_y \chi} \, d\chi \,, \tag{10.2}$$

$$\chi = 2\frac{y}{B}, \; \partial y = \frac{B}{2}\partial\chi, \; \nu_y = k_y \frac{B}{2\pi}, \; \sigma_b^2 = \frac{B^2}{2\lambda_0 R_b} \,, \tag{10.3}$$

$$F_1(\nu_x, \sigma_a) = \int\limits_{-1}^{+1} \cos\left(\frac{\pi}{2}\xi\right) \cdot e^{-j\frac{\pi}{2}\sigma_a^2 \cdot \xi^2} \cdot e^{+j\pi\nu_x \xi} \, d\xi \,, \tag{10.4}$$

$$\xi = 2\frac{x}{A}, \; \partial x = \frac{A}{2}\partial\xi, \; \nu_x = k_x \frac{A}{2\pi}, \; \sigma_a^2 = \frac{A^2}{2\lambda_0 R_a} \,. \tag{10.5}$$

In the following, the antennas are arranged in overall coordinate systems. In this context, the horizontal and vertical slices through the horn antennas, thus the H-and E-planes, parameterized by ϑ_H, ϑ_E, are more feasible than the classical, spherical parameterizing ϑ, φ. The directivity D is not equal to the gain of the antenna. To obtain the latter, the losses of feeding waveguide and horn flaring due to non-infinite conductivity must be considered additionally [31], [32].

10.1 SUM system

More and more, missions of European armed forces have changed from classical homeland defence to peace-keeping missions in conflict zones. Within a variety of tasks, there is usually the necessity to drive convoys or patrols for civil or military supply, security purposes or medical aid and transport. However, such convoys have often been the target of assaults with improvised explosive devices (IEDs), conventional land-mines or suicide bombing persons, which imposes a severe problem for all kind of personnel.

Therefore, the European Defence Agency (EDA) had engaged countermeasures by funding several scientific programs on threat awareness, countermeasures IEDs or landmine detection, which this work is part of. The program, denoted as *Surveillance in an Urban environment using Mobile sensors* (SUM), covers the idea of equipping one or more patrol vehicles with a set of sensors exploiting different physical principles in order to gain detailed insights into the road situation ahead. Providing an added value to a conventional visual camera system, additional measurement data from an infra-red (IR) camera, a millimeter-wave (mmW) radiometer (SUMIRAD and a MMW radar are fused. The results are displayed on a human-machine-interface (HMI). It will assist the vehicle's co-driver to identify suspect objects or persons on or next to the road without forcing the vehicle to stop its cruise.

10.1.1 The consortium for the sensor set

In order to demonstrate the capabilities of such a joint sensor system, a European consortium consisting of GMV[1] (Spain), RMA[2] (Belgium), DLR[3] and TUM[4] (Germany) developed a low-cost demonstration system with partial assignments: Visual and and IR camera as well as the data fusion was provided by RMA, the mmW radiometer by DLR and the mmW radar by TUM. Each sensor was equipped with its own processing unit in order to provide pre-processed data for the data fusion. Furthermore each sensor delivers a figure of merit for each detected object. For the radar sensor, this is derived from the outcome of a constant-false-alarm rate (CFAR) algorithm.

More details on the SUM system itself can be found in [134]. The radiometer is explicated intensively in [135], [136], [137].

---

[1] Grupo Tecnologico e Industrial S.A.
[2] Royal Military Academy
[3] Deutsches Zentrum fuer Luft- und Raumfahrt
[4] Technische Universitaet Muenchen

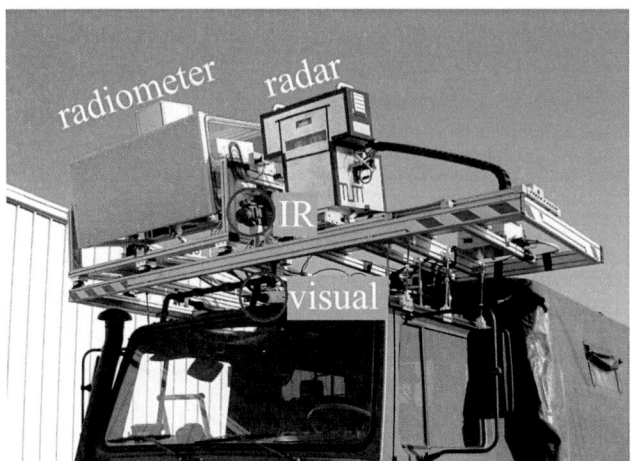

Fig. 10.2 SUM sensor suite mounted on Aluminium test rack on top of test truck: Radiometer for W-band, radar in E-band, cameras for infra-red and visual spectral parts.

10.1.2 Data fusion

The different data sets are then combined by a *fuzzy logic* fed by these figures of merit and a threat level probability is determined for each suspect object. The results can be displayed with a handheld monitor or a tablet PC where the detected object is marked by a coloured bounding box in the visual image. The colour of the bounding box changes with the threat level from blue over green and yellow to red in order to give an intuitive support for the operator while a cruise of the vehicle. More details on the data fusion can be found in [138].

10.1.3 The MMW radar

The radar sensor operates in E-band from 72 GHz up to 79 GHz with variable bandwidths from 1 GHz, standard 3 GHz and even 6 GHz, which provides a range resolution between 15 cm, standard 5 cm and 2.5 cm. The exploited wavelength is around 4 mm. Millimeter-waves still bear a limited penetration capability through e. g. blankets, clothes or awnings, combined with the possibility of high bandwidths for imaging purposes. The system performs with a range of coverage of 100 m.

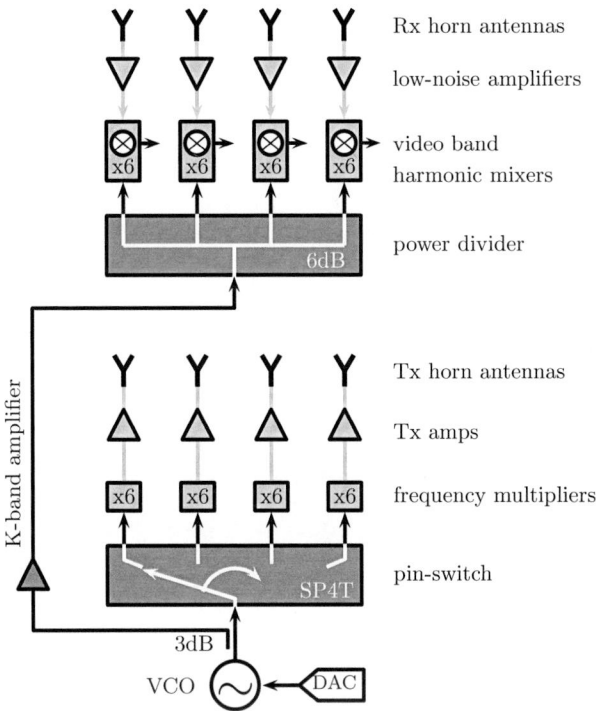

Rx horn antennas

low-noise amplifiers

video band
harmonic mixers

power divider

Tx horn antennas

Tx amps

frequency multipliers

pin-switch

Fig. 10.3 Block diagram of SUM E-band frontend with 4 Tx and 4 Rx elements.

Fig. 10.3 sketches a block diagram of radar frontend: A voltage controlled oscillator VCO, tuned by a 16 Bit digital-to-analog converter DAC provides a modulation ramp in K-band from 12.5 GHz to 13 GHz. The chirp duration was chosen around 10 ms. The DAC drives the VCO via a predistortion look-up table which is stored in the processing unit (see also section 11.2.1). For the transmit path, a single-pole-four-through switch (SP4T) guides the signal to the corresponding Tx path in time multiplex. Each transmit channel is equipped with a frequency multipier of factor six, a power amplifier (PA) and its own pyramidal horn antenna. In K-band section, the local oscillator signal (LO) is split apart from the modulated transmit signal. The LO signal is split further into four equal paths biasing four harmonic mixers with frequency factor six in receiving part. The four receivers in turn have

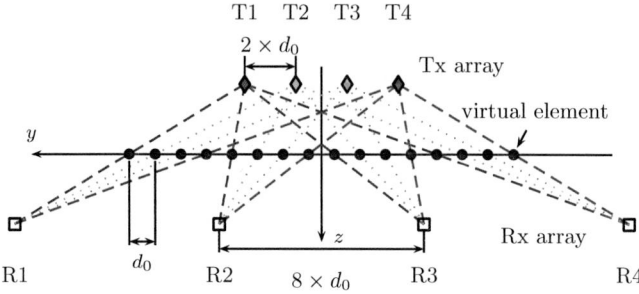

Fig. 10.4 The SUM antenna configuration consists of 4 Tx and 4 Rx, thus 16 virtual elements. $d_0 = 0.8\lambda_0$.

their own pyramidal horn antennas and low-noise-amplifiers (LNA) wich provide constantly four input signals for the harmonic mixers in homodyne principle ($f_{IF} = 0$). Their output signals are amplified, filtered and sorted according to the switching scheme at the SP4T. Then $4 \cdot 4 = 16$ different signals can be obtained. The advantage of harmonic devices offers the possibility to guide all signals in coaxial cables (K-band) before finally changing to retangular waveguides (E-band). The sampling rate can be chosen to 1 MS/s or 2 MS/s.

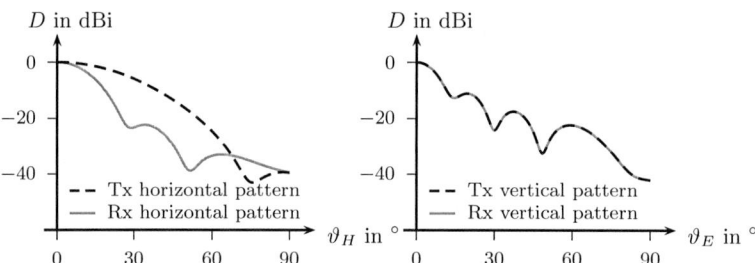

Fig. 10.5 SUM: Normalized directivity antenna pattern according to angles starting at main looking direction (Fig. 10.1). H-plane (left), E-plane (right).

dim. in mm	A	B	a	b	L_{tap}
Tx	6.2	16	3.1	1.55	40
Rx	12.8	16	3.1	1.55	40

Tab. 10.2 SUM aperture dimensions: Horizontal A and vertical B dimensions, taper length L_{tap}, rectangular hollow waveguide dimensions a, b.

10.1.4 Antenna configuration

The polarisation is vertical to ground, thus parallel to incoming wave plane [31], [32]. Fig. 10.5 presents the horizontal and vertical directivity pattern of SUM transmit and receive antennas. Please note, that those pattern were generated by evaluation of *Fresnel integrals*, parameterized by the dimensions of implemented horn antennas: a, b, A, B, L in Fig. 10.1 (equation 10.1) [31], [32]. The virtual array is a uniform linear array ULA, with an element spacing of $d_0 = 0.8\lambda_0$ (Fig. 10.4). Since this element spacing is much larger than the ideal $d_0 = \lambda_0/4$, grating lobes in the half space in front of the radar must occur [31], [32], see section 3.3.1. However, the principle idea for SUM system was to attenuate those by the azimuth

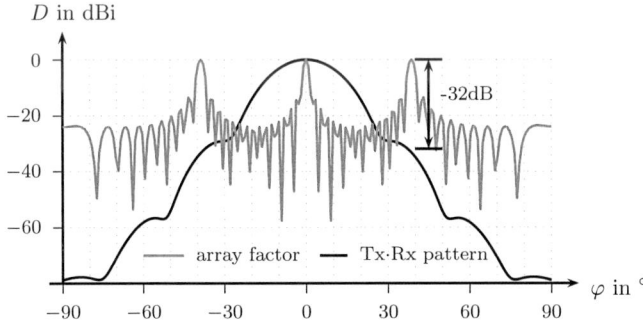

Fig. 10.6 SUM: Normalized array factor and normalized combined antenna pattern (directivity). The grating lobes are attenuated by the antenna pattern.

 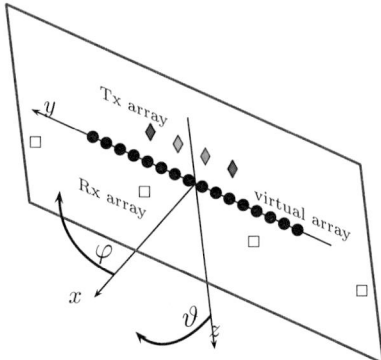

Fig. 10.7 SUM coordinate system: Sensor tower (left) and coordinate system w.r.t. the antenna arrays (right). The main looking direction is the x-axis. The antennas are arranged in y-z-plane. $d_0 = 0.8\lambda_0$.

antenna pattern. Fig. 10.6 demonstrates that the azimuth pattern provides 32 dB attenuation on the grating lobes in this case.

Fig. 10.4 shows the virtual uniform linear array ULA which is formed by four transmitters and four receivers. The antennas are horn antennas milled out of brass. The four transmit antennas are in a single block, whereas the receiving array is split into two halves. By re-arranging the Rx blocks, the different radar operational modes are realized.

Fig. 10.8 SUM: Receiver array configuration for High Resolution Mode and Imaging Mode (top) as well as Surveillance Mode (bottom).

Fig. 10.9 SUM: Transmitter array and system tower without housing.

10.1.5 Operational modes

The setup can be operated in three different modes: 1. Surveillance Mode, which is the standard mode with an angular unambiguity of 70 ° in azimuth, but a low azimuth resolution of 4 °. Here the receiver array is split in the middle and each half is tilted back by 17.5 °. 2. High Resolution Mode. Here the azimuth angular unambiguity is 35 °; however, with 2 ° of angular resolution. 3. Imaging Mode, which exploits methods of Synthetic Aperture Radar (SAR). A lateral synthetic aperture of 1 m on a linear motion slide can be used for azimuth resolution enhancement. The transmit antennas have a 10 dB-beamwidth of 70 ° and illuminate the complete scenery in all operational modes. Fig. 10.8 shows the different Rx antenna configurations for each mode.

By tilting the tower socket (Fig. 10.2, Fig. 10.7), depression angles can be realized, in order to transmit most power to the ground in front of the SUM vehicle. Common depression angles were 8 ° to 10 °.

target	corner as di-hedral	street light as cylinder	grenade shell as cylinder	calibration trihedral
rcs in dBm^2 at $\lambda_0 = 4\,\mathrm{mm}$	≈ 73	≈ 34	≈ 15	8

Tab. 10.3 Radar cross sections of standard targets occuring in SUM project [42].

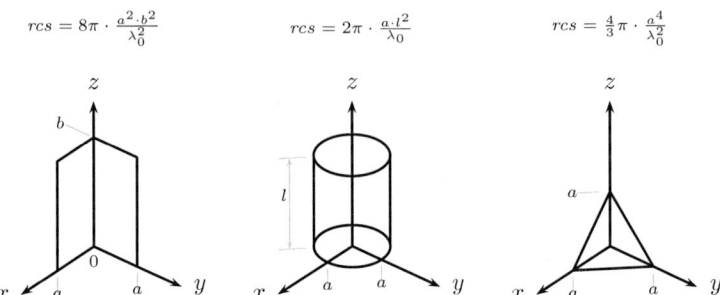

$$rcs = 8\pi \cdot \frac{a^2 \cdot b^2}{\lambda_0^2} \qquad rcs = 2\pi \cdot \frac{a \cdot l^2}{\lambda_0} \qquad rcs = \frac{4}{3}\pi \cdot \frac{a^4}{\lambda_0^2}$$

Fig. 10.10 SUM: Radar cross sections of dihedral and cylinder targets as fundamental target shapes during campaign [42].

10.1.6 Array ambiguities

The SUM carrier, a UNIMOG truck, drove along a road track where different targets had been placed. By comparing two runs on the same targets, once with coverage, once without, the added value of a mmW radar could be demonstrated: The detection rate of the target was equal in both cases. However, the 32 dB grating lobe suppression by the horizontal antenna pattern was not sufficient in this case. During the measurement campaign, the truck passed by street lamps as well as corners of walls with rolling gates

Fig. 10.11 SUM: The sensor carrier truck passes street lamps and wall corners with high radar cross section rcs (red ellipses), thus high reflectivity.

(see Fig. 10.11). With respect to the chosen polarization, those beared a very high *radar cross section*.

Fig. 10.10 shows fundamental target shapes which the order of *radar cross section* can be estimated with, at first instance. The results for approximate dimensions of those unwanted targets as well as for e. g. a grenade shell which was one of the targets searched for, are listed in Tab. 10.3. The assumption of 15 dBm² for the grenade shell is optimistic. However, the burried landmine exhibits a significantly smaller RCS. The overall antenna pattern achieves approximately 32 dB attenuation of the grating lobes. Neverthless, e. g. the wall corner exhibits a *radar cross section* which is larger by the factor of 60 dB compared to grenade shell. Therefore, the wall corner, as well as the street lamp provided a much higher signal response via the grating lobes as the searched targets in the middle of the road ahead (factor 30 dB). By this measurement campaign, it was revealed that the array configuration of Fig. 10.4 had to be revised and adapted.

10.2 Helicopter landing aid radar

In the same context of peace keeping missions which European armed forces are assigned to, not only truck transportation is needed as in case of SUM, but also helicopter lifts. In this project, the focus is set on the landing manoeuvre. Hereby, the very last 10 m (30 ft) are most critical: In dusty or snowy areas the main rotor downwash can push particles aside. Due to their kinetic energy, those particles are reflected at the ground and incline in height. Then they can get intaken again by the main rotor and the procedure starts again. The negative effect for the pilots is that they suffer a total loss of vision. This effect is denoted as *brown-out* in case of dust and *white-out* in case of snow [139], [140], [141]. Since the air pressure altimeters are not reliable at the last 10 m, the only orientation are usually the artificial horizon and eventually a radar altimeter [139], [140], [141]. However, those do not deliver information about possible inclination of ground, parked cars, street lamps, hydrants etc., although those are very dangerous for the helicopter. During the last years, remedys were presented, such as setups with several radar altimeters to estimate the ground inclination [139], [140], [141]. The HeLAR can now be seen as a consequent sequel to those achievements. It shall provide a more detailed picture of ground scenery. Therefore, a frontend setup similar to SUM (Fig. 10.3) had been chosen (Fig. 10.12): A fiths receiver was added for elevation monopulse [47].

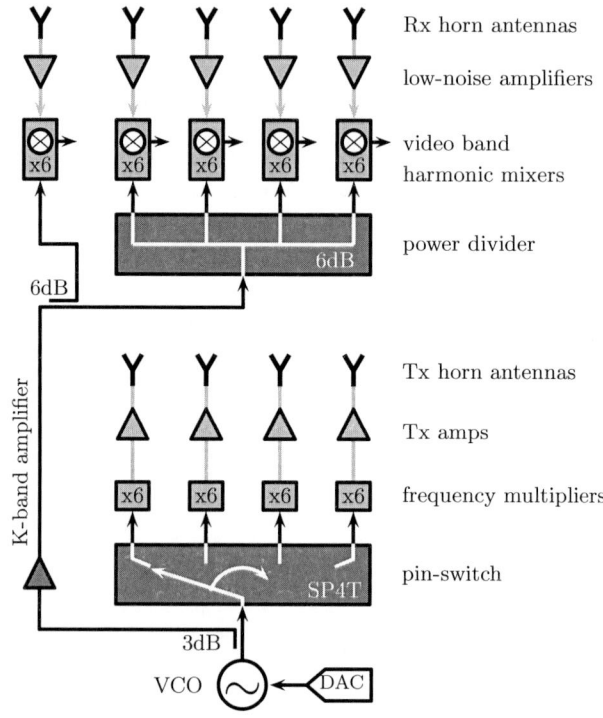

Rx horn antennas

low-noise amplifiers

video band
harmonic mixers

power divider

Tx horn antennas

Tx amps

frequency multipliers

pin-switch

Fig. 10.12 Block diagram of helicopter mmW frontend with 4 Tx and 5 Rx elements.

10.2.1 Array with minimum redundancy

According to SUM, the helicopter landing aid radar HeLAR had also been designed with a ULA endowed (similar to Fig. 10.4) at first instance. However, this was revealed with the insights from SUM measurement campaign. Therefore an array with minimum redundancy was implemented for HeLAR system.

During array redesign, a fundamental distance of d_0 =0.95 mm was chosen, thus d_0 is always smaller than $\lambda_0/4$ over complete bandwidth $\Delta F = [75\,\text{GHz},\ 78\,\text{GHz}]$. The array is an array with *minimum redundancy* of *general type* (compare section 9). It was braced with a factor of three.

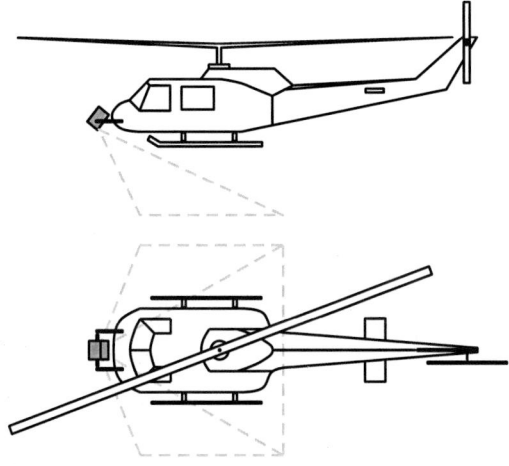

Fig. 10.13 Helicopter lading aid: mounting (helicopter model similar to Bell UH-I, not to scale)

Due to the larger lateral array dimensions, the far-field starts at $2 \cdot \frac{D^2}{\lambda_0} = 4.16\,\text{m}$, which is important for data processing.

10.2.2 Antenna pattern

The single antenna pattern were chosen similarly to SUM sensor. In this case, the four transmit antennas were again manufactured out of one block of brass. Whereas, the receivers were milled as single antennas. In order to

dim. in mm	A	B	a	b	L_{tap}
Tx	5.5	11.6	3.1	1.55	50
Rx	20	11.6	3.1	1.55	50

Tab. 10.4 HeLAR horn antennas: Horizontal A, vertical B dimensions, taper length L_{tap}, rectangular hollow waveguide dimensions a, b.

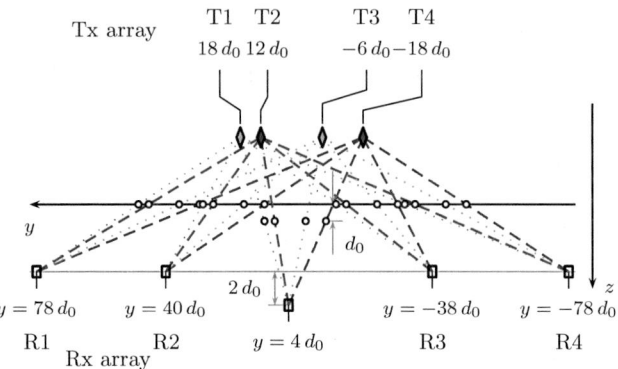

Fig. 10.14 Modified virtual array for HeLAR.

achieve impermeability to splash water, the antenna radome with a rubber inlet serves as labyrinth seal.

10.2.3 Helicopter mount

The best mounting position for such a radar sensor is the nose with the sensor looking slantly to ground (Fig. 10.13). The implemented system features a total weight of approximately 30 kg which was too much for a side or tale mount anyway. Those positions were limited to 10 kg or 6 kg. The sensor is mounted with a cardan gimbal to helicopter. For the first

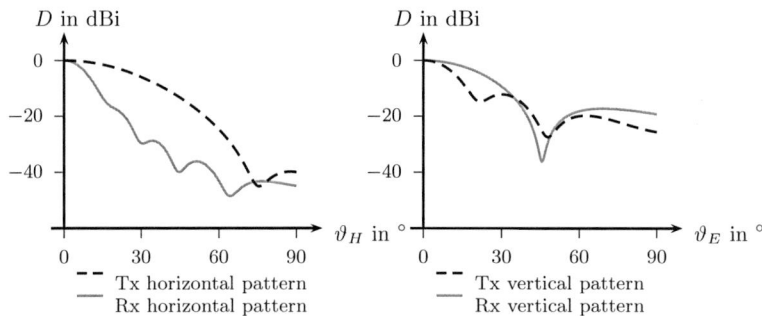

Fig. 10.15 HeLAR: Normalized directivity antenna pattern according to angles starting at main looking direction (Fig. 10.1). *H*-plane (left), *E*-plane (right).

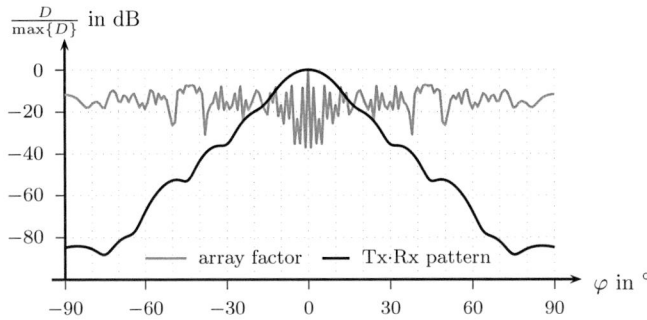

Fig. 10.16 HeLAR: Normalized directivity antenna pattern according to angles starting at main looking direction (Fig. 10.1) with array factor.

tests, this gimbal is fixed, thus the sensor depression angle is fixed, too. Therefore, the pattern were designed for an optimal height of $H = 10\,\mathrm{m}$ (Fig. 10.17).

10.2.4 Tough housing

The use in helicopter poses a challenge in terms of mechanics: Vibration, crash retardation, firmness of screws etc. Therefore a very robust housing was designed. Fig. 10.19 shows the numbering of antennas and finally Fig. 10.20 shows the sensor mounted on (cardan) gimbal for final test.

parameters	H in m	ϑ_0 in °	ϑ_B in °	φ_B in °
	10	37	40	35

Tab. 10.5 Parameters of HeLAR antenna footprint in Fig. 10.15.

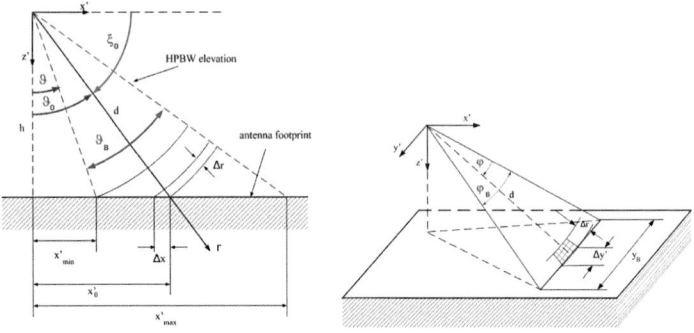

Fig. 10.17 Helicopter lading aid: Slant illumination (left) and lateral (right).

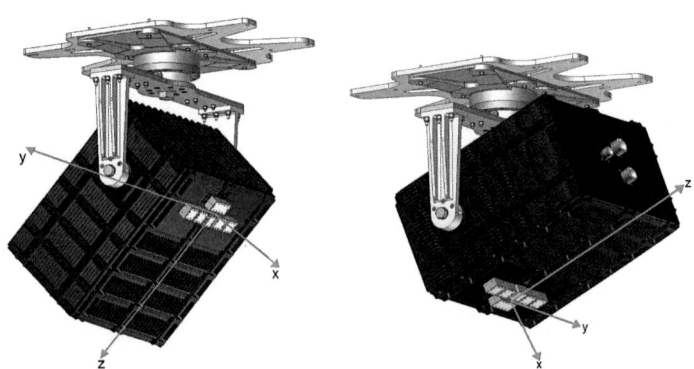

Fig. 10.18 Helicopter lading aid: Two possibilities of mounting.

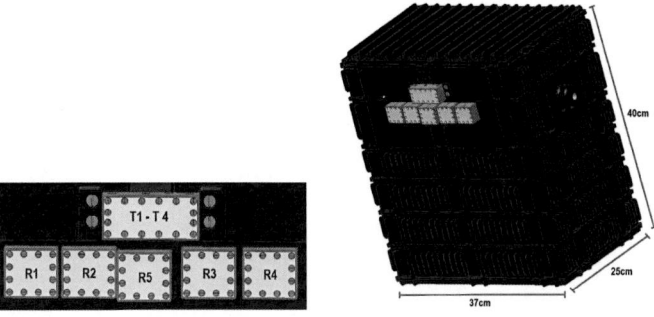

Fig. 10.19 Helicopter lading aid: Antenna numbering (left) and dimensions (right).

Fig. 10.20 Helicopter lading aid during final tests.

11 Correction of modulation and calibration

In chapter 4, types of modulation was introduced. Both implemented systems (chapter 10) exploit linear frequency modulation with the FMCW principle (Fig.10.3, Fig.10.12). Fig.11.1 shows a single channel of the im-

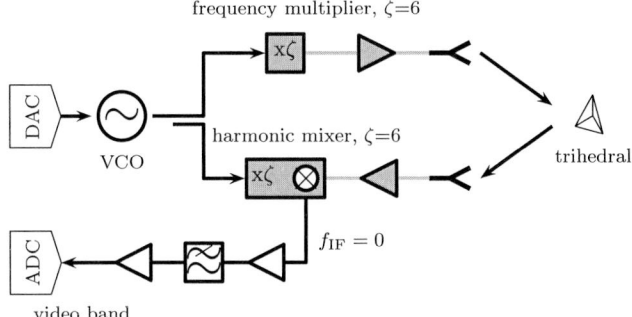

frequency multiplier, ζ=6

Fig. 11.1 Block diagram of one MIMO radar channel and its IF stage.

plemented systems with their IF stage. The following sections shall not only describe distortion effects which occurred but also the countermeasures which were implemented.

11.1 Distortions in radar hardware

11.1.1 Complex signals' amplitude and phase errors

A complex (envelope) signal can be represented by two real components of same amplitude and fixed phase difference of $90°$ [67]:

$$s_I(t) = A \cos\left(\omega_0 t\right) \quad , \tag{11.1}$$

$$s_Q(t) = A \sin\left(\omega_0 t\right) \quad , \tag{11.2}$$

$$\Rightarrow s(t) = s_I(t) + j \cdot s_Q(t) = A\,\mathrm{e}^{+j\,\omega_0 t} \quad . \tag{11.3}$$

An error in amplitude γ and an error in phase ϕ can be modelled by manipulating e. g. the quadrature channel [67]:

$$s_I(t) = A \cos(\omega_0 t) = \frac{A}{2} \left(e^{+j\,\omega_0 t} + e^{-j\,\omega_0 t} \right) , \tag{11.4}$$

$$s_Q(t) = A\,\gamma \sin(\omega_0 t + \phi) = \frac{A\,\gamma}{2j} \left(e^{+j\,\omega_0 t}\, e^{+j\,\phi} - e^{-j\,\omega_0 t}\, e^{-j\,\phi} \right) , \tag{11.5}$$

$$\Rightarrow s(t) = s_I(t) + j \cdot s_Q(t) =$$
$$= A\,e^{+j\,\omega_0 t} \cdot \frac{1}{2} \left(1 + \gamma \cdot e^{+j\,\phi} \right) + A\,e^{-j\,\omega_0 t} \cdot \frac{1}{2} \left(1 - \gamma \cdot e^{-j\,\phi} \right) . \tag{11.6}$$

Now, energy from the desired band at $\delta(\omega - \omega_0)$ is drawn and put into the mirror band at $\delta(\omega + \omega_0)$. The relation of powers in desired band to mirror band is then:

$$\frac{P(\omega - \omega_0)}{P(\omega + \omega_0)} = \left| \frac{1 + \gamma \cdot e^{+j\,\phi}}{1 - \gamma \cdot e^{-j\,\phi}} \right|^2 = \frac{1 + 2\gamma \cos\phi + \gamma^2}{1 - 2\gamma \cos\phi + \gamma^2} . \tag{11.7}$$

By assuming small ϕ and a γ close to one, this can be approximated by [67]:

$$\frac{P(\omega - \omega_0)}{P(\omega + \omega_0)} \approx \frac{4}{\gamma^2 + \phi^2} . \tag{11.8}$$

There are hardware methods to correct the IQ-signal for small values of γ and ϕ [67]. Alternatively, the complex signal can be generated use of *Hilbert* transformation from the real part which secures a $90\,°$ shift between both components at any time (compare to section 4.1.2). However, then only two quadrants of the complex IQ-plane can be described uniquely, which states a problem for complex modulation schemes for communication, such as QAM.

The transient behaviour of IF filter and amplifier, such as loading coupling capacitances etc., can add offset voltages with low frequencies onto the baseband signals [67]. However, they only increase the low frequency parts of the spectrum, hence only corrupt the first range bins (if the range compression is a *Fourier* transformation). Those in turn, are of low interest very often, since they lie in *reactive near-field* where also antenna *cross-talk* takes place.

11.1.2 IF filter and amplifier nonlinearities

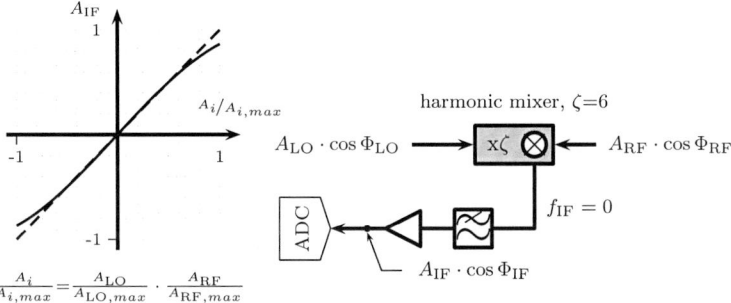

Fig. 11.2 Saturation effect of IF stage (left). Block diagram of IF stage (right).

The chain of harmonic mixer (phase discriminator), IF filter and IF amplifier can also be susceptible for distortions (see Fig. 11.2). The chain is considered to be linear if [67]:

$$A_{IF} \sim A_{RF} \wedge A_{IF} \sim A_{LO} \Rightarrow A_{IF} \sim A_{RF} \cdot A_{LO} , \qquad (11.9)$$

$$\cos \Phi_{IF} \sim \cos (\Phi_{RF} - \zeta \cdot \Phi_{LO}) , \qquad (11.10)$$

is fullfilled. With respect to possible realizations, this is only valid for certain signal power limits and frequency bands defined by the devices' physical limits [37], [142], [143]. With increasing input power, amplifiers and mixers saturate in output power (1dB-compression points) [37], [142]. Besides that, amplifiers can exhibit nonlinear phase responses [37], [142], [143], so can filters [144]. This usually leads to spectral *harmonics* [67]. Fig. 11.3 shows an example: A large wall at ca. 20 m distance was illuminated. The large signal level drove the IF stage into saturation. The red markers point onto the increasing values of 2nd and 3rd harmonic.

At first sight, the saturation curve in Fig. 11.2 can be approximated by a function of uneven symmetry, such as a *sine* function (Fig. 11.2) [67]. However, for small arguments, the *sine* function can be approximated by its argument [43]: $\sin \left(A_i / A_{i,max} \right) \approx A_i / A_{i,max}$. This teaches us, that the signal chain power levels must not reach the saturation limits. This logic and even rather simple requirement actually caused problems for the first indoor test of both radar systems: The SUM radar's baseband amplifiers needed to be trimmed to the lowest possible gain. In case of the helicopter landing aid, even fixed attenuators had to be mounted onto the amplifier inputs.

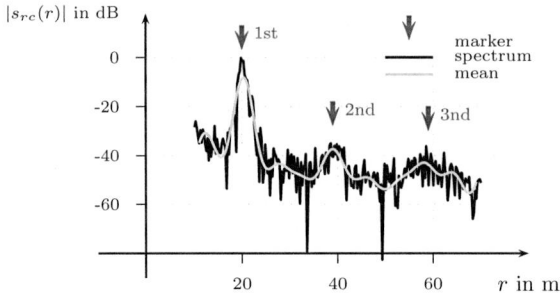

Fig. 11.3 Example of range compressed FMCW radar signal with harmonics, suffering from nonlinear mixer and IF stage (red markers).

Consequently, the analog-to-digital converter must be matched to the power limits, i.e. the digital range (here 16 Bit) is used completely to resolve the wanted power interval (dynamic range). This reduces losses by quantization [67]. As a rule of thumb, the *signal-to-(quantization) noise ratio* gains 6 dB per Bit [104].

11.1.3 AM and FM noise effects

The figures 10.3, 10.12 give a hint of the implemented signal chains. Further Fig. 11.1 focuses on a single MIMO radar channel. There, any device can feature amplitude and phase responses which present themselves as possible *amplitude modulation* (AM) noise, *frequency modulation* (FM) noise or phase noise respectively [63], [67]. All those have in common, that sidebands next

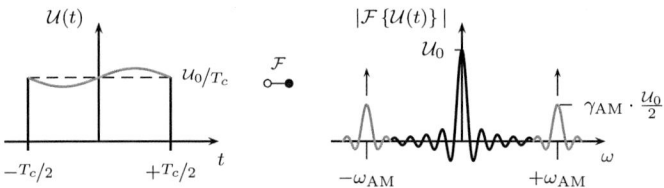

Fig. 11.4 A corrupted envelope function, or amplitude noise, causes sidebands [63].

to the wanted signal parts occur. A simplified model demonstrates the effect of AM noise [63], [104]:

$$s_{\text{AM}}(t) = \mathcal{U}(t) \cdot e^{+j\,\omega_0 t} =$$

$$= \frac{\mathcal{U}_0}{T_c} \cdot \text{rect}\left(\frac{t}{T_c}\right) \cdot [1 + \eta_{\text{AM}} \sin(\omega_{\text{AM}} t)] \cdot e^{+j\,\omega_0 t} =$$

$$= \frac{\mathcal{U}_0}{T_c} \cdot \text{rect}\left(\frac{t}{T_c}\right) \cdot \left[1 + \frac{\eta_{\text{AM}}}{2j}\left(e^{+j\,\omega_{\text{AM}} t} - e^{-j\,\omega_{\text{AM}} t}\right)\right] \cdot e^{+j\,\omega_0 t} =$$

$$= \frac{\mathcal{U}_0}{T_c} \cdot \text{rect}\left(\frac{t}{T_c}\right) \cdot \left[e^{+j\,\omega_0 t} + \frac{\eta_{\text{AM}}}{2j} \cdot e^{+j\,(\omega_0 + \omega_{\text{AM}})t} - \frac{\eta_{\text{AM}}}{2j} \cdot e^{+j\,(\omega_0 - \omega_{\text{AM}})t}\right] .$$

$$(11.11)$$

For a sinusoidal amplitude fluctuation, two sidebands occur (compare to Fig. 11.4). Hereby, η_{AM} can be interpreted as AM *modulation index* [63]. In general, if the amplitude fluctuations were represented by a *Fourier series*, several sidebands would occur. The behaviour in case of FM or phase noise is similar [34], [43], [63], [104]:

$$s_{\text{FM}}(t) = \mathcal{U}(t) \cdot e^{+j\,\omega_0 [1 + \eta_{\text{FM}} \sin(\omega_{\text{FM}} t)]\,t} =$$

$$= \mathcal{U}(t) \cdot e^{+j\,\omega_0 t} \sum_{p=-\infty}^{+\infty} \mathcal{J}_p\left(\eta_{\text{FM}} \cdot \omega_0\right) e^{+j\,p\omega_{\text{FM}} t} . \qquad (11.12)$$

Like in AM case, the frequency or phase modulation causes additional sidebands at $(\omega_0 + p\omega_{\text{FM}})$, weighted by *Bessel functions* $\mathcal{J}_p$ evaluated at the absolute frequency hub $\eta_{\text{FM}} \cdot \omega_0 = \Delta\omega$ [43], [104]. Since both effects, AM and FM noise, result in similar symptoms [145], they will be tackled as one problem (with two facets) in section 11.2. However, it must be mentioned that the use of multipliers and harmonic mixers (Fig. 10.3, Fig. 10.12, Fig. 11.1) within the system setup bares the phase noise of the VCO to be multiplied by six.

11.1.4 Dispersion by transmission lines

The IF signals delivered by the implemented systems feature a bandwidth of maximally $500\,\text{kHz}$. They are guided by coaxial transmission lines excited in fundamental TEM mode, thus well below the cut-off frequency of the next higher TE, TM mode (see [37], [104]). Therefore dispersion is not of interest in this domain. The RF stage rectangular hollow waveguides of both systems are excited by $3\,\text{GHz}$ or even $6\,\text{GHz}$ at centre frequencies of $75\,\text{GHz}$

or above. This gives a relative bandwidth of 4% or 8% respectively. For both cases this is well below the 20% barrier to ultra-wide-band systems [30]. However, dispersion of fundamental TE_{10} (former H_{10}) mode can still reduce the effective linear modulation ramp, which in turn reduces the effective signal bandwidth leads to broadening of the compressed signal [60]. The transmit signal which is guided to the antennas is represented by equations (4.28), (4.31). Its spectrum is the *Fourier* transform:

$$U_{Tx}(f) \sim \int\limits_{-\infty}^{+\infty} \mathcal{U}(t) \cdot e^{+j\, 2\pi\left[\frac{K}{2}\, t^2 - (f - f_0)\, t\right]} dt \,, \tag{11.13}$$

which can be approximated again by the *method of stationary phase* [32], [47], [71], section 4.4.3, appendix A.3:

$$\int\limits_{-\infty}^{+\infty} h(t)\, e^{+j\, \Phi(x)}\, dt \approx \sqrt{\frac{j\, 2\pi}{\Phi''(t_*)}}\, h(t_*)\, e^{+j\, \Phi(t_*)} \,, \tag{11.14}$$

In this case Φ is the phase in equation (11.13):

$$\Phi = 2\pi \left[\frac{K}{2}\, t^2 - (f - f_0)\, t\right] \,, \tag{11.15}$$

$$\frac{\partial \Phi}{\partial t} = 2\pi \left[K t - (f - f_0)\right] \,, \tag{11.16}$$

$$\frac{\partial^2 \Phi}{\partial t^2} = 2\pi K \,. \tag{11.17}$$

where t_* ($\Phi' = 0$) is the *stationary point*: $t_* = (f - f_0)/K$. Inserted into equation (11.13) yields:

$$U_{Tx}(f) \sim U\left(\frac{f - f_0}{K}\right)\, e^{-j\, \frac{\pi}{K}\, (f - f_0)^2} \,. \tag{11.18}$$

The complex propagation coefficient can be defined as $\gamma = \alpha + j\, \beta$ [104]. As α represents the losses, β is the wavenumber [104]. For rectangular hollow waveguide (with length of edges $a = 2\, b$, [37], [104], appendix B.2),

β represents the eigenvalue of the according *Helmholtz* equation [32], [36], [37], [104]:

$$\beta = \sqrt{k_0^2 - \left(\frac{m \cdot \pi}{a}\right)^2 - \left(\frac{n \cdot \pi}{b}\right)^2},$$

$$\beta_{\mathrm{TE}10} = \sqrt{k_0^2 - \left(\frac{\pi}{a}\right)^2} = \frac{2\pi}{c_0} \sqrt{f^2 - f_c^2}, \text{ for TE}_{10} \text{ mode.}$$

(11.19)

Hereby, the *cut-off* wavelength is $\lambda_c = 2\,a$, the cut-off frequency $f_c = c_0/2\,a$ respectively (f_c is the minimum frequency for propagation of this mode) [104].

The resulting signal in spectral domain is a product of equation (11.18) with the spectral response of the waveguide [56], [57], [60]. In order to represent the nonlinear behaviour in spectral domain (appendix B.2), the phase term in equation (11.19) is approximated by a *Taylor* series around f_0 (centre frequency) which is now not aborted after the linear term [57], [60], but after the quadratic one:

$$\beta_{\mathrm{TE}10} \approx$$

$$\approx 2\pi \left[f_0 \frac{\sqrt{1 - (f_c/f_0)^2}}{c_0} + \frac{f - f_0}{c_0 \sqrt{1 - (f_c/f_0)^2}} - \frac{(f_c/f_0)^2 \cdot (f - f_0)^2}{2\,c_0 f_0 \left[1 - (f_c/f_0)^2\right]^{3/2}} \right] =$$

$$= 2\pi \left[\frac{f_0}{v_{ph}} + \frac{f - f_0}{v_g} - \frac{(f - f_0)^2}{2\,a_g} \right],$$

(11.20)

with the substitutions for *phase* and *group velocities* for a rectangular hollow waveguide [104], [146]:

$$v_{ph} = \frac{c_0}{\sqrt{1 - (f_c/f_0)^2}} \qquad \text{phase velocity,} \qquad (11.21)$$

$$v_g = c_0 \sqrt{1 - (f_c/f_0)^2} \qquad \text{group velocity,} \qquad (11.22)$$

$$a_g = \frac{c_0 f_0 \left[1 - (f_c/f_0)^2\right]^{3/2}}{(f_c/f_0)^2} \qquad \text{pseudo group accel.} \qquad (11.23)$$

The term a_g exhibits the unit $1\,\mathrm{m}/s^2$, thus can be interpreted as *accelera-tion*. The spectrum of the transmit signal altered by a rectangular hollow waveguide of length Δr is then:

$$U_{Tx,2} \sim e^{-j\,\frac{\pi}{K}\,(f-f_0)^2} \cdot e^{+j\,2\pi\,f_0\,\frac{\Delta r}{v_{ph}}} \cdot e^{+j\,2\pi\,\frac{\Delta r}{v_g}\,(f-f_0)} \cdot e^{-j\,\pi\,\frac{\Delta r}{a_g}\,(f-f_0)^2} . \quad (11.24)$$

Please note, that the phase term containing the *phase velocity* only provides a constant phase shift, thus does not exert changes on the modulation ramp. Further, the term containing the *group velocity* does not affect the modu-lation ramp either. It provides a linear phase in spectrum, thus results into a delay in time domain, which is expected by a transmission line. How-ever, the quadratic expression with pseudo group acceleration does affect the modulation ramp, hence both exponents together give an altered ramp $\tilde{K}$:

$$-\pi \left(\frac{1}{K} + \frac{\Delta r}{a_g} \right) \cdot (f - f_0)^2 \overset{!}{=} -\pi\,\frac{1}{\tilde{K}} \cdot (f - f_0)^2$$
$$\Rightarrow \quad \frac{\tilde{K}}{K} = \frac{1}{1 + K\,\Delta r/a_g} . \quad (11.25)$$

For typical parameters of the implemented systems like $\Delta F \approx 3\,\mathrm{GHz}$, $T_c \approx 10\,\mathrm{ms}$, $f_0 \approx 75\,\mathrm{GHz}$, this gives:

$$\frac{K}{a_g} \approx 12 \cdot 10^{-9}\,\frac{1}{\mathrm{m}} = 12\,\frac{\mathrm{ppb}}{\mathrm{m}} , \quad \Rightarrow \quad \frac{\tilde{K}}{K} \approx \frac{1}{1 + 12\,\frac{\mathrm{ppb}}{\mathrm{m}} \cdot \Delta r} . \quad (11.26)$$

The order of waveguide lengths in each channel, manufactured for both sys-tems is $1\,\mathrm{m}$, hence the dispersion effect can be neglected for those compact sensors. Besides that, the small remaining effect can be compensated by the resampling technique, demonstrated in section 11.2.2.

11.2 Modulation nonlinearities and calibration

In section 11.1.3, the effects of amplitude or phase fluctuations were demon-strated. Fig. 11.1 appears with several active elements which can poten-tially contribute to an individual amplitude or phase response per radar channel.

11.2.1 VCO predistortion

Hereby, the very first component, the VCO, already contributes to possible phase errors, since a linear tuning voltage usually does not lead to a linear increase in frequency, due to hardware realization with potentially nonlinear elements such as transistors, varactors etc. (compare Fig. 11.5). A popu-

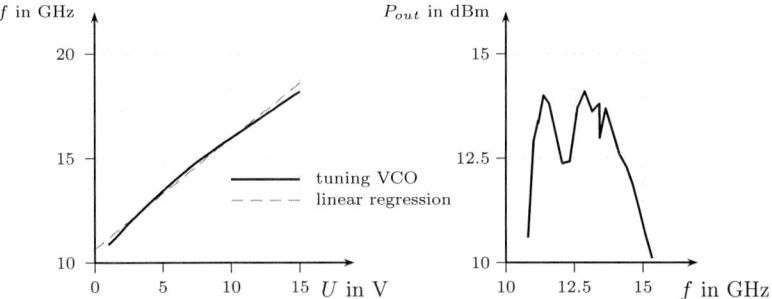

Fig. 11.5 Example of a VCO's nonlinear tuning characteristics (left) and amplitude response (right) (General Microwave V6120A).

lar countermeasure is VCO *predistortion* [147]: First, the tuning voltage to frequency characteristics are measured statically, thus an equidistant frequency grid is chosen and the corresponding tuning voltage is taken. Then this usually nonlinear tuning voltage is reissued to tune the VCO in order to gather a linear frequency output [147]. However, the dynamic or transient behaviour of the VCO is not taken into account. Furthermore, a possible VCO amplitude response (Fig. 11.5) is not compensated by this method. Both lead to remaining phase errors which are to be corrected in the following. Fig. 11.6 demonstrates the range compressed signal after VCO *predistortion* compared to an ideal reference function. The remaining influence of the AM-FM effects is still that large, that the object's response defocuses massively. Hence, more effort must be spent on this issue. The reference function in turn can be regarded as linear phase regression in time domain (Fig. 11.5).

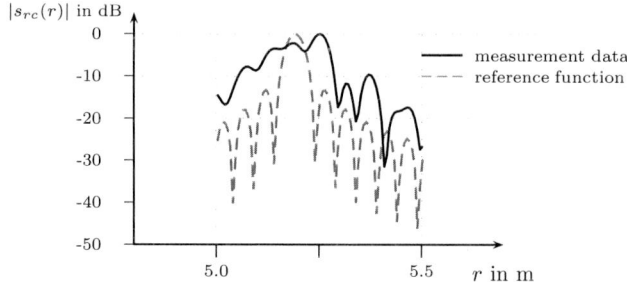

Fig. 11.6 Effect of ramp nonlinearities on range compressed signals: The original measured signal with nonlinearities on the modulation ramp (black) and an ideal reference function at the same nominal position (gray).

11.2.2 Ramp linearization

At first glance, it shall be assumed that all frequency or phase nonlinearities have their seeds in the VCO characteristics (Fig. 11.5). This can be described by an amendment in equations (4.26), (4.29) [65], [148], [149]:

$$f_{Tx}(t) = f_0 + K t + \tilde{\varepsilon}(t) , \qquad (11.27)$$

$$\Rightarrow \ \varphi_{Tx}(t) = 2\pi \left[f_0 t + \frac{K}{2} t^2 + \varepsilon(t) \right] + \phi_0 . \qquad (11.28)$$

Hereby, the phase deviation function $\varepsilon(t)$ is the integral version of $\tilde{\varepsilon}(t)$. The system works with FMCW principle. According to equation (4.49), the phase difference of transmit signal and received echo provided by the mixer (phase discriminator) for each target is then [65], [148], [149]:

$$\varphi_{mx,q}(t) = \varphi_{Tx}(t) - \varphi_{Tx}(t - \tau_q) =$$
$$= 2\pi \left[(f_0 + K t) \tau_q - \frac{K}{2} \tau_q^2 + \varepsilon(t) - \varepsilon(t - \tau_q) \right] . \qquad (11.29)$$

The last term in equation (11.29) must examined in more detail. Thereby, $\varepsilon(t)$ shall be approximated by a *Taylor series*, aborted after the linear term [65], [148], [149]:

$$\varepsilon(t) \approx \varepsilon(t_0) + \frac{\partial\varepsilon}{\partial t}\big|_{t_0} (t - t_0) \ ,$$

$$\varepsilon(t - \tau_q) \approx \varepsilon(t_0 - \tau_q) + \frac{\partial\varepsilon}{\partial t}\big|_{(t_0 - \tau_q)} (t - t_0 + \tau_q) \ , \qquad (11.30)$$

$$\Rightarrow \varepsilon(t) - \varepsilon(t - \tau_q) \approx \frac{\partial\varepsilon}{\partial t}\big|_{t_0} \tau_q \approx \tilde{\varepsilon}(t) \cdot \tau_q \ .$$

Since τ_q is very small, it can be assumed that $\varepsilon(t_0) \approx \varepsilon(t_0 - \tau_q)$, as well as $\partial\varepsilon/\partial t\big|_{t_0} \approx \partial\varepsilon/\partial t\big|_{(t_0 - \tau_q)}$ [65], [148]. Therefore, the phase difference in equation (11.29) is directly proportional to the frequency deviation $\tilde{\varepsilon}(t)$ [65], [148], [149]:

$$\varphi_{mx,q}(t) \approx 2\pi \left[(f_0 + K\, t)\, \tau_q - \frac{K}{2}\, \tau_q^2 + \tilde{\varepsilon}(t) \cdot \tau_q \right] \ . \qquad (11.31)$$

11.2.3 Retrieval of frequency deviation

The frequency deviation $\tilde{\varepsilon}(t)$ can be retrieved by a single isolated target [41], [147], [148], [149]. In terms of hardware, this can be achieved by a pseudo target realized by a delay line and a mixer apart of the remaining system [41], [62], [147], [148], [149] (see Fig. 11.7). Thereby, the ramp signal is branched off directly after the VCO before the multipliers, such that the delay line can be implemented as coaxial cable [41], [147], [148]. The pure nonlinearities can be isolated by another conversion with a reference signal provided by a stable oscillator STALO. Conveniently, the STALO can be an *off-the-shell* quartz oscillator. Then the delay line should be trimmed in order to cause a nominal distance frequency equal to the reference frequency (*homodyne* principle [66]). Then the output signal of the second mixer is a complex voltage directly proportional $\exp(+j\, 2\pi\, \tilde{\varepsilon}(t)\, \tau_d)$. It can be exploited as input signal for a controller [41] or, after analog-to-digital conversion, for a digital signal processor DSP [148], [149]. Those in turn are to retune the VCO. Now a linearized ramp is provided. In case of an ideal multiplier, the transmitted ramp is also linear, thus the signal is corrected for all range bins with the drawback of increased hardware effort [41], [62], [147], [148], [149]. There are also similar approaches with direct digital synthesizers DDS [150].

Alternatively to a hardware solution, the signal of a single isolated target can be determined by a test measurement: Ideally this is a point scatterer

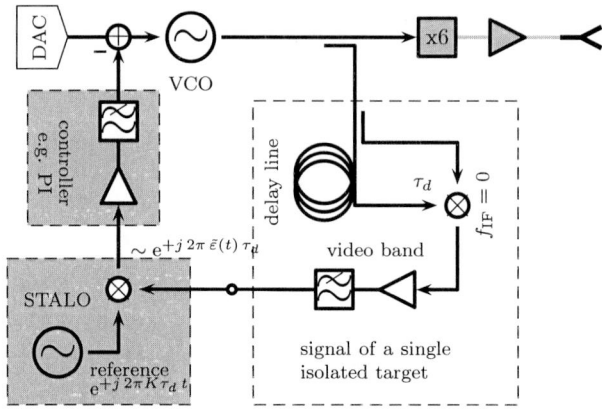

Fig. 11.7 Hardware ramp linearization [62], [147], [148].

like a trihedral reflector at known distance to the radar in an anechoic environment. However, this is not feasible for field measurements. Therefore, there are to be chosen: Looking directions offering a minimum of reflections and an electrically large trihedral whose reflected signal dominates the measured signal. The FMCW signal is *Fourier* transformed, the trihe-

RCS of trihedral

$$rcs = \tfrac{4}{3}\,\pi \cdot \tfrac{l^4}{\lambda^2}$$

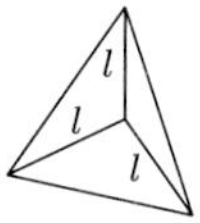

Fig. 11.8 The radar-cross section RCS of a trihedral reflector [42].

dral target is isolated by a window function (bandpass) in spectrum and inversely *Fourier* transformed. Then the retrieved time domain signal features a similar phase like in equation (11.31). Hereby, a trade-off must be concluded: A narrow band filter in frequency domain would exclude signal contributions from other scatterers sited in range bins close by. However, this results in a broad impulse response in time domain which in turn favours *Gibb's phenomenon* [56] until it even outreaches the phase deviation effects. This effect is reduced by use of a broadband filter (thus broad in spectral domain) with the drawback of the requirement to have an isolated target. The big advantage to the proposed hardware solution is the fact that in case of a test measurement, all distortion effects of transmitter and receiver stage can be collected and projected on individual frequency deviations $\tilde{\varepsilon}_i(t)$ per MIMO channel. Therefore, the principal setup of one radar

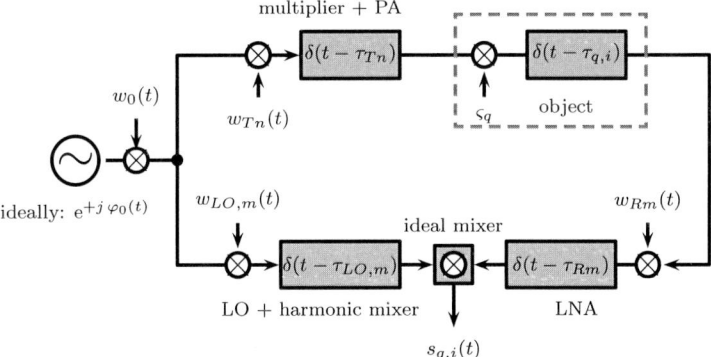

Fig. 11.9 Simplified model representing delays and potential AM or FM noise contributions per radar channel.

channel has to be examined in more detail. Fig. 11.9 sketches a simplified model of possible fluctuations and distortions: The ideal signal is combined with an amplitude and phase response, indicated by the complex factor $w_0(t)$. Possible amplitude responses by the multipliers, amplifiers (power amplifier PA, low-noise amplifiers LNA), as well as for the harmonic mixers are represented by complex functions $w_{Tn}(t)$, $w_{Rm}(t)$, $w_{LO,m}(t)$. The different delays caused by concentrated elements, transmission lines and flight time to the object, are modelled by individual concentrated blocks. Since the harmonic mixers in LO-paths are not linked to the multipliers in transmit path by means of amplitude control etc., different phase deviations must be assumed for each transmit and LO-path. This results in individual amplitude responses, as documented in Fig. 11.10. Equation (11.29) must be extended to obtain the phase deviations for each channel i:

$$\tau_{ges,q,i} = \tau_{Tn} + \tau_{q,i} + \tau_{Rm} - \tau_{LO,m} \,, \tag{11.32}$$

$$\varphi_{q,i}(t) = 2\pi \left[(f_0 + K_i t)\, \tau_{ges,q,i} - \frac{K_i}{2}\, \tau_{ges,q,i}^2 \right] + \delta\varphi_i(t) \,, \tag{11.33}$$

$$\delta\varphi_i(t) = 2\pi \left[\varepsilon_{LO,m}(t - \tau_{LO,m}) - \varepsilon_{Tn,Rm} \left(t - \tau_{Tn} - \tau_{q,i} - \tau_{Rm} \right) \right] \,. \tag{11.34}$$

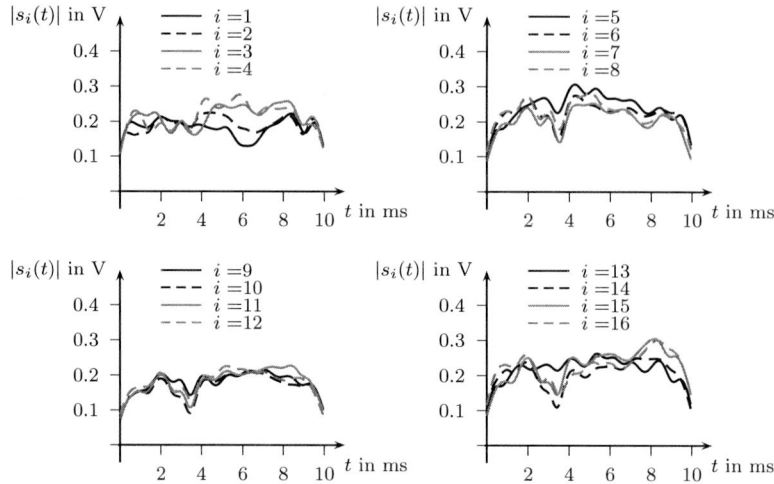

Fig. 11.10 Isolated trihedral target at 4.5 m recorded by helicopter landing aid radar setup: Absolute values or envelope functions in time domain after spectral windowing and inverse *Fourier* transformation, channels $i = 1 \ldots 16$. Parameters: $\Delta F = 3$ GHz, $T_c = 10$ ms, spectral window $\Delta r = \pm 0.45$ m around target, Kaiser window with $\beta = 3$.

11.2.4 Internal transmission line lengths

First of all, systematic and time-invariant effects must be separated from unknown deterministic and statistic processes. One time-invariant and systematic effect are possibly the different internal effective electric lengths per radar channel. In the following, a test measurement recorded by the helicopter landing aid radar shall serve as example. The unwrapped phase of an isolated target signal is collected in a matrix $\boldsymbol{P}$, whereby the MIMO channels' phases over time serve as columns, hence the column numbering corresponds to index i:

$$\boldsymbol{P} = \left[\ldots, \ \angle \{s_i(t)\}, \ \ldots \right] . \tag{11.35}$$

Ideally, the isolated target signal should bear a constant frequency, thus a linear phase, which is not the case for the measured signal. Therefore, it is feasible to introduce reduced ramps K_i per channel. By calculus of *linear regression* (*method of least-squares, general inverse matrix, Gauss method*) a linear function equal to the first part of equation (11.33) can be obtained [43].

$$\varphi_i(t) = 2\pi\,K\,\tau_{ges,q,i}\,t + 2\pi\,\left(f_0\,\tau_{ges,q,i} - \frac{K}{2}\,\tau_{ges,q,i}^2 \right) = m_i\,t + \phi_i\ . \quad (11.36)$$

Please note, that a possible coupling between several antennas would have a minor effect on the modulation but mainly affect the complex (radiation) impedance at feed input [31], [32]. A priori, each channel could exhibit a (generally reduced) ramp K_i due to the effects introduced at the beginning of this chapter. However, it shall be assumed that all MIMO channels i feature the same chirp rate K. Differences are then mapped onto the *round-trip delays* $\tau_{ges,q,i}$. Hereby, the quadratic term with respect to τ in equation (11.36) affects only the constant phase ϕ_i for each channel. The searched parameters m_i, ϕ_i are posed then to a matrix $\boldsymbol{M}$:

$$\boldsymbol{M} = \begin{bmatrix} m_1 & \dots & m_i & \dots & m_{NM} \\ \phi_1 & \dots & \phi_i & \dots & \phi_{NM} \end{bmatrix}\ . \quad (11.37)$$

If $\boldsymbol{T}$ holds a discrete time vector and ones as columns:

$$\boldsymbol{T} = \begin{bmatrix} 0 & \dots & t & \dots & T_c \\ 1 & \dots & 1 & \dots & 1 \end{bmatrix}^{\mathrm{T}} = [\boldsymbol{t},\,\boldsymbol{1}]\ , \quad (11.38)$$

the *linear regression problem* can be postulated [43]. Its solution is the so-called *general inverse matrix* [43]:

$$\min \|\boldsymbol{T}\,\boldsymbol{M} - \boldsymbol{P}\|_2^2\quad ,\quad \Rightarrow\quad \boldsymbol{M} = \left(\boldsymbol{T}^{\mathrm{T}}\,\boldsymbol{T}\right)^{-1}\,\boldsymbol{T}^{\mathrm{T}}\,\boldsymbol{P}\ . \quad (11.39)$$

The first row of $\boldsymbol{M}$ contains all slopes $m_i = (2\pi K\,\tau_{ges,q,i})$, whereas the second row strings all start phases ϕ_i. Both can be exploited to impose reference functions $s_{ref,i}(t)$:

$$s_{ref,i}(t) = \mathrm{e}^{+j\,(m_i\,t + \phi_i)}\ . \quad (11.40)$$

The individual phase deviation per channel $s_{\delta\varphi,i}(t)$ can then be gathered by interferometry:

$$s_{\delta\varphi,i}(t) = \frac{s_i(t)}{|s_i(t)|} \cdot s^*_{ref,i}(t) = e^{+j\,\delta\varphi_i(t)} \text{ , for } i \in [1;\, NM] \text{ ,} \qquad (11.41)$$

$$\delta\varphi_i(t) = 2\pi \left[\varepsilon_{LO,m}(t - \tau_{LO,m}) - \varepsilon_{Tn,Rm}(t - \tau_{Tn} - \tau_{q,i} - \tau_{Rm})\right] \text{ .} \qquad (11.42)$$

Fig. 11.9 and equation (11.34) lead to a phase deviation depending on two different functions: The first one, $\varepsilon_{LO,m}$, incorporates the deviations of local oscillator path, whereas the second one, $\varepsilon_{Tn,Rm}$, represents the combined deviations of transmitter and corresponding receiver. However, the first is always the same for all targets, whereas the second depends on the *round-trip delay*, thus the radial range of object. In the following, a strategy similar to [65] is proposed. Fig. 11.11 shows a phase deviation example recorded with the helicopter landing aid radar. The observed target is

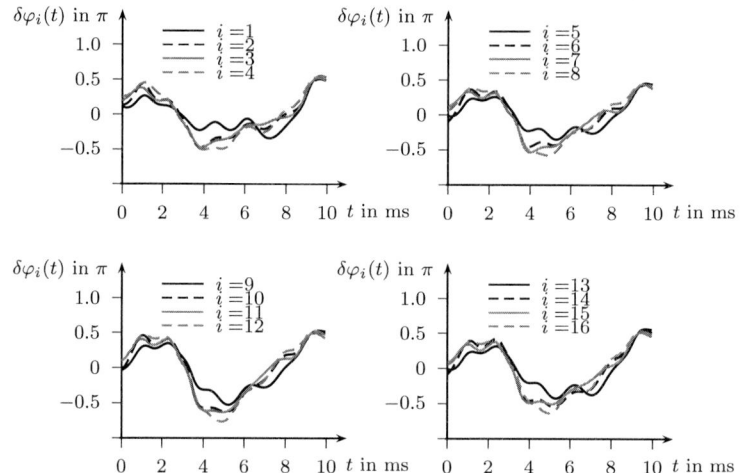

Fig. 11.11 Example for individual phase deviations from an ideal linear slope $2\pi K \tau_{ges,q,i} \cdot t$ (landing aid radar): Trihedral at 4.5 m with internal transmission line lengths, $rcs = 2\pi\,\mathrm{m}^2$, $\Delta F = 3\,\mathrm{GHz}$, $T_c = 10\,\mathrm{ms}$, spectral window $\Delta r = \pm 0.45\,\mathrm{m}$ around target, Kaiser window with $\beta = 3$.

a trihedral reflector at 4.5 m under $0\,^\circ$ azimuth and $90\,^\circ$ elevation with $\Delta F = 3\,\mathrm{GHz}$, $T_c = 10\,\mathrm{ms}$. The trihedral is constructed with a characteristic length $l = 7\,\mathrm{cm}$ (Fig. 11.8) which features a radar cross-section (RCS) of $rcs = 2\pi\,\mathrm{m}^2 \simeq 8\mathrm{dBm}^2$ at $\lambda_0 \approx 4\,\mathrm{mm}$ (Fig. 11.8) [42]. The *Kaiser window* exhibited $\beta = 3$ as envelope parameter in spectral domain at $\Delta r = \pm 0.45\,\mathrm{m}$ around the target. The corresponding frequency deviation is then:

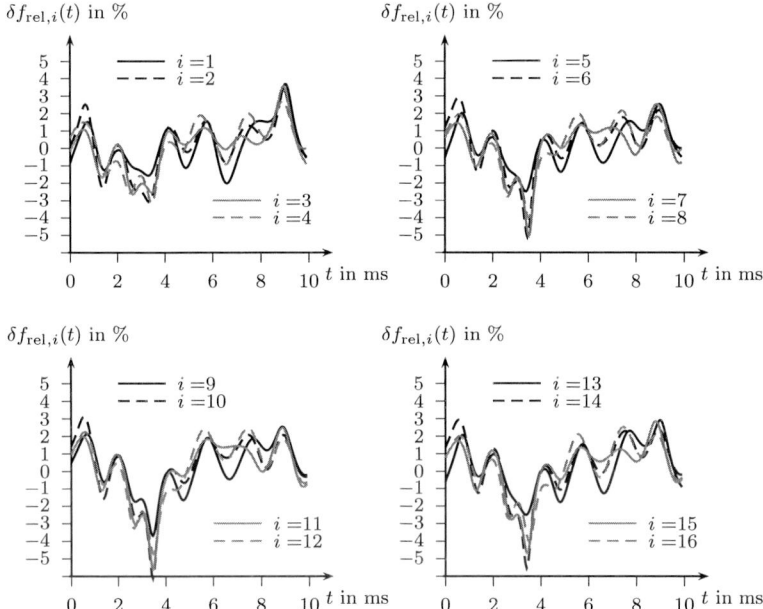

Fig. 11.12 Example for individual frequency deviations relative to ideal $K\tau_{ges,q,i}$ (landing aid radar): Trihedral at 4.5 m with internal transmission line lengths, $rcs = 2\pi\,\mathrm{m}^2$, $\Delta F = 3\,\mathrm{GHz}$, $T_c = 10\,\mathrm{ms}$, spectral window $\Delta r = \pm 0.45\,\mathrm{m}$ around target, Kaiser window with $\beta = 3$.

$$\delta f_i(t) = \frac{1}{2\pi}\frac{\partial \delta\varphi_i(t)}{\partial t}\ .\tag{11.43}$$

If the absolute frequency deviation is divided by the frequency provided by the linear regression, this will give the relative frequency deviation (Fig. 11.12).

$$\delta f_{\mathrm{rel},i}(t) = \frac{\delta f_i(t)}{K \, \tau_{ges,q,i}} = 2\pi \, \frac{\delta f_i(t)}{m_i(t)} = \frac{1}{m_i(t)} \, \frac{\partial \delta \varphi_i(t)}{\partial t} \, . \tag{11.44}$$

Fig. 11.12 demonstrates that already a relative frequency deviation of $\pm 5\%$ is responsible for a defocusing effect like in Fig. 11.6. A maximum peak detection in all radar channels shall be carried out in spectral domain. The results m_i from the linear regression calculus are manipulated to provide the measured distances (per channel i) to the target including the internal lengths of transmission lines:

$$m_i \stackrel{!}{=} 2\pi \, K \tau_{ges,q,i} = \frac{4\pi}{c_0} \, K \, r_i \, , \quad \Rightarrow \quad r_i = \frac{c_0}{4\pi} \cdot \frac{m_i}{K} \, . \tag{11.45}$$

Fig. 11.13 compares the results. The linear regression discloses again the fundamental convolution character of the virtual array: N electrical lengths for Tx and M electrical lengths for Rx lead to $N \cdot M$ combinations. Fig. 11.14 presents the linear regression results minus the radial distance to the trihedral reflector ($\delta r_i = r_i - r_{0q}$), hence shows the internal electrical lengths and reveals that there are four different electrical lengths in Tx part and five in Rx (in this case), which are combined systematically. Those

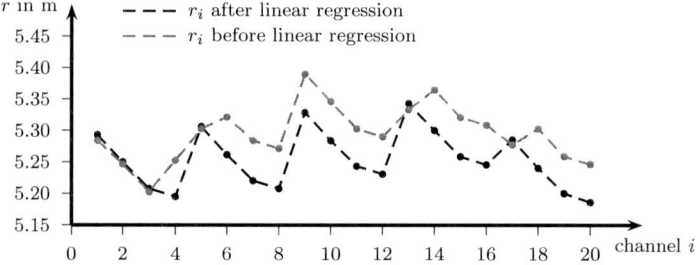

Fig. 11.13 Comparison of range measurement to trihedral target (including internal transmission lines) provided by linear regression (black) and after maximum peak detect in defocused spectrum (gray) (helicopter landing aid). Nominal target distance $r_{0q} = 4.5\,\mathrm{m}$, $rcs = 2\pi\mathrm{m}^2$.

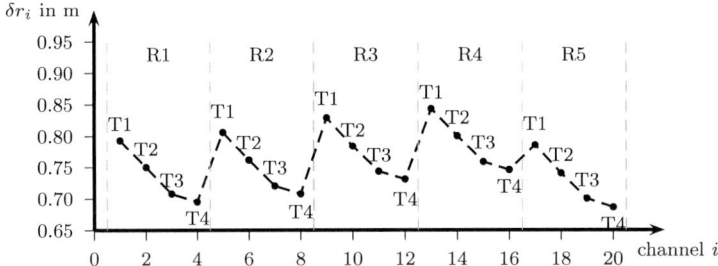

Fig. 11.14 The internal electrical lengths can be regarded as convolution between Tx lengths and Rx lengths. The active Tx and Rx elements are indicated (helicopter landing aid).

shall be represented by δT_n for the transmitters and δR_m for the receiving and local oscillator path:

$$\delta T_n = \frac{c_0}{2}\,\tau_{Tn} \ , \ n \in [1;\,N] \ , \tag{11.46}$$

$$\delta R_m = \frac{c_0}{2}\,(\tau_{Rm} - \tau_{LO,m}) \ , \ m \in [1;\,M] \ , \tag{11.47}$$

$$\Rightarrow \quad \delta r_i = \frac{c_0}{2}\,(\tau_{Tn} + \tau_{Rm} - \tau_{LO,m}) = \delta T_n + \delta R_m \ , \ i \in [1;\,NM] \ . \tag{11.48}$$

The terms δT_n, δR_m are now aligned in vector $\boldsymbol{\delta L}$. The measured internal lengths are collected in $\boldsymbol{\delta r}$. Equation (11.48) can then be re-written as minimum search in matrix vector notation:

$$\min\left\{\left\|\boldsymbol{A}_{sys}\,\boldsymbol{\delta L} - \boldsymbol{\delta r}\right\|_2^2\right\} \ , \tag{11.49}$$

$$\min\left\{\left\|\begin{bmatrix}\mathbf{1}_{N\times1}\,\boldsymbol{p}_1^{\mathrm{T}} \ , & \boldsymbol{E}_N \\ \vdots & \vdots \\ \mathbf{1}_{N\times1}\,\boldsymbol{p}_m^{\mathrm{T}} \ , & \boldsymbol{E}_N \\ \vdots & \vdots \\ \mathbf{1}_{N\times1}\,\boldsymbol{p}_M^{\mathrm{T}} \ , & \boldsymbol{E}_N\end{bmatrix}\begin{bmatrix}\vdots \\ \delta Rm \\ \vdots \\ \delta Tn \\ \vdots\end{bmatrix} - \begin{bmatrix}\vdots \\ \delta r_i \\ \vdots\end{bmatrix}\right\|_2^2\right\} \ . \tag{11.50}$$

The vector $\mathbf{1}_{N \times 1}$ is a column containing N ones, whereas $\boldsymbol{p}_m$ is a column of $(M-1)$ zeros, however a single one entry at mth position is picked out. $\boldsymbol{E_N}$ is a $N \times N$ unity matrix (eye).

$$\boldsymbol{A}_{sys} \in \{0, 1\}^{NM \times (M+N)} \ , \boldsymbol{\delta L} \in \mathbb{R}^{(N+M) \times 1} \ , \boldsymbol{\delta r} \in \mathbb{R}^{(NM) \times 1} \ ,$$

$$\mathbf{1}_{N \times 1} = \begin{bmatrix} 1 \\ \vdots \\ 1 \end{bmatrix} \in \{1\}^{N \times 1} \ , \boldsymbol{p}_m = \begin{bmatrix} 0 \\ \vdots \\ 1 \\ 0 \\ \vdots \end{bmatrix} \text{mth;} \in \{0, 1\}^{M \times 1} \ . \tag{11.51}$$

Equation 11.50 can formally be solved by the *method of least-squares* again [43]. However, instead of $(N + M)$, $\boldsymbol{A}_{sys}$ is of rank $(N + M - 1)$, hence $\boldsymbol{A}_{sys}^{\mathrm{T}} \boldsymbol{A}_{sys}$ is singular and cannot be inverted. This can be demonstrated by a few fundamental operations on the $(M + N)$ columns of $\boldsymbol{A}_{sys}$: The very first column contains N ones.:

$$\boldsymbol{A}_{sys} = \begin{bmatrix} \mathbf{1}_{N \times 1}, & \mathbf{0}_{N \times (M-1)}, & \boldsymbol{E_N} \\ & \vdots & \vdots \\ \mathbf{0}_{M(N-1) \times 1}, & \mathbf{1}_{N \times 1} \, \tilde{\boldsymbol{p}}_m^{\mathrm{T}}, & \boldsymbol{E_N} \\ & \vdots & \vdots \end{bmatrix} \ , \tag{11.52}$$

with $\tilde{\boldsymbol{p}}_m \in \{0, 1\}^{(M-1) \times 1} \ , m \in [2; \ M] \ .$

Now the last N columns are subtracted from the very first one, whereas the remaining $(M-1)$ columns are added. This gives a first column $(\boldsymbol{A}_{sys}(:, 1)$, as *Matlab* notation) containing only zeros, which is possible for all N, M and sets the column rank and therefore the total rank to $(N + M - 1)$.

$$\Rightarrow \boldsymbol{A}_{sys}(:, 1) = \begin{bmatrix} \mathbf{1}_{N \times 1} - \boldsymbol{E_N} \, \mathbf{1}_{N \times 1} + \mathbf{0}_{N \times (M-1)} \\ \vdots \\ \mathbf{0}_{N \times 1} - \boldsymbol{E_N} \, \mathbf{1}_{N \times 1} + \mathbf{1}_{N \times 1} \, \tilde{\boldsymbol{p}}_m^{\mathrm{T}} \, \mathbf{1}_{(M-1) \times 1} \\ \vdots \end{bmatrix} = \mathbf{0}_{NM \times 1} \ . \tag{11.53}$$

Similar considerations can be engaged for row manipulations on A_{sys}. Analogously, it can be observed that the first N rows span a linear subspace of dimension N. However, the following blocks of N rows only increase the rank by one. Therefore, the rank $(N + M - 1)$ emerges again. Now, for solving the minimum search of equation (11.49), one degree of freedom has to be set to fixed length. W.l.o.g. δR_M is chosen. The corresponding column of A_{sys} scaled by δR_M is subtracted from equation (11.49), thus the corresponding column in A_{sys} as well as the entry in δL are left away, which yields a modified minimum search:

$$\min \left\{ \left\| \tilde{A}_{sys} \, \tilde{\delta L} - b_{\delta r} \right\|_2^2 \right\} \text{, with } b_{\delta r} = \delta r - \begin{bmatrix} 0_{M(N-1) \times 1} \\ 1_{N \times 1} \end{bmatrix} \delta R_M \text{,} \tag{11.54}$$
$$\tilde{A}_{sys} \in \{0, 1\}^{NM \times (M+N-1)} \text{, } \tilde{\delta L} \in \mathbb{R}^{(N+M-1) \times 1} \text{.}$$

The *least squares* solution for the remaining internal lengths is then [43]:

$$\tilde{\delta L} = \left(\tilde{A}_{sys}^{\mathrm{T}} \tilde{A}_{sys} \right)^{-1} \tilde{A}_{sys}^{\mathrm{T}} b_{\delta r} \text{.} \tag{11.55}$$

By re-inserting the degree of freedom δR_M into $\tilde{\delta L}$, δL is obtained again. The remaining residual can be determined by [43]:

$$res = b_{\delta r} - \tilde{A}_{sys} \, \tilde{\delta L} = \delta r - A_{sys} \, \delta L \text{.} \tag{11.56}$$

A feasible choice for the helicopter landing aid radar was $\delta R_5 = 0.3\,\mathrm{m}$. Then this yields Tab. 11.1. The remaining residual of 1.8 mm originates not only in measurement errors but also in possible boresight errors, hence yaw or pitch errors (azimuth or elevation) between the sensor and the test target. Those in turn would lead to small range deviations. In terms of

δR_m	δR_1	δR_2	δR_3	δR_4	δR_5
value in m	0.3084	0.3209	0.3437	0.3588	**0.3000**
δT_n	δT_1	δT_2	δT_3	δT_4	$\| res \|_2$
value in m	0.4850	0.4408	0.3998	0.3869	1.8 mm

Tab. 11.1 Internal effective electrical lengths of helicopter landing aid frontend.

mathematics equal to range cell migration, however within one range cell (see also equation (5.16)). Regarding Tab. 11.1 with respect to calibration, it might be more convenient to relate all transmitter and receiver lengths to a reference like e.g. δR_M, δT_N. This is listed in Tab. 11.2. Now, the parameters τ_{Tn}, $(\tau_{Rm} - \tau_{LO,m})$ in equations (11.46), (11.47) can be determined by the internal effective electrical lengths. However, they serve as arguments distributed over two different phase deviation functions in equation (11.33), which prevents a straight forward solution.

But actually, the first part in equation (11.33), $\varepsilon_{LO,m}(t - \tau_{LO,m})$, is the same for all targets and only depends on the receiver path. Just as the internal transmission line lengths, those can be compensated by dividing two signals or multiplying one signal with the complex conjugate of another.

Therefore, two deviation signals $s_{\delta\varphi,1,i}(t)$, $s_{\delta\varphi,2,i}(t)$ featuring the same trihedral at alternative radial distances shall be compared to each other. An interferometric approach cancels the local oscillator dependence, similar to the approach presented in [65] where the Tx characteristics become extinct:

$$
s_{\delta\varphi,i}(t) = s_{\delta\varphi,1,i}(t) \cdot s^*_{\delta\varphi,2,i}(t) = \mathrm{e}^{+j\,(\delta\varphi_{1,i}(t) - \delta\varphi_{2,i}(t))} =
$$
$$
= \mathrm{e}^{-j\,2\pi\,[\varepsilon_{Tn,Rm}(t - \tau_{Tn} - \tau_{1,i} - \tau_{Rm}) - \varepsilon_{Tn,Rm}(t - \tau_{Tn} - \tau_{2,i} - \tau_{Rm})]}\ . \tag{11.57}
$$

δR_m	δR_1	δR_2	δR_3	δR_4	δR_5
value	δR_5+ 8.40 mm	δR_5+ 20.9 mm	δR_5+ 43.7 mm	δR_5+ 58.8 mm	δR_5
δT_n	δT_1	δT_2	δT_3	δT_4	$\delta R_5 + \delta T_4$
value	δT_4+ 98.1 mm	δT_4+ 54.0 mm	δT_4+ 12.9 mm	δT_4	$\delta r_{20} = 0.6863$ m

$$\|\boldsymbol{res}\|_2 = 1.8\,\mathrm{mm}$$

Tab. 11.2 Internal effective electrical lengths of the helicopter landing aid frontend related to δR_5 or δT_4.

The substitution $\tilde{t} = t - \tau_{Tn} - \tau_{2,i} - \tau_{Rm}$ applied to equation (11.57) gives:

$$s_{\delta\varphi,i}(\tilde{t}) = e^{+j\,2\pi\,\left[\varepsilon_{Tn,Rm}(\tilde{t}) - \varepsilon_{Tn,Rm}\left[\tilde{t} - (\tau_{1,i} - \tau_{2,i})\right]\right]} \approx$$
$$\approx e^{+j\,2\pi\,\tilde{\varepsilon}_{Tn,Rm}(\tilde{t})\cdot(\tau_{1,i} - \tau_{2,i})} = e^{+j\,2\pi\,\tilde{\varepsilon}_i(\tilde{t})\cdot(\tau_{1,i} - \tau_{2,i})} . \tag{11.58}$$

Now the frequency deviation $\tilde{\varepsilon}_{Tn,Rm}(t) = \tilde{\varepsilon}_i(t)$ is formally:

$$\tilde{\varepsilon}_{Tn,Rm}(t) = \tilde{\varepsilon}_i(t) = \frac{1}{2\pi\,(\tau_{1,i} - \tau_{2,i})} \cdot \arctan\left(\frac{\Im\left\{s_{\varepsilon,i}(t)\right\}}{\Re\left\{s_{\varepsilon,i}(t)\right\}}\right) =$$
$$= \frac{c_0}{4\pi\,(r_{1,i} - r_{2,i})} \cdot \arctan\left(\frac{\Im\left\{s_{\varepsilon,i}(t)\right\}}{\Re\left\{s_{\varepsilon,i}(t)\right\}}\right) . \tag{11.59}$$

The a-priori known *flight times* $\tau_{1,i}$, $\tau_{2,i}$ are supposed to be purely free-space transitions and can be determined by a distance measurement from radar frontend to object by e. g. tape measure, laser distance measure etc. The time shift formally yields new frequency deviation functions. This retrieved and time-shifted functions per channel $\tilde{\varepsilon}_i(t)$ serve then as fundament for the following filter techniques. The phase deviation can be related to the total phase hub [41], [151], [152]: The phase deviation is $2\pi\tilde{\varepsilon}_i \cdot \Delta\tau_{12,i}$ with $\Delta\tau_{12,i} = (\tau_{1,i} - \tau_{2,i})$. The total phase hub per chirp can be written as $2\pi\Delta F \cdot \Delta\tau_{12,i}$. The normalized phase deviation can be interpreted as a normalized relative *time variation* for each MIMO channel i:

$$\delta t_{rel,i} = \frac{\delta\varphi_i(t)}{2\pi\,KT_c\,\Delta\tau_{12,i}} = \frac{2\pi\,\tilde{\varepsilon}_i \cdot \Delta\tau_{12,i}}{2\pi\,\Delta F \cdot \Delta\tau_{12,i}} = \frac{\tilde{\varepsilon}_i}{\Delta F} . \tag{11.60}$$

In the following section, it can be found that a time deviation δt_i can be derived from $\delta t_{rel,i}$ by:

$$\delta t_i = \delta t_{rel,i} \cdot T_c = \frac{\tilde{\varepsilon}_i}{K} . \tag{11.61}$$

Among others, also in [41] a figure of merit for the phase and frequency deviations (per channel i) was introduced: The so-called *linearity* $\mathcal{L}_i$, which is the maximum absolute value of $\delta t_{rel,i}$ [41]:

$$\mathcal{L}_i = \max\left\{\left|\frac{\tilde{\varepsilon}_i}{K}\right|\right\} = \max\left\{|\delta t_{rel,i}|\right\} . \tag{11.62}$$

Just for the sake of completeness, with help of equations (11.30), (11.58), the relative frequency deviation is then:

$$\delta f_{rel,i}(t) = \frac{1}{2\pi K \Delta \tau_{12}} \frac{\partial \delta \varphi_i(t)}{\partial t} = \frac{1}{K} \frac{\partial \tilde{\varepsilon}_i(t)}{\partial t} = \frac{\partial \delta t_i}{\partial t}. \tag{11.63}$$

11.3 Data interpolation and resampling

One drawback of first retrieving the error functions $\tilde{\varepsilon}_i(t)$ and using them e. g. for a dispersive software filter, is a possibly high computational load [65]. An alternative method for linearizing the modulation charcteristics is *resampling* [147], [148], [149] [153], [154]. Here, the basic idea is not to sample the signals in equidistant time steps but at equidistant phase steps defined by a reference signal [147], [148], [149] [153], [154]. Those equidistant phase increments lead to a non-equidistant time base [147], [148], [149] [153], [154], which defines intermediate time steps where the signal is to be interpolated or re-sampled. The before-mentioned reference signal can be generated either by a hardware setup like in Fig. 11.7 or, like in this case, by methods introduced in section 11.2.3. Fig. 11.15 demonstrates exemplarily the need of a correction process: First of all, it seems that the MIMO channels exhibit different behaviour in time and frequency domain compared to each other. Furthermore, the spectra of the same MIMO channels change over range.

Three test measurements with a single isolated and static target (trihedral) shall be recorded:

$$s_{1,i}(t) = U_{1,i} \cdot e^{+j\,2\pi\,(Kt+f_0)\,\tau_{ges,1,i}} \cdot e^{-j\,\pi K \tau_{ges,1,i}^2} \cdot e^{+j\,\delta\varphi_{1,i}(t)}, \tag{11.64}$$

$$s_{2,i}(t) = U_{2,i} \cdot e^{+j\,2\pi\,(Kt+f_0)\,\tau_{ges,2,i}} \cdot e^{-j\,\pi K \tau_{ges,2,i}^2} \cdot e^{+j\,\delta\varphi_{2,i}(t)}, \tag{11.65}$$

$$s_{3,i}(t) = U_{3,i} \cdot e^{+j\,2\pi\,(Kt+f_0)\,\tau_{ges,3,i}} \cdot e^{-j\,\pi K \tau_{ges,3,i}^2} \cdot e^{+j\,\delta\varphi_{3,i}(t)}. \tag{11.66}$$

Those three signals $s_1(t)$, $s_2(t)$, $s_3(t)$ are recorded under $0\,°$ azimuth and elevation angle with a trihedral reflector at distances of $4.5\,\mathrm{m}$, $5.5\,\mathrm{m}$, $6.5\,\mathrm{m}$. Later, it will be proofed that only two signals are needed for a correct resampling process. However, the third signal provides an alternative signal for proofing a chosen way of processing to be valid generally. Again $\tau_{ges,q,i}^2 \approx \tau_{0q}^2$ can be applied if feasible. Fig. 11.15 illustrates that both signals are defocused massively by nonlinearities.

As already suggested, the deterministic effects like internal effective electric lengths δT_n, δR_m, the phase deviation of local oscillator path or different amplitude coefficients can be compensated by dividing two signals. All further radar signals in this section are pre-processed by dividing them by

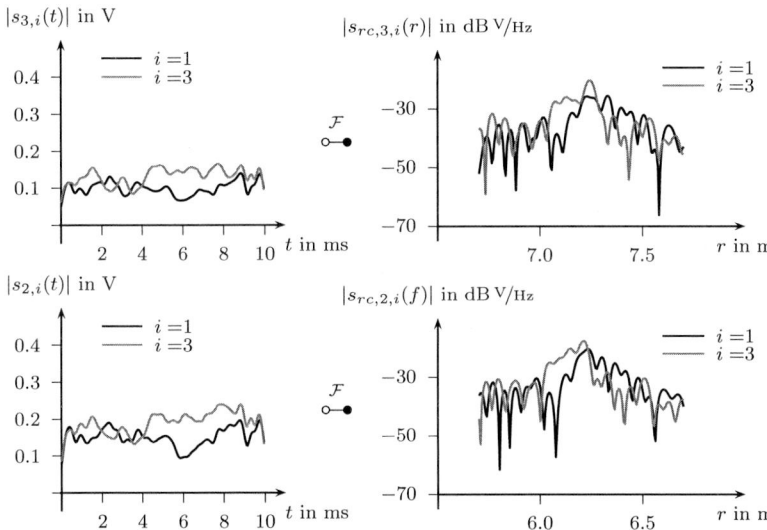

Fig. 11.15 Example for two isolated but defocused signals recorded by helicopter landing aid radar. Top row: Trihedral target at 6.5 m, $rcs \approx 2\pi \, \text{m}^2$ (with internal lengths $+0.7\,\text{m}$) and the absolute value of its inverse Fourier transform. Bottom row: Analogously trihedral target at 5.5 m with internal lengths. Both signal spectra are cut out by a Kaiser window of shape parameter 3 as window of $\pm 1\,\text{m}$ around the nominal radial distance. Both cases are illustrated for their MIMO channels $i = 1$, $i = 3$.

$s_1(t)$. In case of an arbitrary radar measurement divided by $s_1(t)$, it can be obtained:

$$s_i(t) = \frac{s_{mx,i}(t)}{s_{1,i}(t)} =$$

$$= \sum_{q=1}^{Q} \frac{U_{q,i}}{U_{1,i}} \cdot \mathrm{e}^{+j\,2\pi\,(Kt+f_0)\,(\tau_{q,i}-\tau_{1,i})} \cdot \mathrm{e}^{-j\,\pi K\left(\tau_{q,i}^2-\tau_{1,i}^2\right)} \cdot \mathrm{e}^{+j\,2\pi\,\bar{\varepsilon}_i(\bar{t})\,(\tau_{q,i}-\tau_{1,i})} \cdot$$

$$\text{(11.67)}$$

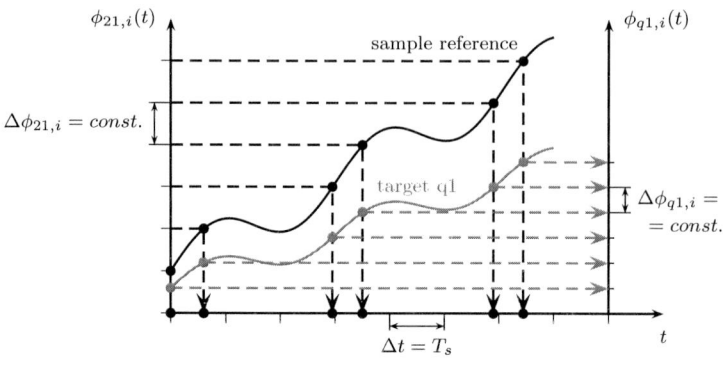

• new sampling time steps

Fig. 11.16 The resampling principle: Instead of sampling along equidistant time steps, sampling along equidistant phase increments. A reference phase $\phi_{21,i}$ defines a non-equidistant sampling scheme in time-domain. The same scheme provides equidistant phase increments for alternative target phase $\phi_{q1,i}$

The general phase terms can be represented slightly shorter by:

$$\phi_{q1,i}(t) = 2\pi \left(\tau_{q,i} - \tau_{1,i}\right) \cdot \left[Kt + \tilde{\varepsilon}_i(\tilde{t})\right] + \phi_{const,q1,i} \ . \tag{11.68}$$

If the two test signals are pre-processed in same manner, this yields the *sample reference signal* $s_{sr}(t)$:

$$s_{sr,i}(t) = s_{21,i}(t) = \frac{s_{2,i}(t)}{s_{1,i}(t)} =$$

$$= \frac{U_{2,i}}{U_{1,i}} \cdot e^{+j\,2\pi\,(Kt+f_0)\,(\tau_{2,i}-\tau_{1,i})} \cdot e^{-j\,\pi K\left(\tau_{2,i}^2 - \tau_{1,i}^2\right)} \cdot e^{+j\,2\pi\,\tilde{\varepsilon}_i(\tilde{t})\,(\tau_{2,i}-\tau_{1,i})} \ . \tag{11.69}$$

The alternative *sample reference signal* (for demonstration purpose) is then:

$$s_{sr,2,i}(t) = s_{31,i}(t) = \frac{s_{3,i}(t)}{s_{1,i}(t)} = \frac{U_{3,i}}{U_{1,i}} \cdot e^{+j\,2\pi\,(Kt+f_0)\,(\tau_{3,i}-\tau_{1,i})} \cdot$$
$$\cdot e^{-j\,\pi K\left(\tau_{3,i}^2 - \tau_{1,i}^2\right)} \cdot e^{+j\,2\pi\,\tilde{\varepsilon}_i(\tilde{t})\,(\tau_{3,i}-\tau_{1,i})} \ . \tag{11.70}$$

Up to now, the phase has been considered as a function of time in our deliberations. Thereby, it shall be required that the frequency deviation effects are rather small compared to the modulation itself. Fig. 11.12 proofs this for the implemented systems. (Nevertheless, as already seen, those deviations end in large effects.) However, since the linear frequency modulation usually dominates its deviation effects, the channels' unwrapped phases can not only be considered as *continuous functions* but even as *strictly monotonic increasing* (in *up-chirp* case) or *strictly monotonic decreasing* (in *down-chirp* case) (see also [154]). In this case, the relation of time and phase can be deemed as *bijective* [154]. Then, equation (11.68) shall be solved for t (with $q = 2$), which leads to the actual time base (in *sample reference signal*):

$$t_{re,i} = \frac{\phi_{21,i} - \phi_{const,21,i}}{2\pi\,K\Delta\tau_{21,i}} - \frac{\tilde{\varepsilon}_i}{K} = \frac{\phi_{21,i}}{2\pi\,K\Delta\tau_{21,i}} - \frac{\tilde{\varepsilon}_i}{K} + \frac{\tau_{1,i} + \tau_{2,i}}{2} - \frac{f_0}{K} =$$
$$= \frac{\phi_{12,i}}{2\pi\,K\Delta\tau_{12,i}} - \frac{\tilde{\varepsilon}_i}{\Delta F}\,T_c + \frac{\tau_{1,i} + \tau_{2,i}}{2} - \frac{T_c}{B_{rel}}\,,$$
$$\text{with } \Delta\tau_{21,i} = \tau_{2,i} - \tau_{1,i}\,,\ \frac{f_0}{K} = \frac{T_c}{B_{rel}}\,,\ B_{rel} = \frac{\Delta F}{f_0}\,.$$

$$(11.71)$$

The unique transformation between time and phase is valid globally, thus for $t \in [0; T_c]$, in case of *strictly monotonic* phase functions ϕ_i. Please note, that the same approach still can be applied in sections in case of phase functions which are just *strictly monotonic in sections* [154].

Now, the wanted mathematical expression for the trend of $\phi_{21,i}$ can be chosen, inserted into equation (11.3) and $t_{re,i}$ defines new individual (resampling) time steps per channel i. If the not mandatorily equidistant time base $t_{re,i}$ put into equation (11.68), this yields:

$$\phi_{q,i}(t = t_{re}) = \cdots = \frac{\tau_{q,i} - \tau_{1,i}}{\tau_{2,i} - \tau_{1,i}}\left(\phi_{21,i} - \phi_{const,21,i}\right) + \phi_{const,q1,i} =$$
$$= \cdots = \frac{\tau_{q,i} - \tau_{1,i}}{\tau_{2,i} - \tau_{1,i}}\,\phi_{21,i} - \pi K\left(\tau_{q,i} - \tau_{1,i}\right)\cdot\left(\tau_{q,i} - \tau_{2,i}\right)\,.$$

$$(11.72)$$

Of course, the matter of choice for $\phi_{21,i}(t)$ is a linear function. However, any other strictly monotonic ramp form, such as nonlinear schemes like suggested in section 4.5.2 would be thinkable. The constant term $\phi_{const,21,i}$ will be of special interest for coherency between all MIMO channels in the following. The second summand in second line of equation (11.72) is again a term quadratic in τ, which can be neglected for near-range applications but is listed here for the sake of completeness. Equation (11.72) demonstrates that the general radar signal was successfully disburdened from non-linearities for all ranges [147], [148], [149] [153], [154]. Therefore, this method bears the same result as a range-dependent, thus dispersive, filter in frequency domain [147], [148], [149] [153], [154]. Fig. 11.16 sketches a graphic solution of the principle procedure for the new time base. A desired linear phase function would look like:

$$\phi_{21,i}(t) = 2\pi K \Delta \tau_{21,i} t + \phi_{const,21,i} . \tag{11.73}$$

Inserted into equation gives:

$$t_{re,i} = \cdots = t - \frac{\tilde{\varepsilon}_i(\tilde{t} = t)}{K} = t - \delta t_i . \tag{11.74}$$

This shows, that after determining the channels' frequency deviation functions $\tilde{\varepsilon}_i$, the new time base can be determined directly.

In case of a discrete time base t, the chirp length T_c was split equidistantly in $(P-1)$ steps, providing P sample points. A possible linear phase function providing P sampling points for a new time base could then look like:

$$
\begin{aligned}
\phi_{21,i}(p) &= 2\pi K \Delta \tau_{21,i} \frac{T_c}{P-1} p + \phi_{const,12,i} = \\
&= 2\pi \Delta F \Delta \tau_{21,i} \frac{p}{P-1} + \phi_{const,12,i} , \quad p \in [0; \, P-1] .
\end{aligned} \tag{11.75}
$$

This gives the individual discrete re-sampling time base per channel:

$$
\begin{aligned}
t_{re,i} &= \frac{T_c}{P-1} p - \frac{\tilde{\varepsilon}_i \left(\tilde{t} = T_c \, p/P-1 \right)}{K} = \\
&= T_c \left[\frac{p}{P-1} - \frac{\tilde{\varepsilon}_i(\tilde{t})}{\Delta F} \right] = T_c \left[\frac{p}{P-1} - \delta t_{rel,i} \right] .
\end{aligned} \tag{11.76}
$$

This helps to define the new time base, the re-sampling can take place which in turn automatically fulfills equation (11.75) and gives a linearized signal. Fig. 11.17 shows an example for the *relative time variation* $\delta t_{rel,i}$. It can be concluded that a very little variation of a few per mill ($\mathcal{L} = 3.6 \cdot 10^{-3}$) cause a defocusing effect like in Fig. 11.15, which matches observations of Piper [152]. The bandwidth is $\Delta F =$ 3 GHz for the transmit signal. The multipliers (Fig. 11.9) exhibit the factor 6, thus the VCO needs to provide a bandwidth of 500 MHz. According to Fig. 11.17, the maximum

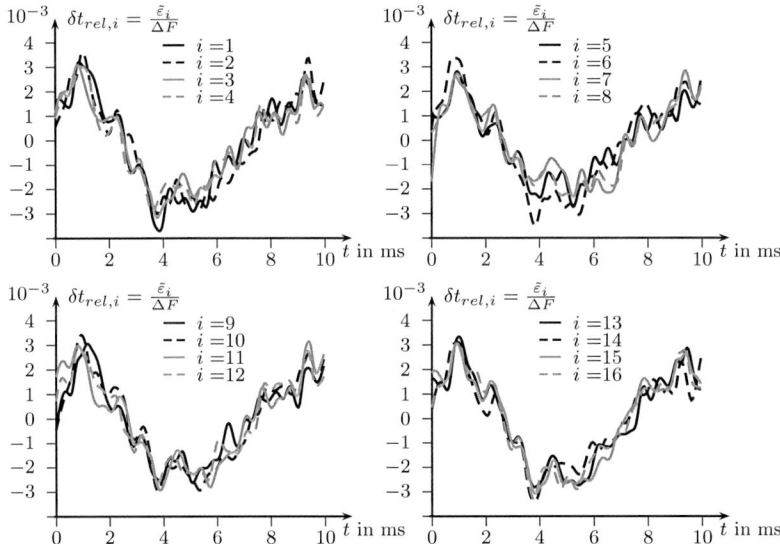

Fig. 11.17 Example for time variation, hence frequency deviation per channel i divided by bandwidth $\Delta F = 3$ GHz of a sample reference signal generated by two signals: trihedral reflector at 5.5 m and 4.5 m, $p \in [0; 4999]$. Provided linearity $\mathcal{L} = 3.6 \cdot 10^{-3}$.

frequency deviation is $\mathcal{L}_i \cdot \Delta F = 10.8$ MHz, thus traced back to a deviation of 1.8 MHz at the *voltage controlled oscillator* (VCO). Its *tuning sensivity* is approximately 400 MHz/V, hence the *frequency deviation* can be mapped to a *voltage deviation* of maximally 4.5 mV, whereas the total tuning voltage range is approximately 1.25 V (between 3 V and 4.5 V) [155]. For the sake of completeness, the *pushing sensivity* is maximally 250 kHz/V [155], which describes unwanted tuning characteristics via voltage ripple on the VCO's power supply [155]. This is one of the effects which were collected to a system wide or global AM or FM noise in Fig. 11.9. The new time base now requires signal values at eventually very different time steps. Those can be presented by either a very high oversampling rate in data acquisition, zero-padding in spectral domain with inverse Fourier transformation or interpolation in time domain. We concentrate on the latter possibility, interpolation in time domain. The first method would require a high hardware effort with the drawback of lacking an a-posteriori application, whereas the second method bears a high computational effort due to the

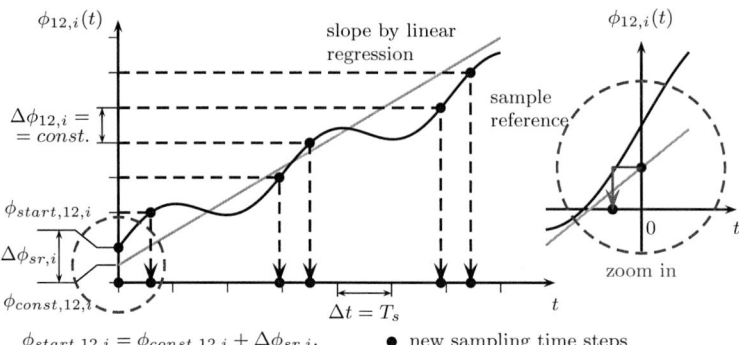

$\phi_{start,12,i} = \phi_{const,12,i} + \Delta\phi_{sr,i},$ • new sampling time steps

Fig. 11.18 A reference phase $\phi_{21,i}$ defines a non-equidistant sampling scheme in time-domain. A crucial point is the start phase which can differ from start phase delivered by linear regression (left). The choice of $\phi_{const,21,i}$ can cause negative time steps (right).

fact that also unnecessary sampling values are computed in addition. Both methods are rather greedy compared to an interpolation operation in time domain which is only evaluated at precisely defined time steps. The mathematical operation can be formulated as a convolution with an interpolation kernel [34], [51], [61]:

$$s_{sr,i}(t_{re,i}) = s_{sr,i}(t) * \left[w(t) \cdot \frac{\sin\left(\frac{\pi}{T_s}t\right)}{\frac{\pi}{T_s}t} \right]_{t_{re,i}} . \tag{11.77}$$

Hereby, the window function $w(t)$ shall limit the unlimited *sinus cardinalis function* in equation (11.77) to a feasible length which is a compromise between computational load and accuracy [61]. This window function $w(t)$ and the *sinus cardinalis* together from the so called interpolation kernel [34], [61]. Of course, there are several other methods for interpolation such as *splines* etc. The reader shall be relegated to literature here, e. g.[156], [157], [158], [159], [160], [161]. However, a crucial point and absolute requirement is, that the sampling theorem is fulfilled during data acquisition [61]. This secures, that any intermediate signal value can be generated by means of interpolation [51], [53], [54], [56], [162].

Nevertheless, there is a drawback in choosing the expression in equation (11.75) for $\phi_{21,i}$ in order to obtain $t_{re,i}$ in equation (11.76). If the actual measured phase is larger than the linear regression, thus $\tilde{\varepsilon}_i(t = 0) > 0$, then equation (11.74) leads to samples to be interpolated in the past, hence negative $t_{re,i}$ occur. Fig. 11.18 illustrates the effect. In this case the signal must be extrapolated to the past, which can lead to complications or additional computational load, depending on the selected method. Therefore a feasible objective should be $t_{re,i} \geq 0$ for all channels i. A possible remedy is to modify equation (11.73):

$$\phi_{21,i}(t) = 2\pi\, K\Delta\tau_{21,i}\, t + \phi_{start,21,i} =$$
$$= 2\pi\, K\Delta\tau_{21,i}\, t + \phi_{const,21,i} + \Delta\phi_{sr,i} \,, \tag{11.78}$$

$$\Delta\phi_{sr,i} = \frac{\delta\varphi_i(t=0)}{2\pi K\Delta\tau_{21,i}} \,, \quad \Rightarrow \quad t_{re,i} = t - \frac{1}{K}\left[\tilde{\varepsilon}_i(\tilde{t}) - \tilde{\varepsilon}_i(0)\right] . \tag{11.79}$$

The amendment in equation 11.79 indicates that the new time base is basically affected by the frequency deviation with its start value subtracted.

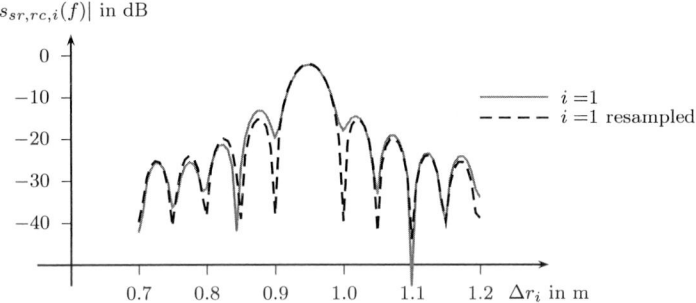

Fig. 11.19 Spectra for original and resampled sample reference signal $s_{sr,i}(t)$ (equation (11.69)). Parameters: $\Delta F = 3\,\text{GHz}$, $\Delta r_{21} \approx 1\,\text{m}$.

Fig. 11.19 demonstrates the spectrum of channel $i = 1$ helicopter landing aid before and after the resampling process at $s_{sr,1}(t)$. Hereby, the differences are rather small due to the effect that the difference in radial distance of the test trihedral target is only 1 m. However, the resampling of an alternative signal with the same modified time base gives Fig. 11.20: Here, although the difference in radial distance is only 2 m, the remaining modulation nonlinearities already defocus the original signal. The resampled version is close to si-function shape. The reduced sidelobe levels mirror in reduced amplitude deviation in time-domain (compare with section 11.1.3).

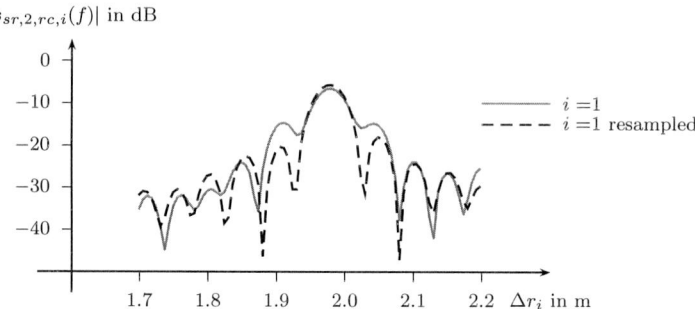

$|s_{sr,2,rc,i}(f)|$ in dB

Fig. 11.20 Spectra for original and resampled signal $s_{sr,2,i}(t)$ (equation (11.67)). Parameters: $\Delta F = 3\,\mathrm{GHz}$, $\Delta r_{31} \approx 2\,\mathrm{m}$.

The possible amplitude responses per channel i (Fig. 11.9) should be cancelled automatically by the quotients in equation (11.67):

$$\frac{U_{q,i}}{U_{1,i}} \approx \frac{\sigma_q}{\sigma_1} \,, \quad \frac{U_{2,i}}{U_{1,i}} \approx \frac{\sigma_2}{\sigma_1} \,, \tag{11.80}$$

which should give the same amplitudes for all channels already in time domain, hence the substitution of LO deviation characteristics (equation (11.67)) could provide an amplitude calibration at the same time. However, the possible amplitude responses of any involved hardware device (Fig. 11.9) can still vary massively over time, which settles in the remaining side-lobe levels of the resampled signals. The amplitude responses per channel do not change significantly by the linear resampling process. Section 11.3.4 shall present further improvement of the amplitude behaviour in time domain by sidelobe reduction in frequency domain. Nevertheless, the very different channels' absolute values in the spectral maxima equalize significantly by the operation of equation (11.67) or (11.69) respectively. The resampling method does not affect this process negatively (Fig. 11.22). Fig. 11.22 presents a variation of 1.5 dB over the channels i. This stands in harsh contrast to the amplitude variation of ca. 6 dB or more before dividing each signal with $s_1(t)$ and resampling it. The amplitude variation is a crucial point for the later evaluation by minimum redundancy radar processing (compare with section 9). A similar consideration can be imposed for the phases in those spectral maxima (Fig. 11.23). Fig. 11.23 shows the phases before and after the resampling method applied to *sampling reference signal* along all channels in the spectral maximum.

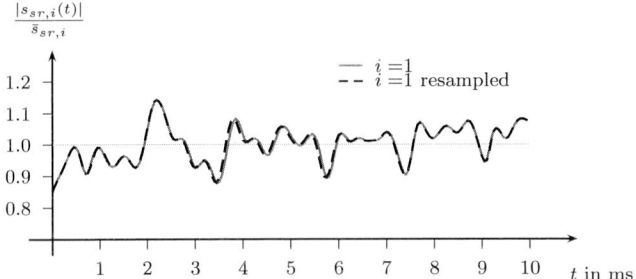

Fig. 11.21 Relative amplitudes in time domain before and after resampling process for a sample reference signal. $\Delta F = 3\,\mathrm{GHz}$, trihedral $rcs \approx 2\pi\,\mathrm{m}^2$, channel $i = 1$, signal amplitude $|s_{sr,i}(t)|$ was normalized by its mean value $\bar{s}_{sr,i}$.

First, the range compressed *sampling reference signal* before re-sampling shall be examined: By recalling section 4.4.3, the phase along all channels can be expressed by:

$$\varphi_{sr,i}(\Delta r_{21}) = \varphi_{sr,Tn,Rm}(r_{02} - r_{01}) =$$

$$= \phi_{Tn} + \frac{4\pi}{c_0}f_0\left[\delta_i^2(r_{02}) - \delta_i^2(r_{01}) - \boldsymbol{d}_i \cdot [\boldsymbol{e}_r(\vartheta_2, \varphi_2) - \boldsymbol{e}_r(\vartheta_1, \varphi_1)]\right]\ .$$

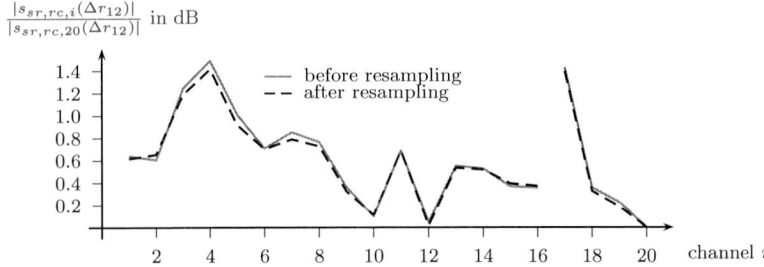

Fig. 11.22 Range compressed sample reference signal: Amplitudes relative to channel 20 over channel numbering within the spectral maximum at $\Delta r_{12} \approx 1\,\mathrm{m}$ before and after resampling process. $\Delta F = 3\,\mathrm{GHz}$, trihedral $rcs \approx 2\pi\,\mathrm{m}^2$. Please note, that the channels 1-16 and 17-20 are azimuth processed separately.

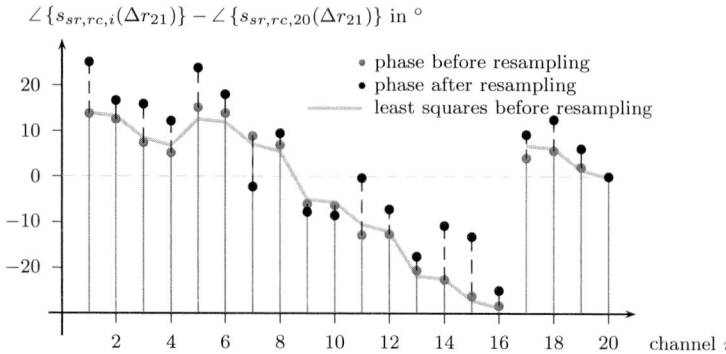

Fig. 11.23 Range compressed sample reference signal: Phases relative to channel 20 over channel numbering within the spectral maximum at $\Delta r_{12} \approx$ 1 m before and after resampling process. $\Delta F = 3$ GHz, trihedral $rcs \approx 2\pi\,\mathrm{m}^2$. Please note, that the channels 1-16 and 17-20 are azimuth processed separately.

$$\tag{11.81}$$

The incorporated variables shall be retrieved again by regression calculus over all channels. Therefore the term

$$\left[\delta_i^2(r_{02}) - \delta_i^2(r_{01})\right] = \frac{d_{Tn}^2 + d_{Rm}^2}{4}\left(\frac{1}{r_{02}} - \frac{1}{r_{01}}\right) = \frac{d_{Tn}^2 + d_{Rm}^2}{4}\,\xi_{21}\ ,$$

$$\tag{11.82}$$

describes the difference between two different phase front curvatures. For the following calculus, the variable ξ_{21} as function of r_{01}, r_{02} was introduced. Since the difference Δr_{21} is known from the linear regression over phases (as demonstrated in section 11.2.4): $\Delta r_{21,i} = (m_{21,i}\cdot c_0)/(4\pi K)$, ξ_{21} will provide a second independent equation which presents the opportunity to calculate r_{01}, r_{02} exactly out of the measured data beyond the necessity of an exact measurement by e.g. a laser range detector etc. However, the index i indicates that the linear regression result can exhibit slightly different values for $\Delta r_{21,i}$. At first sight this is encountered by performing the mean value over these Δr_{21}. However, it is symptomatic for the independent resampling process along each channel. This will be solved by a more sophisticated regression calculus later on. Then the exact distances r_{01}, r_{02} can be determined by solving a quadratic equation:

$$\Delta r_{21} = r_{02} - r_{01} \, , \quad \xi_{21} = \frac{1}{r_{02}} - \frac{1}{r_{01}} \, , \tag{11.83}$$

$$\Rightarrow r_{01}^2 + \Delta r_{21} r_{01} + \frac{\Delta r_{21}}{\xi_{21}} = 0 \, ,$$

$$\Rightarrow r_{01,a,b} = -\frac{\Delta r_{21}}{2} \underset{(-)}{+} \sqrt{\left(\frac{\Delta r_{21}}{2}\right)^2 - \frac{\Delta r_{21}}{\xi_{21}}} \, , \quad r_{01} > 0 \, , \tag{11.84}$$

$$\Rightarrow r_{02} = r_{01} + \Delta r_{21} \, .$$

The trihedral test target was mounted at same height as the landing aid radar sensor for recording the *sample reference*, thus it was ensured that $\vartheta_1 = \vartheta_2 = 90°$ (see coordinate system in section 10.2, Fig. 10.18). Since the landing aid radar's virtual elements are distributed in yz-plane, it can then be written:

$$\boldsymbol{d}_i \cdot [\boldsymbol{e}_r(\vartheta_2, \varphi_2) - \boldsymbol{e}_r(\vartheta_1, \varphi_1)] = \boldsymbol{d}_i \cdot \boldsymbol{e}_y \left[\sin\varphi_2 - \sin\varphi_1\right] \approx$$
$$\approx \boldsymbol{d}_i \cdot \boldsymbol{e}_y \left[\varphi_2 - \varphi_1\right] = \boldsymbol{d}_i \cdot \boldsymbol{e}_y \Delta\varphi_{21} \, , \text{ for small } \varphi_1, \varphi_2 \, . \tag{11.85}$$

This describes the possible shift in azimuth during the data recording. The four constant phase terms $\phi_{T1}, \ldots, \phi_{T4}$ serve as *slack variables* to overcome possible discontinuities caused by mapping the phase onto the *principle value* interval $[-\pi, +\pi]$. It is essential to chose *slack variables* for the transmitters, not for the receivers: The difference of curvature is uniquely defined by the height of phase steps between outer and inner elements, such as from virtual element 4 to 5. *Slack variables* for the receivers would introduce the degree of freedom to shift complete blocks of outer and inner element phases relative to each other. This would provide ambiguities of choosing two curvatures. Therefore, r_{01}, r_{02} could even be negative. Now, the regression can be formulated:

$$\min \|\boldsymbol{A}_{sr}\, \boldsymbol{x}_{sr} - \boldsymbol{b}_{sr}\|_2^2 \, , \tag{11.86}$$

$$A_{sr} = \begin{bmatrix} E_N, & \frac{4\pi}{c_0} f_0 \frac{d_{Tn}^2 + d_{Rm}^2}{4}, & -\frac{4\pi}{c_0} f_0 \, d_i \cdot e_y \\ \vdots & \vdots & \vdots \end{bmatrix} \in \mathbb{R}^{(NM) \times (N+2)} \, ,$$

(11.87)

$$x_{sr} = \begin{bmatrix} \phi_{T1} & \cdots & \phi_{T4}, & \xi_{21}, & \Delta\varphi_{21} \end{bmatrix}^{\mathrm{T}} \in \mathbb{R}^{(N+2) \times 1} \, ,$$

(11.88)

$$b_{sr} = \begin{bmatrix} \cdots & \varphi_{sr,i}(\Delta r_{21}) & \cdots \end{bmatrix}^{\mathrm{T}} \in \mathbb{R}^{(NM) \times 1} \, .$$

(11.89)

The matrix A_{sr} is of rank $(N + 2)$, thus $A_{sr}^{\mathrm{T}} A_{sr}$ is also of rank $(N + 2)$ and can be inverted [48]. The *least squares* solution is again [43]:

$$x_{sr} = \left(A_{sr}^{\mathrm{T}} A_{sr} \right)^{-1} A_{sr}^{\mathrm{T}} b_{sr} \, .$$

(11.90)

The remaining residual is then [43]:

$$res_{sr} = b_{sr} - A_{sr} \, x_{sr} \, .$$

(11.91)

The results for the *sample reference* before the resampling are listed in Tab. 11.3 and plotted in Fig. 11.23. If the same regression procedure (equation

ϕ_{T1} in °	ϕ_{T2} in °	ϕ_{T3} in °	ϕ_{T4} in °	r_{02} in m	r_{01} in m	$\Delta\varphi_{21}$ in °
3.51	3.96	3.11	4.70	5.40	4.45	-0.15

$$\|res_{sr}\|_2 = 0.11 \text{ rad}$$

Tab. 11.3 Results of linear regression over phases of range compressed *sample reference signal* before resampling process.

(11.90)) is imposed after the so far proposed resampling method, the results degrade massively: The calculated ranges are mapped to wrongly 7.64 m and 6.70 m, the residual's *Euclidean* norm increases to 0.39 rad. Further, if the set of parameters in Tab. 11.3 is applied to phases after the resampling process b_{re}, an alternative residual of *Euclidean* norm 0.57 rad can be obtained: $res_{alt} = b_{re} - A_{sr}x_{sr}$. In Fig. 11.23, the larger variation in phase can be observed.

The reason for this loss of coherency is due to the independent linear regression over phases in time domain which is performed independently in each channel. Although each channel must exhibit slightly different $\Delta r_{21,i}$

(compare to section 4.4.3), the corresponding mathematical conditions must be enforced during the resampling procedure. Otherwise, it is possible that slightly different slopes might occur during the linear regression. Furthermore, possible constant phase errors were not taken into account. This leads to the conclusion that the proposed procedure must be adapted to produce fixed phase conditions between the MIMO channels. Furthermore,

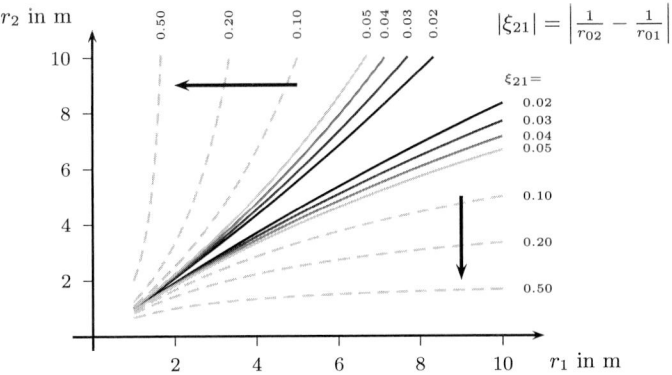

Fig. 11.24 Absolute value of ξ_{21}: Below 0.02 (polygon in the middle), the degrees of freedom for linear regression must be reduced by ξ_{21}. For $|\xi_{21}| \geq 0.02$ ansatz from equation (11.90) yields reliable results (black arrows).

the position of calibration targets should be adapted to match a certain region of parameter ξ_{21}. For $\xi_{21} \in [0.02; 0.5]$ (see Fig. 11.24), the difference of both phase front curvatures is that large, the linear regression over lateral phase delivers reliable values.

For values $\xi_{21} < 0.02$, the curvatures have flattened so much, so that the linear regression on phase does not deliver reliable values anymore. The phase terms of Tx elements are more dominant. The linear regression tries to compensate that by delivering too small values of r_{01}, r_{02}, thus to large curvatures. Therefore in case of $\xi_{21} < 0.02$, it is recommended to set ξ_{21} fixed (known from e.g. tape measurement), subtract the fifth column multiplied by this ξ_{21} from both subtrahends in equation (11.86) and solve the linear regression for the Tx phases and an azimuth offset $\Delta\varphi_{21}$. The degrees of freedom

11.3.1 Coherent resampling

For a coherent resampling scheme, we must step back to the equations (11.35) till (11.39), which provide a linear regression over the unwrapped phases of each channel's sample reference signal:

$$
T = \begin{bmatrix} 0 & 1 \\ \vdots & \vdots \\ t & 1 \\ \vdots & \vdots \\ T_c & 1 \end{bmatrix} = [\boldsymbol{t}, \boldsymbol{1}] \ , \quad \boldsymbol{P} = \begin{bmatrix} \ldots , & \angle\{s_{sr,i}(t)\} \ , & \ldots \end{bmatrix} \ . \tag{11.92}
$$

$$
\boldsymbol{T} \in \mathbb{R}^{L \times 2} \ , \quad \boldsymbol{P} \in \mathbb{R}^{L \times (NM)} \ , \tag{11.93}
$$

$$
\min \|\boldsymbol{T} \cdot \boldsymbol{M} - \boldsymbol{P}\|_2^2 \ , \quad \Rightarrow \quad \boldsymbol{M} = \left(\boldsymbol{T}^{\mathrm{T}} \boldsymbol{T}\right)^{-1} \boldsymbol{T}^{\mathrm{T}} \boldsymbol{P} \ , \tag{11.94}
$$

$$
\boldsymbol{M} = \begin{bmatrix} m_1 & \cdots & m_i & \cdots & m_{NM} \\ \phi_1 & \cdots & \phi_i & \cdots & \phi_{NM} \end{bmatrix} \in \mathbb{R}^{2 \times (NM)} \ . \tag{11.95}
$$

The measured range differences $\Delta r_{121,i}$ can be extracted from the first row of $\boldsymbol{M}$:

$$
\Delta r_{21,i} = \frac{m_i \cdot c_0}{4\pi K} \ \wedge \ \Delta r_{21,i} = \left(\Delta r_{21} + \frac{d_{Tn}^2 + d_{Rm}^2}{4} \xi_{21} - \boldsymbol{d}_i \cdot \boldsymbol{e}_y \, \Delta\varphi_{21} \right) \ . \tag{11.96}
$$

For the example of 5.5 m and 4.5 m *sample reference signal* recorded by the landing aid radar, a mean value of $\Delta r_{21} = 0.9474$ m at a *standard deviation* [43] of 0.57 mm can be obtained. Since the bandwidth of the exploited signal bears a 3 GHz bandwidth, the *range cell* is of length 5 cm. Therefore, no range cell migration takes place [61], [71]. Fig. 11.25 demonstrates that no systematic relationship between individual $\Delta r_{21,i}$ and the actual differences in path lengths could be extracted. Nevertheless, the magnitude of phase deviation in the preceding section might persuade the reader that the limit of measurement accuracy is touched here. Consequentially, instead of individual slopes m_i, a standard slope shall be applied to all channels: $m = 4\pi K \Delta r_{21}/c_0$, with Δr_{21} as mean value. The corresponding approximation for the signal phases could then look like:

$$
\ldots (Kt + f_0) \cdot \Delta r_{21,i} \cdots \approx \ldots Kt \cdot \Delta r_{21} + f_0 \cdot \Delta r_{21,i} \ldots \tag{11.97}
$$

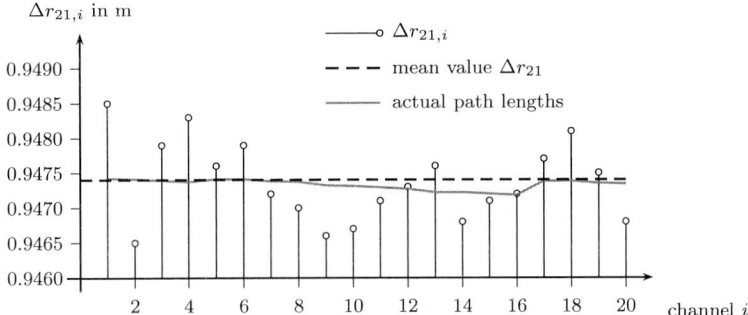

Fig. 11.25 Linear regression: Results for $\Delta r_{21,i}$ compared to its mean value and the actual path lengths matching the measurement parameters $r_{0,1} \approx 4.5\,\text{m}$, $r_{0,2} \approx 5.5\,\text{m}$ as well as a slight azimuth deviation of $-0.15°$. *Standard deviation* of $\Delta r_{21,i}$ at 0.57 mm.

This means, that the pure phase due to the round trip delay between radar and the object is of course dominated by the mean value Δr_{21}. However, if all slopes get equalized, thus all MIMO channels bear the same *beat frequency*, a special focus must be set on the static phase relations between the different MIMO channels in order to enable a correct angular processing. Therefore, the constant phases in M's second row have to be adapted slightly, which leads to a refreshed approach to equation (11.94), whereby all known parameters including the *beat frequency* are put to P's side. In general, the subtraction of chirp values is often denoted as *deramping* [71], [61]:

$$\min \left\| \begin{bmatrix} 1 \\ \vdots \\ 1 \end{bmatrix} \begin{bmatrix} \phi_1 & \cdots & \phi_i & \cdots & \phi_{NM} \end{bmatrix} - \left(P - m \begin{bmatrix} 0 \\ \vdots \\ t \\ \vdots \\ T_c \end{bmatrix} \begin{bmatrix} 1 & \cdots & 1 \end{bmatrix} \right) \right\|_2^2 = ,$$

$$= \min \left\| \mathbf{1}_{L \times 1} \begin{bmatrix} \phi_1 & \cdots & \phi_i & \cdots & \phi_{NM} \end{bmatrix} - P_{deramp} \right\|_2^2 , \tag{11.98}$$

$$\Rightarrow \quad \begin{bmatrix} \phi_1 & \cdots & \phi_i & \cdots & \phi_{NM} \end{bmatrix} = \frac{1}{L} \begin{bmatrix} 1 & \cdots & 1 \end{bmatrix} P_{deramp} . \tag{11.99}$$

The solution of the *least squares problem* gives an equivalent expression to mean value of each channel's *deramped* phase. The *frequency deviation* functions can then be derived from the deviation of *deramped phases* from their *mean value*:

$$\tilde{\varepsilon}_i(t) = \frac{(\angle\{s_{sr,i}(t)\} - \phi_i) \cdot c_0}{4\pi\,\Delta r_{21}} = \frac{\delta\varphi_i(t) \cdot c_0}{4\pi\,\Delta r_{21}} \; . \tag{11.100}$$

Please note, that the fundamental shape of the *frequency deviation* is determined by the deviation from the unified linear slope m in each channel. Possible constant phase errors in ϕ_i lead to different start values of $\tilde{\varepsilon}_i$, thus constant frequency offsets. But those in turn are subtracted during the resampling process (equation (11.74)).

Now, the same linear regression ansatz performed in spectral maximum of the sample reference signal (equation (11.86)), can be applied to the constant phases $[\ldots, \phi_i, \ldots]$:

$$\min \|\boldsymbol{A}_{sr}\,\boldsymbol{x}_c - \begin{bmatrix} \phi_1 & \cdots & \phi_i & \cdots & \phi_{NM} \end{bmatrix}^\mathrm{T}\|_2^2 \; ,$$
$$\boldsymbol{res}_c = \begin{bmatrix} \phi_1 & \cdots & \phi_i & \cdots & \phi_{NM} \end{bmatrix}^\mathrm{T} - \boldsymbol{A}_{sr}\,\boldsymbol{x}_c \; . \tag{11.101}$$

In case of the second sample reference $s_{sr,2,i}(t)$, the same procedure yields

ϕ_{T1} in $^\circ$	ϕ_{T2} in $^\circ$	ϕ_{T3} in $^\circ$	ϕ_{T4} in $^\circ$	r_{02} in m	r_{01} in m	$\Delta\varphi_{21}$ in $^\circ$
-69.90	-69.54	-70.46	-68.87	5.37	4.42	-0.15

$$\|\boldsymbol{res}_c\|_2 = 0.10 \text{ rad}$$

Tab. 11.4 Results of linear regression over constant phase terms ϕ_i derived from *sample reference* $s_{sr,i}(t)$, equation (11.98).

Tab. 11.5. The indices c shall denote coherent resampling. However, the matrix $\boldsymbol{A}_{sr}$ stays unchanged. The results in $\boldsymbol{x}_c$ are comparable to those

ϕ_{T1} in °	ϕ_{T2} in °	ϕ_{T3} in °	ϕ_{T4} in °	r_{03} in m	r_{01} in m	$\Delta\varphi_{31}$ in °
-113.0	-111.5	-113.1	-111.8	6.41	4.43	-0.29

$$\|\boldsymbol{res}_c\|_2 = 0.10 \text{ rad}$$

Tab. 11.5 Results of linear regression over constant phase terms ϕ_i derived from 2nd *sample reference* $s_{sr,2,i}(t)$, equation (11.98).

in $\boldsymbol{x}_{sr}$ (see Tab. 11.3). By adding the new insights, the actual *sample reference signal*'s phase must be updated to:

$$\phi_{21,i}(t) = \frac{4\pi}{c_0}\left(Kt + \tilde{\varepsilon}_i\right)\Delta r_{21} + \phi_i =$$

$$= \frac{4\pi}{c_0}\left(Kt + \tilde{\varepsilon}_i\right)\Delta r_{21} + \phi_{const,21,i} + res_{c,i} + \phi_{Tn} \,, \tag{11.102}$$

$$\text{with } \phi_{const,21,i} = \frac{4\pi}{c_0}\left[f_0 \cdot \Delta r_{21,i} - \frac{K}{c_0}\Delta r_{21}\cdot(r_{01} + r_{02})\right] \,.$$

Please note, that the second summand of $\phi_{const,21,i}$ is again the remaining quadratic term which is kept for the sake of completeness. Now, the coherent resampling process is performed analogously to equation (11.79) by individual interpolation in each channel:

$$\phi_{21,i}(t_{re,i}) = \phi_{21,i}\left(t - \frac{\tilde{\varepsilon}_i}{K} + \frac{\tilde{\varepsilon}_i(0)}{K}\right) =$$

$$= \frac{4\pi}{c_0}\left[Kt + \tilde{\varepsilon}_i(0)\right]\Delta r_{21} + \phi_{const,21,i} + res_{c,i} + \phi_{Tn} = \tag{11.103}$$

$$= \frac{4\pi}{c_0}Kt\,\Delta r_{21} + \phi_{const,21,i} + res_{c,i} + \phi_{Tn} + \Delta\phi_{sr,i} \,.$$

However, this proofs that the phase relations between the different channels are slightly corrupted by the constant phase terms $(res_{sr,i} + \phi_{Tn} + \Delta\phi_{sr,i})$. This can be corrected by a simple multiplication by a complex term. Additionally, an amplitude correction term can be derived from spectral domain (Fig. 11.22), hence from the range compressed signals: One channel is posed as amplitude reference. All channels are then divided by their ab-

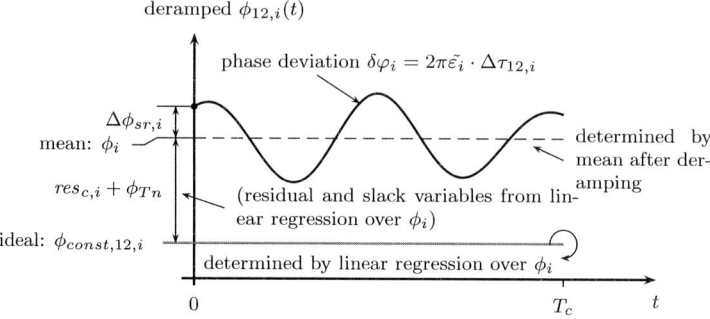

Fig. 11.26 Principle composition of phases and frequency deviation in case of coherent resampling, sketched at deramped phases.

solute amplitude and multiplied by the reference amplitude. Phase and amplitude correction composed together, yields a complex correction factor which is applied either before or after the resampling process:

$$s_{corr} = \frac{|s_{sr,rc,ref}(\Delta r = \Delta r_{21})|}{|s_{sr,rc,i}(\Delta r = \Delta r_{21})|} \cdot \mathrm{e}^{-j\,(res_{c,i}+\phi_{Tn}+\Delta\phi_{sr,i})} \ . \qquad (11.104)$$

The corrected resampled signals are then:

$$s_{re,i}(t) = s_{sr,i}(t = t_{re,i}) \cdot s_{corr} \ . \qquad (11.105)$$

Now, in order to survey the MIMO channels phase relation, thus the effectivity of proposed processing steps, the start and stop phases, thus the phase relations between the channels at $t = 0$ and $T = T_c$ were examined. Therefore, the start phases were related to one channel e. g. $i = 20$, as well as the stop phases. If the trend over i is the same at start and end of the measurement, the MIMO channels are likely to be synchronous. Fig. 11.27 shows the difference of stop and start phase after resampling $P = 5000$ points in each channel. The largest difference is -1.8 °. However, for confirming coherency, the phase behaviour over time must also satisfy the same linear slope. Therefore, the resampled phases are examined again with respect to frequency deviations. The result is sketched in Fig. 11.28. If the resampled reference signal is deramped again it can be found that $\mathcal{L}_i$ shrinked to $3.3 \cdot 10^{-4}$, which is one order of magnitude lower than before. Hence, the maximum frequency deviation for the transmit signal is now lower than 0.99 MHz, thus lower than 165 kHz traced back to the VCO (fac-

11.3. Data interpolation and resampling

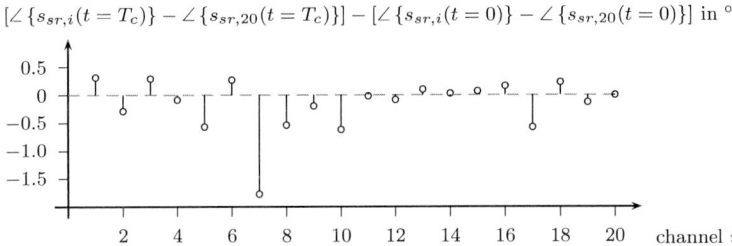

$[\angle\{s_{sr,i}(t=T_c)\} - \angle\{s_{sr,20}(t=T_c)\}] - [\angle\{s_{sr,i}(t=0)\} - \angle\{s_{sr,20}(t=0)\}]$ in $°$

Fig. 11.27 Difference of stop and start phases of *sample reference* after the coherent resampling in time domain. $\Delta F = 3\,$GHz, trihedral *rcs* $\approx 2\pi\,$m^2. $P = 5000$.

tor 6). Transformed into a voltage deviation (tuning sensivity 400 MHz/V), gives 0.4 mV [155]. The total tuning voltage range of 1.25 V was driven by a 16 Bit digital-to-analog converter (DAC) with an output interval from 3 V to 4.5 V. Then the 0.4 mV correspond to only 20 DAC steps (out of 65536), which not only demonstrates the quality of accuracy but also shows that the numerical limits were reached.

The static phase differences between the channels after coherent resampling features the possibility of another phase correction factor: The ideal version of $s_{1,i}(t)$, adapted appropriately to equation (11.97). A multiplication of $s_{re,i}(t)$ with the afore mentioned would not only allow to revoke a possible curvature of $s_{1,i}(t)$ but also reveals the shift by exact range r_{01}. The possible azimuth shift $\Delta\varphi_{21} = (\varphi_2 - \varphi_1)$ shall be allocated completely to $s_{1,i}(t)$, since its values are well below the intended system azimuth resolution of approximately $1\,°$. The required parameters are provided by the linear regression in equation (11.101).

$$s_{re,i}(t) = s_{sr,i}(t = t_{re,i}) \cdot s_{corr} \cdot s_{1,ideal,i}(t)\,,$$
$$\text{with } s_{1,ideal,i}(t) = \mathrm{e}^{+j\,\frac{4\pi}{c_0}\,Kt\,\Delta r_{01}} \cdot \mathrm{e}^{+j\,\frac{4\pi}{c_0}\,f_0\cdot\left[\delta_i^2(r_{01}) - d_i\cdot e_y\,(-\Delta\varphi_{21})\right]}\,.$$

$$(11.106)$$

Up to now, individual time bases $t_{re,i}$ securing coherency between the MIMO channels, phase and amplitude correction factors, as well as compensation of $s_{1,i}(t)$ have been derived from *sample reference signal* $s_{sr}(t)$. However, the quality of proposed method can be tested by applying it to 2nd sample reference signal $s_{sr,2}(t)$.

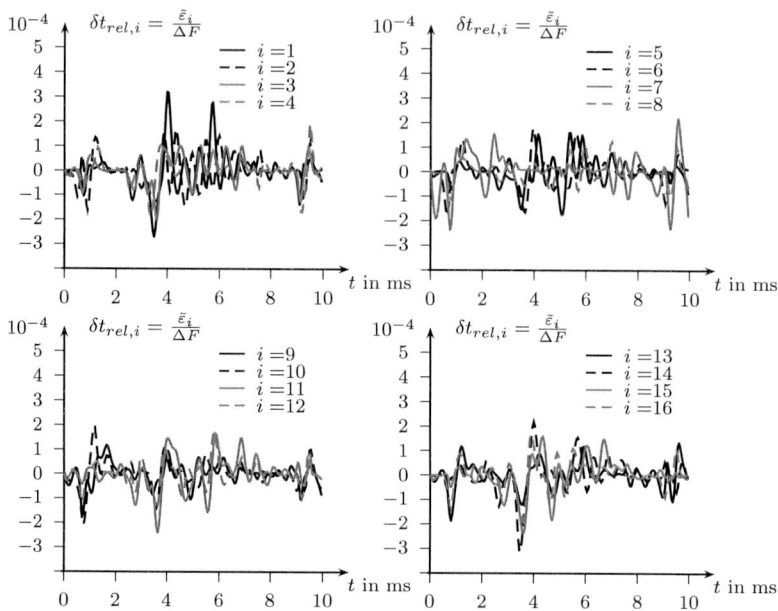

Fig. 11.28 Example for time variation, thus frequency deviation per channel i divided by bandwidth $\Delta F = 3\,\mathrm{GHz}$ of the interpolated *sample reference signal*. Provided linearity $\mathcal{L} = 3.3 \cdot 10^{-4}$.

First the amplitude correction of equation (11.106) is tested: Fig. 11.29 demonstrates that the proposed amplitude correction works successfully for the *sample reference* itself. The slight remaining discrepancies are owed to the fact, that the range compression by *Fourier transformation* was carried out as FFT, thus discretely. Depending on the discretization in range bins, the exact maxima of compressed signals are not meet exactly if it is sighted e. g. at the edge between two range bins. This can be encountered by a high zero-padding length in time domain, which interpolates the exact values in spectrum. However, applied to the 2nd *sample reference signal* $s_{sr,2}(t)$, the amplitudes can still vary around $\pm 1.5\,\mathrm{dB}$. Fig. 11.30 compares the amplitude correction factors derived from *sample reference signal* $s_{sr,i}(t)$ and 2nd *sample reference signal* $s_{sr,2,i}(t)$. It must be stated that there are still discrepancies in amplitude correction.

As next step, the phase relation are tested: If equation (11.105) is executed with *sample reference* and again a linear regression of

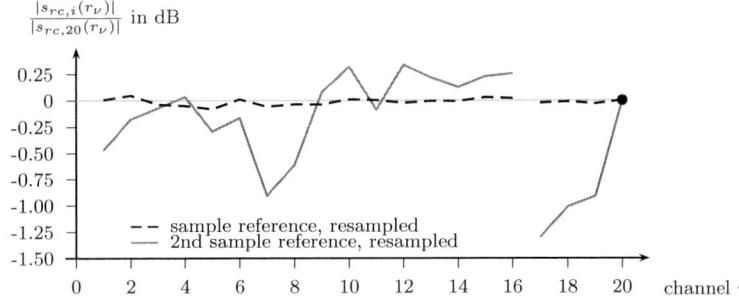

Fig. 11.29 Range compressed signal after resampling and correction: Amplitudes relative to channel 20 over channel numbering within the spectral maximum. Please note, that the channels 1-16 and 17-20 are azimuth processed separately.

$\min \| \boldsymbol{A}_{sr}\, \boldsymbol{x}_{sr} - \boldsymbol{b}_{sr} \|_2^2$ is performed on the phases in spectral maximum, the results of Tab. 11.6 will obtained. The *Euclidean* norm $\| \boldsymbol{res}_{sr} \|_2 = 0.0022$ rad proofs a nearly perfect result. Now, the parameters of Tab. 11.6 are used to form $s_{1,ideal,i}(t)$ for equation (11.106). The nearly ideal result is documented in Fig. 11.31. However, the effect on the 2nd sample reference signal is not satisfying: The linear regression over the spectral maxima's phases delivers misleading parameters: $\Delta\varphi_{31} = -0.12°$, $r_{03} = 5.67\,\text{m}$, $r_{01} = 4.42\,\text{m}$, $\| \boldsymbol{res}_{sr,2} \|_2 = 0.4383\,\text{rad}$. Though, by comparing Tab. 11.4 and

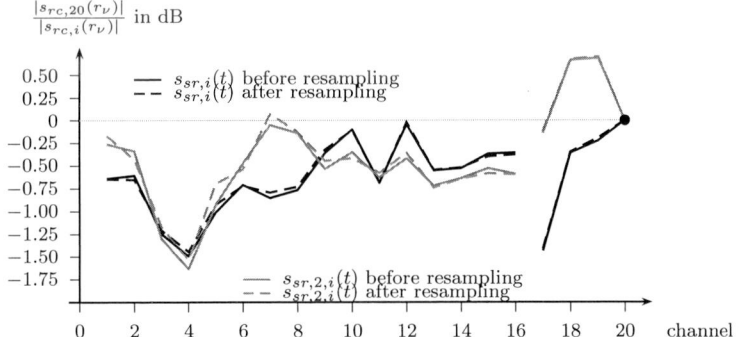

Fig. 11.30 Amplitude correction factors before and after resampling process. Furthermore, comparison of amplitude correction for *sample reference signal* $s_{sr,i}(t)$ and 2nd *sample reference signal* $s_{sr,2,i}(t)$. Please note, that the channels 1-16 and 17-20 are azimuth processed separately.

ϕ_{T1} in °	ϕ_{T2} in °	ϕ_{T3} in °	ϕ_{T4} in °	r_{02} in m	r_{01} in m	φ_2 in °
-8.24	-8.23	-8.26	-8.24	5.37	4.42	0.00

$$\|\boldsymbol{res}_{sr}\|_2 = 0.0022 \text{ rad}$$

Tab. 11.6 Results of linear regression over phases of range compressed *sample reference signal* after resampling, correction and compensation by $s_{1,ideal}(t)$.

11.5, another value for r_{03} was expected: r_{03} =6.41 m. If a desired phase trend is set up and the actual phases are subtracted, the *Euclidean* norm of the result delivers even 0.5491 rad. A possible explanation of this effect is, that the frequency deviation $\tilde{\varepsilon}_i$ contains also statistical contributions which differ from measurement to measurement, hence a further examination of $\tilde{\varepsilon}_i$ is required.

11.3.2 Separation of phase deviation and phase noise

The advantage of generating two *sample references*, bears now the possibility to compare each derived frequency deviation or both $\delta t_{rel,i}$ respectively.

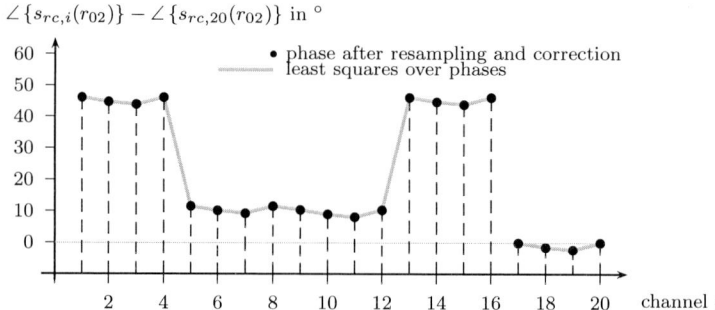

Fig. 11.31 Range compressed resampled *sample reference signal* with correction of amplitude and curvature (equation (11.106)): Phases relative to channel 20 over channel numbering within the spectral maximum at $\Delta r_{02} \approx 5.37$ m. Please note, that the channels 1-16 and 17-20 are azimuth processed separately.

$\angle \{s_{rc,i}(r_{03})\} - \angle \{s_{rc,20}(r_{03})\}$ in $°$

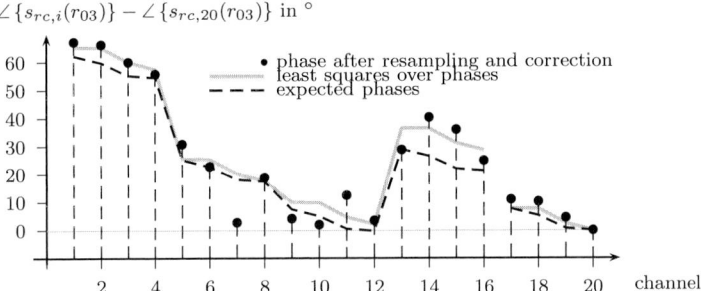

Fig. 11.32 Range compressed resampled 2nd *sample reference signal* with correction of amplitude and curvature (equation (11.106)): Phases relative to channel 20 over channel numbering within the spectral maximum at $\Delta r_{03} \approx 6.41$ m. Please note, that the channels 1-16 and 17-20 are azimuth processed separately.

Although the frequency deviations vary only little from *sample reference* to *sample reference*. However, the preceding sections proved that already little deviations can have tremendous effects. Therefore, the regression calculus analogously to equation (11.54) can be exploited again to extract fundamental frequency deviations for each transmit and each receiver element, and separating them from superimposed noise effects:

$$\min \left\{ \|\tilde{\boldsymbol{A}}_{sys} \, \tilde{\boldsymbol{X}}_\varepsilon - \tilde{\boldsymbol{F}}\|_2^2 \right\} \, , \quad \tilde{\boldsymbol{F}} = \begin{bmatrix} \vdots \\ \tilde{\varepsilon}_i(t) \\ \vdots \end{bmatrix} - \begin{bmatrix} \mathbf{0}_{M(N-1)\times 1} \\ \mathbf{1}_{N\times 1} \end{bmatrix} \cdot \tilde{\varepsilon}_{RM}(t) \, ,$$

$$\tilde{\boldsymbol{X}}_\varepsilon = \begin{bmatrix} \tilde{\varepsilon}_{R1}(t) \\ \vdots \\ \tilde{\varepsilon}_{R(M-1)}(t) \\ \vdots \\ \tilde{\varepsilon}_{Tn}(t) \\ \vdots \end{bmatrix} \, , \text{ with } \tilde{\varepsilon} \text{ as rows,}$$

$$\tilde{\boldsymbol{X}}_\varepsilon \in \mathbb{R}^{((M-1)+N)\times P} \, , \quad \tilde{\boldsymbol{F}} \in \mathbb{R}^{MN\times P} \, .$$

$$(11.107)$$

Again a degree of freedom must be chosen due to rank reasons: The deviation of receiver M, $\tilde{\varepsilon}_{RM}$ is set to half of $\tilde{\varepsilon}_{MN}$. The minimization is again solved by the general inverse [43], the matrix $\tilde{\boldsymbol{X}}_\varepsilon$ is completed to $\boldsymbol{X}_\varepsilon$ by inserting $\tilde{\varepsilon}_{RM}$. The regression frequency deviation $\tilde{\varepsilon}_r$ per channel i and the corresponding residual matrix $\boldsymbol{Res}_\varepsilon$ representing the noise contribution per channel i are then:

$$\tilde{\varepsilon}_r = \boldsymbol{A}_{sys}\,\boldsymbol{X}_\varepsilon \ , \quad \boldsymbol{Res}_\varepsilon = \tilde{\boldsymbol{A}}_{sys}\,\tilde{\boldsymbol{X}}_\varepsilon - \tilde{\boldsymbol{F}} = \begin{bmatrix} \vdots \\ \delta\tilde{\varepsilon}_i(t) \\ \vdots \end{bmatrix} \ . \tag{11.108}$$

The previous frequency deviation functions can then be expressed by the sum of a regressed function and noise:

$$\tilde{\varepsilon}_i(t) = \tilde{\varepsilon}_{r,i}(t) + \delta\tilde{\varepsilon}_i(t) \ , \text{ with } \tilde{\varepsilon}_{r,i}(t) = \tilde{\varepsilon}_{r,Tn}(t) + \tilde{\varepsilon}_{r,Rm}(t) \ . \tag{11.109}$$

If the regression frequency deviations $\tilde{\varepsilon}_{r,i}$ derived from both *sample reference* functions are compared to each other, the differences shrink significantly (Fig. 11.33). This can be further compared by executing the *Fourier* transformation on both types of $\tilde{\varepsilon}_{r,i}$.

Now, the regression frequency deviation functions $\tilde{\varepsilon}_{r,i}$ are put into equation (11.79) to form a regressed time base. However, if it founds the

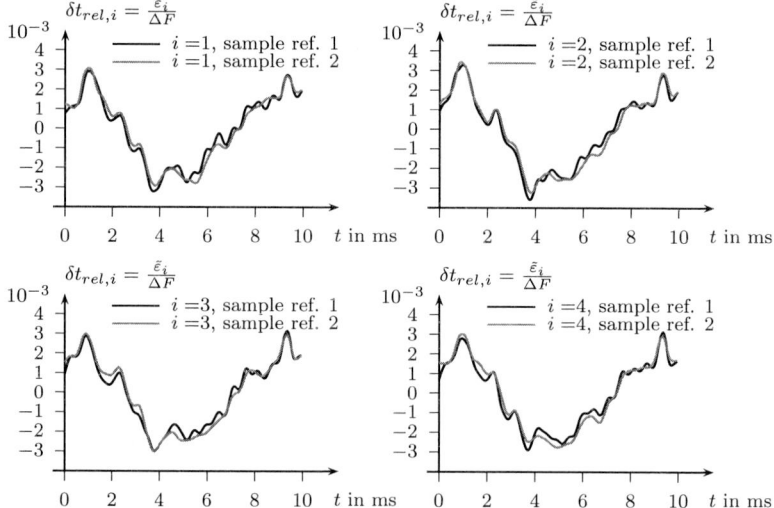

Fig. 11.33 Comparison of relative time deviations delivered by regression calculus on both sample reference signals. Channels $i = 1 \ldots 4$.

resampling of both *sample references*, phase and amplitude correction and compensation with ideal $s_1(t)$, the linear regression over constant phases ϕ_i (equation (11.101)) will provide wrong values in both cases: In case of $s_{sr}(t)$, r_{02} =4.87 m instead of 5.37 m, $\Delta\varphi_{21} = 0.015°$ instead of $0°$ at $\|res_c\|$ =0.081 rad. In case of $s_{sr,2}(t)$, the processing provides r_{03} =5.12 m instead of 6.41 m, $\Delta\varphi_{31} = -0.11°$ instead of expected $-0.14°$ at $\|res_c\|$ =0.21 rad. So the intention of reducing possible phase noise effects, have even decreased the performance of the complete process. Therefore another step back to the principle time base consideration in equation (11.79) must be done. There, the very first value of $\tilde{\varepsilon}_i$ is subtracted to avoid possible negative time steps to be interpolated right at the beginning. Unfortunately, a closer look to the start values of the frequency deviations, the start values of the regression frequency deviations of both *sample references* make clear that they are the parts which differ most, thus are especially affected by swing-in effects due to the chosen window functions. Therefore, the start values are more or less characteristic for each measurement, hence leave their mark on the time base, which was intended to be used for all measurements of one series.

One might assume that resampling is not feasible for coherent MIMO radars. However, the regressed frequency deviations equalize after those swing-in effects (Fig. 11.33). Furthermore, the best coherency between MIMO channels is obtained when the phase relations delivered by regression over all constant phases Φ_i (equation (11.101)) is maintained during the complete process from $t = 0$ to $t = T_c$. However, the time base from equation (11.79) does not secure this. There, it is still possible that one channel needs a step interpolated in near future, while another channels steps back a little in time, because the first frequency deviation minus its start value is negative, while the other one is still positive. In this case the phases between the channels start to shear. Even if the phase is increased continuously afterwards, each such transition contributes the breaking up the phase relations.

A logical consequence is, only admitting interpolation steps forward in time, not back, compared to the original time step. How can this be achieved? First, only regression frequency deviations are taken. By this, it the usage gets more universal, with the drawback, that little phase errors $\delta\tilde{\varepsilon}_i(t)$ must be accepted for each measurement. Then instead of the start values, the individual maximum values of $\tilde{\varepsilon}_{r,i}$ are subtracted in each channel. In terms of mathematics, this starts again at:

$$
\phi_{21,i}(t) = \frac{4\pi}{c_0} \left[Kt + \tilde{\varepsilon}_i(t) \right] \Delta r_{21} + \phi_i =
$$
$$
= \frac{4\pi}{c_0} \left[Kt + \tilde{\varepsilon}_{r,i}(t) + \delta\tilde{\varepsilon}_i(t) \right] \Delta r_{21} + \phi_{const,21,i} + \phi_{Tn} + res_{c,i} \; .
$$
(11.110)

The altered time base shall be:

$$
t_{re,i} = t - \frac{1}{K} \left[\tilde{\varepsilon}_{r,i}(t) - \tilde{\varepsilon}_{r,i,max} \right] \; .
$$
(11.111)

If equation (11.113) is inserted into equation (11.110), it can be obtained:

$$
\phi_{21,i}(t = t_{re,i}) = \frac{4\pi}{c_0} Kt\,\Delta r_{21} + \phi_i + \frac{4\pi}{c_0} \left[\delta\tilde{\varepsilon}_i(t) + \tilde{\varepsilon}_{r,i,max} \right] \Delta r_{21} \; .
$$
(11.112)

The individual maxima $\tilde{\varepsilon}_{r,i,max}$ still cause phase errors. For $s_{sr}(t)$ this yields $r_{02} = 5.23\,\mathrm{m}$, $\Delta\varphi_{21} = -0.01°$, $\|res_c\| = 0.045\,\mathrm{rad}$, whereas for $s_{sr,2}(t)$, $r_{03} = 6.03\,\mathrm{m}$, $\Delta\varphi_{31} = -0.13°$, $\|res_c\| = 0.122\,\mathrm{rad}$. The parameters r_{02}, r_{03} responsible for the correct curvature are still processed in wrong manner. The reason can be found in the individual phase errors caused by $\tilde{\varepsilon}_{r,i,max}$. Please note, that the applied individual phase correction factor was adapted to: $\exp\left[-j\,(\phi_{Tn} + res_{c,i}) \right]$.

The mentioned individual phase error, can be encountered by subtracting the maximum of all maxima, $\max_i \{\tilde{\varepsilon}_{r,i,max}\}$, thus by choosing the following time base:

$$
t_{re,i} = t - \frac{1}{K} \left[\tilde{\varepsilon}_{r,i}(t) - \max_i \{\tilde{\varepsilon}_{r,i,max}\} \right] \; .
$$
(11.113)

Equation (11.112) can then be rewritten to:

$$
\phi_{21,i}(t = t_{re,i}) =
$$
$$
= \frac{4\pi}{c_0} Kt\,\Delta r_{21} + \phi_i + \frac{4\pi}{c_0} \delta\tilde{\varepsilon}_i(t)\,\Delta r_{21} + \frac{4\pi}{c_0} \max_i \{\tilde{\varepsilon}_{r,i,max}\} \Delta r_{21} \simeq
$$
$$
\simeq \frac{4\pi}{c_0} Kt\,\Delta r_{21} + \phi_i + \frac{4\pi}{c_0} \delta\tilde{\varepsilon}_i(t)\,\Delta r_{21} \; .
$$
(11.114)

Now, the phase error caused by $\max \{\tilde{\varepsilon}_{r,i,max}\}$ is the same for all channels i and can thus be cancelled. The only remaining phase error is caused by

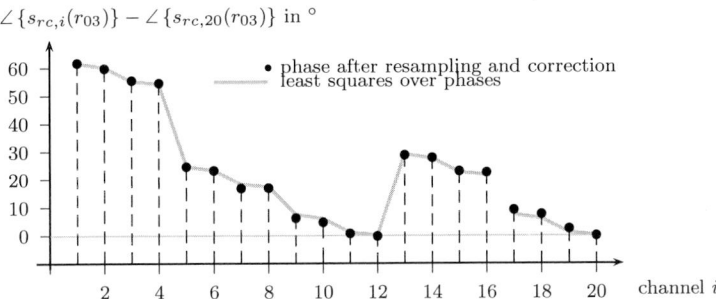

$\angle\{s_{rc,i}(r_{03})\} - \angle\{s_{rc,20}(r_{03})\}$ in $^\circ$

- phase after resampling and correction
- least squares over phases

Fig. 11.34 Range compressed resampled 2nd *sample reference signal* with correction of amplitude and curvature (equation (11.106)): Phases relative to channel 20 over channel numbering within the spectral maximum at $\Delta r_{03} \approx 6.41$ m. Please note, that the channels 1-16 and 17-20 are azimuth processed separately.

$\delta\tilde{\varepsilon}_i(t)$, which is zero mean due to the regression calculus, however the total effect from $t = 0$ to $t = T_c$ cannot be neglected a priori, which will be examined briefly at the end of this section.

Generally, the phase error due to the start value does not occur anymore. Therefore the complex correction factor (entry i) can be adapted to:

$$s_{corr,i} = \frac{|s_{sr,rc,ref}(\Delta r = \Delta r_{21})|}{|s_{sr,rc,i}(\Delta r = \Delta r_{21})|} \cdot e^{-j\,(res_{c,i}+\phi_{Tn})} \ . \tag{11.115}$$

The lateral results for $s_{sr}(t)$ cannot be separated from Fig. 11.31. The parameters correspond to Tab. 11.6, though with $\|\boldsymbol{res}_c\|_2 = 0.007$ rad. In case of processing $s_{sr,2}(t)$ the expected values from Tab. 11.5 are delivered: $r_{03} = 6.42$ m, $\Delta\varphi_{31} = -0.14^\circ$, $\|\boldsymbol{res}_c\|_2 = 0.067$ rad. This proofs, that the chosen method is feasible applied on recorded data a-posteriori.

11.3.3 Correction of statistic frequency deviation

With linear regression, the deterministic deviation can be separated from its statistical part. The first can be saved in memory and re-used for each data set, whereas the latter is unique per measurement run. Fig. shows the statistical part of equation (). If measured signals are resampled according to equation (11.113), ((11.114)), the spectra are corrected and yield results like in Fig. 11.36. Nevertheless, the bleak linearity of magnitude $\mathcal{L} = 1 \cdot 10^{-3}$ causes again incoherency between channels, thus equation (11.114) will provide targets in same range bin, however with phase errors

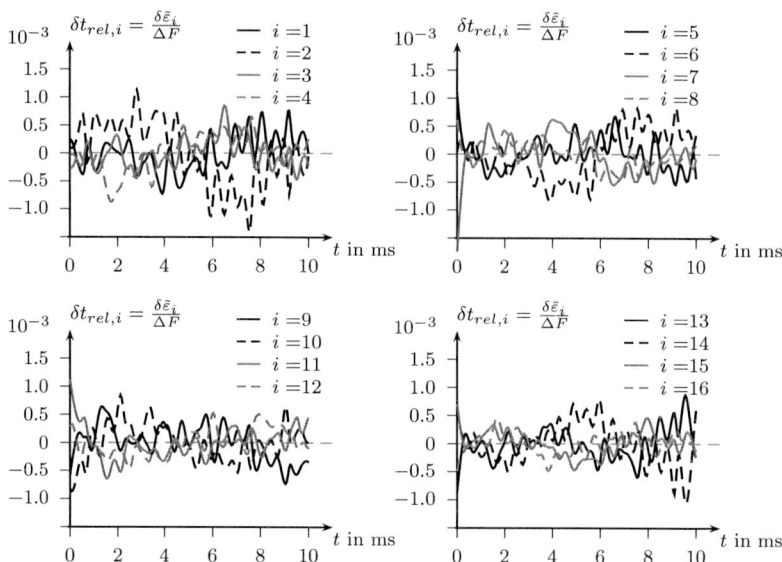

Fig. 11.35 Example for statistic time variation, thus statistic frequency deviation per channel $\delta\bar{\varepsilon}_i$ divided by bandwidth $\Delta F = 3\,\text{GHz}$ of a sample reference signal generated by two signals: trihedral reflector at 5.5 m and 4.5 m. Remaining linearity $\mathcal{L} = 1 \cdot 10^{-3}$.

like in equation (8.22) for targets which are significantly beyond the range of sample reference area. Angular processing becomes corrupted though.

The corruption term in equation (11.114) generally depends on the distance (τ_q), thus differs for each range cell:

$$\delta s_{q,i}(t) = \mathrm{e}^{+j\,2\pi\delta\bar{\varepsilon}_i(t)\,(\tau_{0q}-\tau_{01})} \ . \tag{11.116}$$

The corresponding spectrum can be expressed by:

$$\delta S_{q,i}(f) = \frac{1}{T_c} \int_{T_c} \mathrm{e}^{+j\,2\pi\delta\bar{\varepsilon}_i(t)\,(\tau_{0q}-\tau_{01})}\, \mathrm{e}^{-j\,\omega t}\, dt = \ldots$$

$$= \frac{1}{T_c} \int_{T_c} \mathrm{e}^{+j\,\frac{2\pi}{K}\,[(f_{r,q}-f_{r,1})\,\delta\bar{\varepsilon}_i(t)-Kft]}\, dt \ , \tag{11.117}$$

$$\text{with } f = K\tau \ .$$

The according sample reference signal or measurement signal can be expressed accordingly:

$$s_i(t) = \sum_{q=1}^{Q} \frac{U_{q,i}}{U_{1,i}} \cdot e^{+j\left(\frac{4\pi}{c_0} Kt\,\Delta r_{q1} + \phi_i\right)} * \delta s_{q,i}(t) . \tag{11.118}$$

In next instance, it shall be assumed that $\delta\tilde\varepsilon_i(t)$ is known. Then a range (run time) depended compensation filter can be defined:

$$\delta s_{\varepsilon,i}(t) = e^{-j\,2\pi\delta\tilde\varepsilon_i(t)\,(\tau - \tau_{01})} = e^{-j\,\frac{2\pi}{K}\,\delta\tilde\varepsilon_i(t)\,(f - f_{r,1})} . \tag{11.119}$$

The corresponding spectrum is then:

$$\delta S_{\varepsilon,i}(f) = \frac{1}{T_c} \int_{T_c} e^{-j\,2\pi\delta\tilde\varepsilon_i(t)\,(\tau - \tau_{01})}\, e^{-j\,\omega t}\, dt = \dots$$

$$= \frac{1}{T_c} \int_{T_c} e^{-j\,\frac{2\pi}{K}\,[(f - f_{r,1})\,\delta\tilde\varepsilon_i(t) + Kf\,t]}\, dt , \tag{11.120}$$

$$\text{with } f = K\tau .$$

Now, equation (11.120) is multiplied onto equation (11.117). This yields:

$$\delta S_{q,i}(f) \cdot \delta S_{\varepsilon,i}(f) = \frac{1}{T_c} \int_{T_c} e^{-j\,\frac{2\pi}{K}\,(f - f_{r,q})\,\tilde\varepsilon_i(t)} \cdot e^{-j\,4\pi ft}\, dt , \tag{11.121}$$

$$\delta S_{q,i}(f) \cdot \delta S_{\varepsilon,i}(f) \to \delta(f = 0) \text{ for } f = f_{r,q} \wedge T_c \to +\infty .$$

For the overall signal, this means no shift in channels occurs anymore:

$$s_i(t) = \sum_{q=1}^{Q} \frac{U_{q,i}}{U_{1,i}} \cdot e^{+j\left(\frac{4\pi}{c_0} Kt\,\Delta r_{q1} + \phi_i\right)} . \tag{11.122}$$

By a closer look on equation (11.120), it can be concluded that equivalently the time base can be altered by $\delta\tilde\varepsilon_i(t)$. The measurement signal must then be resampled again. Nevertheless, it had been left open where the knowledge about $\delta\tilde\varepsilon_i(t)$ is retrieved from.

Up to now, all signal information could be gathered from calibration measurements by post-processing methods. However, here comes the point, where a hardware setup like in Fig. 11.7 exhibits the big advantage, to acquire the individual statistical frequency deviation per measurement run.

In post-processing case, a measurement must be resampled as proposed in equation (11.113) though, then a significant target must be isolated by a window function in spectral domain, transformed into time domain again, where a linear regression calculus on the unwrapped phases can be performed again, providing $\delta\tilde{\varepsilon}_i(t)$.

Instead of a linear regression calculus, a quadratic regression could be imposed in case of moving platform (as this is the case for the two implemented systems). This quadratic regression would deliver a possible Doppler shift which causes a quadratic phase (compare to section 4.4.3).

Alternatively, denoising, compressed sensing, thus optimization techniques are examined as remedy.

11.3.4 Mismatched filter by windowing

In section 4.5.2 possibilities in sidelobe reduction in range compression were discussed. In FMCW principle, it is possible to apply a *Taylor* window in time domain before Fourier transformation, thus range compression. The answer is a *mismatched filter* response [47], [56], [59], [67]. Fig. 11.36 demonstrates that the size of range bins increases. The sidelobes are reduced significantly, though. For details on the window functions please see also appendix C.3.

Similar to regression over frequency deviation, regression over amplitude in range compressed maxima could also be considered. However, there, the remaining noise contribution is still that large, that no proper decomposition in transmitter and receiver amplitudes could be found.

11.3.5 Range dependent curvature filter

After resampling, the measurement signals were compensated in amplitude, constant phases, the azimuth deviation $\Delta\varphi_{21}$ as well as the curvature of $s_1(t)$ by equation (11.106). By applying a *Taylor* window in time domain, sidelobe reduction in range domain can be achieved. The only disturbance left is then the remaining curvature of measurement signal. This can be compensated by a range dependent filter in spectral domain after range compression, according to equation 4.73.

$$s_{re,rc,plane}(r) = s_{re,rc}(r) \cdot e^{-j\frac{4\pi}{c_0}f_0 \cdot \frac{d_{Tn}^2 + d_{Rm}^2}{4r}} \quad . \tag{11.123}$$

By the substitution:

$$f = K\tau = K\frac{2r}{c_0} \text{ , for monostatic radar,} \tag{11.124}$$

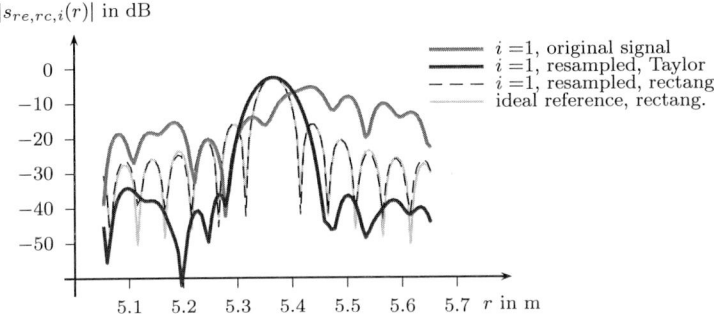

Fig. 11.36 Comparison of spectra and defocus: Original signal of trihedral reflector at ca. 5.35 m, signal after coherent resampling and additional correction process, effect of *Taylor window*, as well as ideal reference function. Parameters: Bandwidth $\Delta F = 3\,\mathrm{GHz}$, the original signal was shifted by 80 cm to balance internal lengths, its amplitude was increased by 15dB to have all signals in same power level.

it can be easily changed between frequency and range representation. The output of equation (11.123) is a plane wave representation for ranges.

11.3.6 Summary of proposed pre-processing

This section sketched the effort which is needed to provide data of good quality, thus equal amplitude and synchronous phase for subsequent azimuth processing. All in all two signals with isolated targets are needed to form the *sample reference signal* which in turn provides all correction or compensation factors, as well as the frequency deviation functions and the most important, the time base.

Whereas Fig. 11.38 sketches a very complex procedure to obtain a new time base and all needed correction parameters, Fig. 11.39 demonstrates that those in turn can be used in straight forward manner to provide measurements deliberated from phase shifts by internal transmission line lengths, phase shifts between channels, the main amplitude errors and modulation nonlinearities. Additionally, improved sidelobe suppression and de-curving phase fronts in order to obtain plain wave signals for all ranges, was demonstrated. Furthermore, all these methods were tested and proofed on real measurement data.

Fig. 11.37 discloses that the complete calibration and correction procedure concentrates in four signals which can be saved after test measurement and loaded for pre-processing of measurement data.

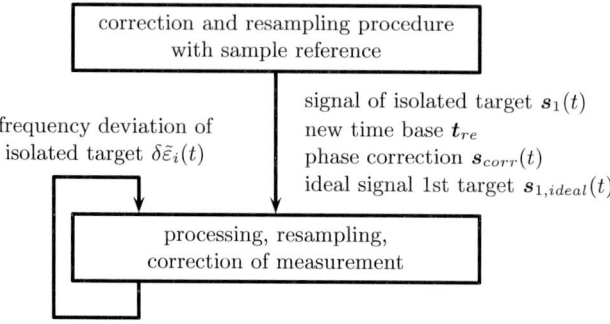

Fig. 11.37 Interface between calibration procedure and measurement pre-processing: Only four fundamental signals need to be exchanged.

For detailed correction, the isolation of a single target and retrieval of $\delta\tilde{\varepsilon}_i(t)$, is demonstrated in Fig. 11.40.

11.3.7 Further methods

The classical approach would be a dispersive filter [65]: Since the phase deviation depends on the round trip delay, the spectrum of the corrupted signal is divided in sub-parts. Each sub-part is corrected with an adapted correction function [65]. This is equivalent to a frequency dependent, thus dispersive, filter in frequency domain [65]. The drawback is a high computational load [65].

Beyond hardware linearization, dispersive filters or re-sampling, there are several other (related or hybrid) methods, such as expanding the phase errors or frequency deviation with help of Fourier series [145], modelling the modulation errors as sinusoidal error [151], [152], modelling the phase errors as polynomial and estimating the coefficients which is optimized in an recursive loop [154].

In [65], similar to the proposed methods, the signal is normalized by a test signal to get rid of (in this case) all transmit effects. Then the phase errors as well as the *residual video phase* are pre-processed with RVP method [64]. The remaining phase error is then independent of τ, can be estimated and corrected in a recursive loop [65].

Furthermore, the chirp length can be gated in time domain in order to optimize the chirp length and to reduce the spectral width [163] (with help of [151]).

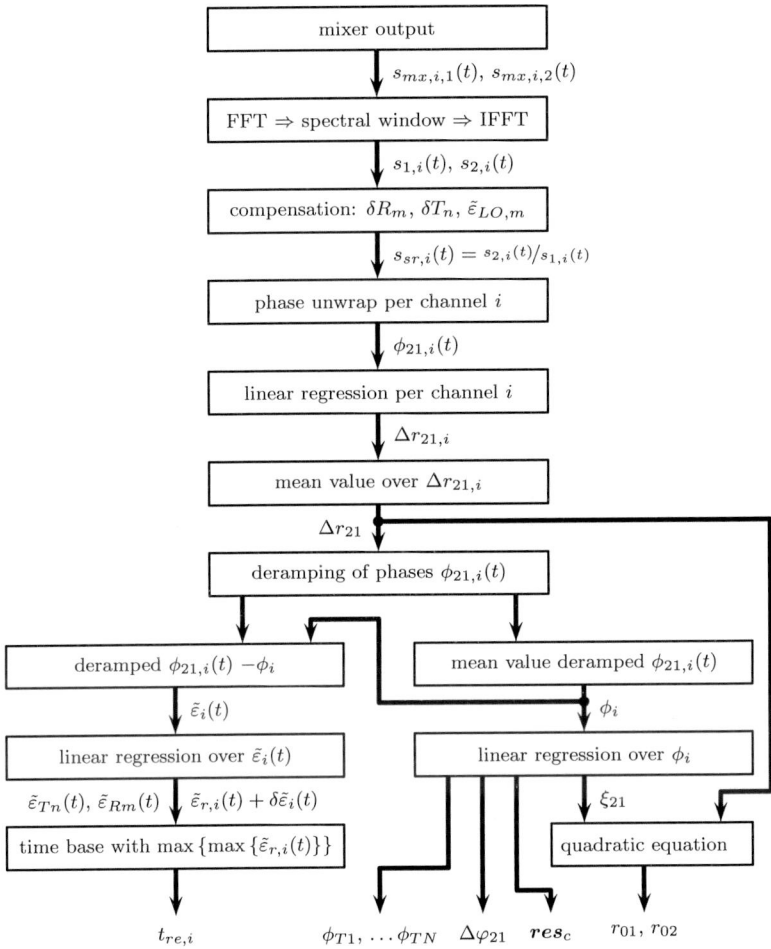

Fig. 11.38 Block diagram of *sample reference* process providing all compensation factors as well as the time base.

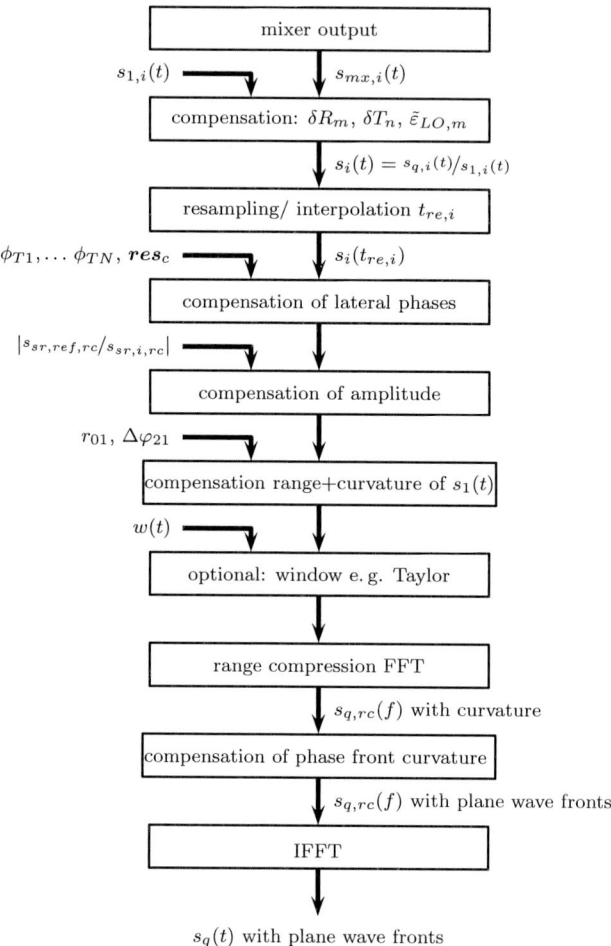

Fig. 11.39 Block diagram of correction and resampling of measured data (only deterministic parts).

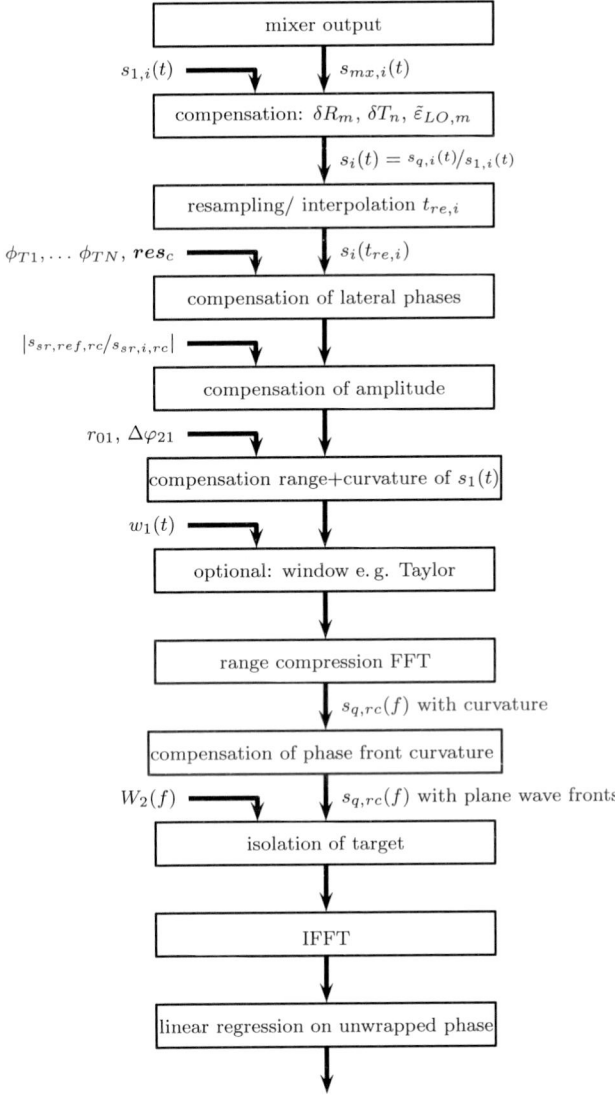

Fig. 11.40 Block diagram of first run on correction and resampling of measured data of isolated target providing statistical frequency deviation $\delta\bar{\varepsilon}_i(t)$.

12 Angular processing results

12.1 Linear array

Fig. 12.1 Test scenario: Trihedral of $l = 7$ cm in front of helicopter landing aid.

For testing of fundamental system function, simple scenarios where set up, such as a single trihedral in the lab's corridor. At first instance, beamformer results $S_{bf}(\Psi) = a_\nu R_n a_\nu^{\mathrm{H}}$ normalized by their maxima shall be compared to each other. Hereby, the beamformer approach on the sparsely filled covariance matrix without *augmentation* shall be denoted as *sparse beamformer*, whereas the approach based on the completely filled covariance matrix by copies shall be depicted by *augmented beamformer*. The results will be compared to a fictive array with Γ fully individual elements (in case of helicopter landing aid radar $\Gamma = 79$). This will be denoted as *ideal, full beamformer*.

Fig. 12.2 shows the result of such a sparse beamformer approach on only $\Gamma_u = 16$ distributed according to Fig. 10.14 (MR radar of *general* MR type). In pre-processing the steps summarized in Fig. 11.39 had been performed. Although, no statistical frequency deviation had been corrected, the sparse beamformer result is very close to ideal signal performance.

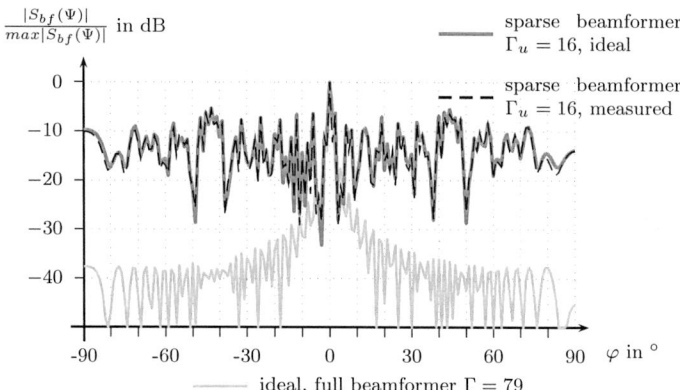

Fig. 12.2 Comparison of sparse beamformer results derived from measured data (dashed, black) and ideal signal (gray), both with sparse $\Gamma_u = 16$ elements. Contrary to that: Response of full beamformer derived from dense $\Gamma = 79$ elements (lightgray). Target: Trihedral with $l = 7\,\text{cm}$ at 6.5 m in front of helicopter landing aid.

Although, the signal belongs to a *coherent signal scenario*, the classical *augmentation* technique (section 5.4.2) can be applied here, since it is a single target scene, hence no mixed terms can occur in angular processing. Fig. 12.3 demonstrates that the *augmented beamformer* is superior to *sparse beamformer* by approximately 10 dB (Fig. 12.3). Nevertheless, there is still a discrepancy in performance between *augmented* and *ideal, full beamformer* which is founded in complex errors such as remaining amplitude noise or statistical frequency deviation in combination with noise contribution which propagate within the *augmentation procedure* (see section 8).

As next steps, the eigenvectors shall be examined: As proposed for the *coherent signal scenario* in section 5.3.3, there should be a single dominant eigenvector representing the weighted sum of directional vectors [45]. This can indeed be found in Fig. 12.4.

If sparse signals are reconstructed by means of compressed sensing CS, an example of *reduced matching pursuit* is demonstrated in Fig. 12.5 and 12.6. The fist demonstrates the outcome for a single target scenario at $r = 8\,\text{m}$, $\varphi = 4\,°$. The second one is at the same range cell, but with two targets $\varphi_1 = 0\,°$, $\varphi = 4\,°$. Originally, these were two single target measurements, which were added for generating a scenario of two identical targets. Nevertheless, both figures show feasible results.

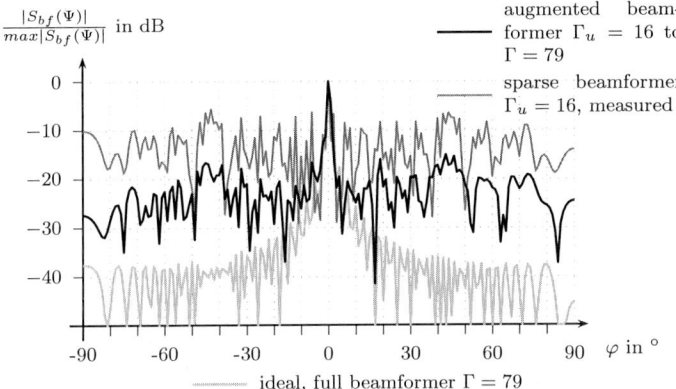

Fig. 12.3 Comparison of sparse beamformer results derived from measured data (gray) with augmented beamformer out of the same data (black), the first with sparse $\Gamma_u = 16$ elements, the latter with $\Gamma = 79$. Contrary to that: Response of full beamformer derived from dense $\Gamma = 79$ elements (lightgray). Target: Trihedral with $l = 7$ cm at 6.5 m in front of helicopter landing aid.

As next step, the forecasted danger of aliasing effects in CS shall be examined (compare to section 9.5). For the helicopter landing aid radar, those could be found e. g. for targets separated by $2°$ in azimuth, which is a larger spacing than the expected azimuth resolution of $1°$ to $1.4°$. In order to demonstrate those effects not only measurement data were examined but also a step back to simulation was done. Basically, both show the same result. An example can be seen in Fig. 12.7: The coherent sum of grating lobes is larger than the true target responses. In this case e. g. a *matching pursuit algorithm* would select to wrong basis function. For the *basis pursuit*, it would mean that there are cross-polytopes belonging to grating lobe responses meeting the L_1-norm more efficiently than for the true target response.

Since those results are less satisfying, the *adapted spatial smoothing algorithm* is also taken into account.

12.2 Cascaded linear array

Starting from the aliasing problem documented in Fig. 12.7, the *spatial smoothing* can be a remedy, which can be seen in Fig. 12.8. A single ques-

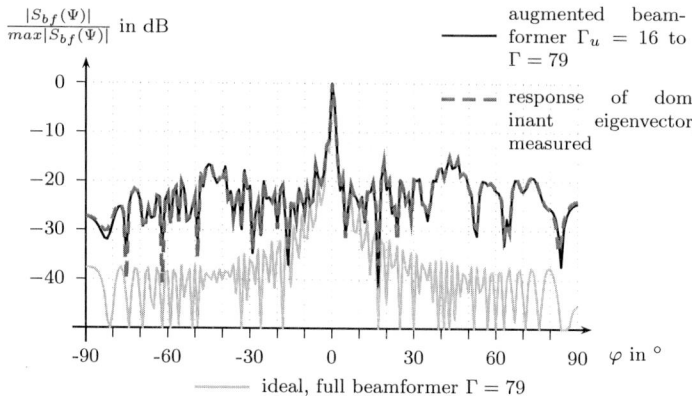

Fig. 12.4 Comparison of augmented beamformer's dominant eigenvector (gray) with augmented beamformer out of the same data (black, dashed), $\Gamma = 79$. Contrary to that: Response of full beamformer derived from dense $\Gamma = 79$ elements (lightgray). Target: Trihedral with $l = 7\,\text{cm}$ at $6.5\,\text{m}$ in front of helicopter landing aid.

tion remains. Does the noise floor increase for targets with higher difference in azimuth angles. But this is not the case, as demonstrated in Fig. 12.9. Both figures were generated as proposed in section 7.1.2: In total, only 16 sparse virtual elements are shifted to 16 sparse lateral positions. Nevertheless, this gives a synthetic uniform virtual element distribution of 156 different virtual elements separated by $\lambda_0/4$. Out of those 60 different sparse columns for matrix D in *spatial smoothing* can be selected. Therefore, in spite of a coherent scenario and in spite of very sparse arrays, the aliasing effects could be suppressed successfully and the minimum redundancy methods could be adapted to coherent signal scenarios for coherent MIMO systems.

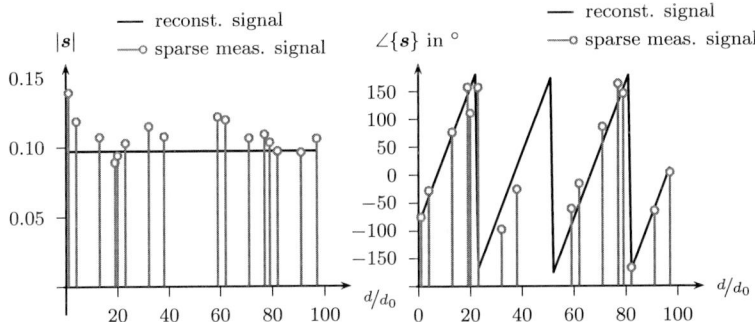

Fig. 12.5 In maximum of range compressed signal: Sparse signal of 16 virtual elements (out of 97 grid points) and estimated full signal, reconstructed by a single atom $a_\nu(\varphi = 4°)$ and its complex weight $0.9556 \exp(+j\,148.2°)$ with *reduced matching pursuit*. The scenario was a single trihedral at $r = 8\,\text{m}$ and $\varphi = 4.5°$. The reconstructed signal contains 80% of total energy in this range cell.

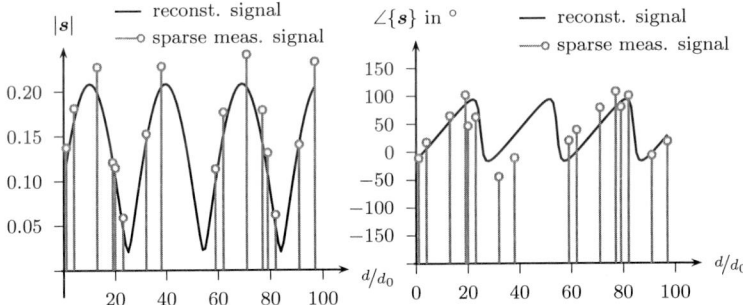

Fig. 12.6 In maximum of range compressed signal: Sparse signal of 16 virtual elements (out of 97 grid points) and estimated full signal, reconstructed by just two atoms $a_\nu(\varphi = 0°)$, $a_\nu(\varphi = 4°)$ and their complex weights $1.126 \exp(+j\,38.89°)$, $0.9241 \exp(+j\,152.7°)$ with *reduced matching pursuit*. The scenario were two trihedral targets at $r = 8\,\text{m}$, $\varphi = 0°$ and $\varphi = 4.5°$. The reconstructed signal contains again 80% of total energy in this range cell.

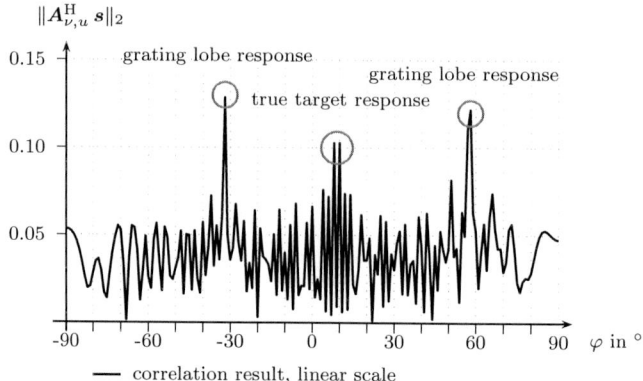

Fig. 12.7 In maximum of range compressed signal: The correlation of sparse signal with all sparse basis functions is the decision fundament not only for *reduced matching pursuit* but for CS in general. *General* MR virtual arrays can bear aliasing effects. The correct targets are $\varphi_1 = 8\,^\circ$, $\varphi_2 = 10\,^\circ$. Highest peaks belong to coherent superposition of grating lobes.

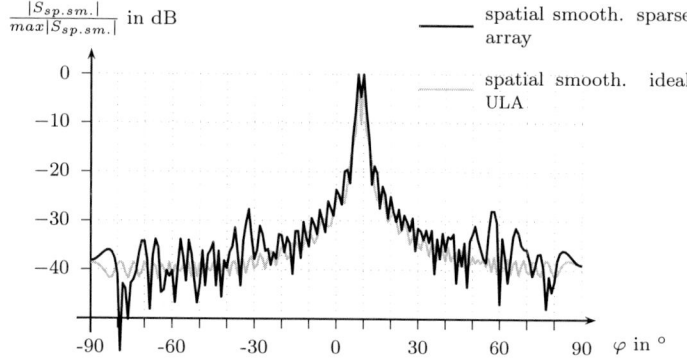

Fig. 12.8 Comparison of simulation results: *Adapted smoothing algorithm* applied to 16 element sparse virtual array (*general* MR Fig. 6.22), shifted laterally to 16 positions, giving $L_s = 60$ columns for matrix $\boldsymbol{D}$ (see section 7.1.2) in black, contrary to the ideal spatial smoothing provided by a ULA. According to Fig. 12.7 the targets are at $\varphi_1 = 8\,^\circ$, $\varphi_2 = 10\,^\circ$.

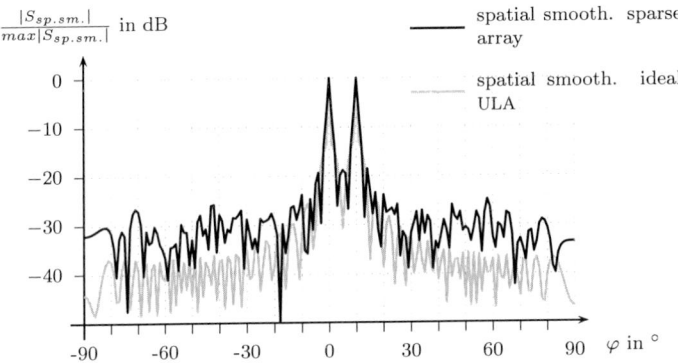

Fig. 12.9 Comparison of simulation results: *Adapted smoothing algorithm* applied to 16 element sparse virtual array (*general* MR Fig. 6.22), shifted laterally to 16 positions, giving $L_s = 60$ columns for matrix $\boldsymbol{D}$ (see section 7.1.2) in black, contrary to the ideal spatial smoothing provided by a ULA. The targets are at $\varphi_1 = 0\,^\circ$, $\varphi_2 = 10\,^\circ$.

13 Conclusion

Starting from general electromagnetics and scattering theory, the reader was guided to the field of *coherent* MIMO setups and *virtual arrays*. The fundamental convolutional character of virtual arrays could be demonstrated and proved. Then several methods of modulation were examined. The linear frequency modulation in had been chosen and a detailed signal model was constructed. This model not only contained the classical phase relations for range and azimuth of targets but also remaining phase front curvature, quadratic parasitic terms from modulation and phase noise. This signal model was then connected to angular processing methods, especially by examination of covariance matrix. There, both fundamental signals scenarios, the *uncorrelated* and the *coherent* one were clearly distinguished. As main topic of the thesis, the field of minimum redundancy arrays, which is classically feasible for the *uncorrelated* signal case, was successfully combined with virtual array techniques which in turn is a typically *coherent* signal scenario. Those ideas were not only developed further by cascading virtual arrays but also adapted to find ways of *coherent* spatial processing (*adapted spatial smoothing*). As own field, *compressed sensing* methods were taken for the same purpose. Additionally, not only a signal model was imposed but also description of noise and error propagation were found.

After parts of theory, two different implemented systems incorporating the proposed methods were introduced briefly. Thereby, also practical challenges were explored, matched to a signal model and solved as far as possible. Among others, topics like amplitude noise, saturation effects, transmission line dispersion and, most important, phase deviations in modulation were investigated. Possible technical solutions were not only suggested but even proven with measurement data. Special focus was taken onto the method of resampling FMCW signals. In this context, deterministic phase deviation due to mis-driving the signal source and statistical phase noise could be distinguished. The suggested method was able to improve the quality of measured data enormously. At the very end, the thesis was closed with a mixture of measured and simulated data, proving and summing the preceding chapters. The field of minimum redundancy could be connected successfully with *coherent* MIMO arrays. An extensive appendix provides the reader all necessary facts for deeper understanding of this thesis.

Fazit

Ausgehend von der Theorie der elektromagnetischen Wellenausbreitung und deren Streuung, wurde der Leser an das Gebiet der *kohärenten* MIMO Gruppenantennen, bzw. *virtueller* Gruppenantennen, herangeführt. Der grundlegende Faltungscharakter von *virtuellen* Gruppenantennen konnte gezeigt werde. Danach wurden mehrere Methoden der Modulation untersucht. Es wurde eine lineare Frequenzmodulation gewählt. An Hand derer wurde ein detailliertes Signalmodell aufgestellt. Dieses Modell enthielt nicht nur die Phasenbeziehungen für Entfernung und Winkelablage der Ziele, sondern auch die verbleibende Krümmung der Phasenfront, quadratische Fehlerterme auf Grund der Modulation und Phasenrauschen. Dieses Signalmodell wurde dann in der Winkelschätzung untersucht. Besonderes Augenmerk lag hier auf der Kovarianzmatrix. Dabei wurden die beiden grundsätzlichen Fälle *unkorreliertes* und *kohärentes* Signalszenario konsequent unterschieden. Danach wurde das Kernthema der Arbeit, sprich das Gebiet von Gruppenantennen mit minimaler Redundanz (im Auftreten der Elementabstände), welches klassisch zu *unkorrelierten* Signalen passt, erfolgreich mit *virtuellen* Gruppenantennen verbunden. Diese wiederum gehören eigentlich zu einem *kohärenten* Signalszenario. Diese Ideen wurden dann nicht nur weiter entwickelt, um Antennengruppen zu kaskadieren, sondern auch angepasst, um Wege für eine kohärente Winkelprozessierung zu finden (*adapted spatial smoothing*). Als eigenständiges aber hoch aktuelles Thema wurden hierzu auch Methoden zu *Compressed Sensing* als Lösungsansatz verwendet. Zusätzlich wurde nicht nur ein Signalmodell aufgestellt, sondern auch Beschreibungen für Systemrauschen und Fehlerfortpflanzung gefunden.

Nach theoretischen Abschnitten, wurden kurz zwei Systeme erläutert, die die vorgeschlagenen Gruppenantennen aufweisen und am Institut entwickelt wurden. Dabei wurden auch die praktischen Herausforderungen untersucht, mit dem Signalmodell in Verbindung gebracht und, so weit möglich, gelöst. Unter anderem waren Themen wie Amplitudenrauschen, Sättigungseffekte, Leitungsdispersion und, am wichtigsten, Phasenabweichungen von der Modulation Gegenstand der Untersuchung. Mögliche technische Lösungen wurden nicht nur vorgeschlagen, sondern auch mit Messdaten bewiesen. Spezieller Augenmerk wurde auf die Methode *Neuabtastung* von FMCW-Signalen gelegt. In diesem Zusammenhang konnte zwischen deterministischen Phasenfehlern auf Grund ungenauer Durchstimmvorgänge der Signalquelle und statistischem Phasenrauschen unterschieden werden. Die vorgestellt Methode war in der Lage, die Signalqualität signifikant zu verbessern. Die Arbeit wurde mit einer Mischung aus gemessenen und prozessierten, sowie simulierten Daten abgeschlossen. Diese belegen und akkumulieren die Erkenntnisse der vorangegangenen Kapitel. Zusammenfassend konnte das Feld Antennengruppen mit minimaler Redundanz (bzgl. Häufigkeit von Elementabständen) erfolgreich mit kohärenten MIMO-Antennengruppen verbunden werden. Ein ausführlicher Anhand bietet dem Leser alle notwendigen Fakten zum tieferen Verständnis der Arbeit.

A Mathematical methods

A.1 Functional analysis

This shall be a short introduction to the denoted topic, hence the prior condition is that the reader is familiar with mathematical terms like *group*, *ring* or *field*. Otherwise the reader shall be relegated to [43].

A.1.1 Norms in vector spaces

Norms of vectors $\boldsymbol{x} = [\ldots, x_\gamma, \ldots]$ over the field $\mathbb{F}^\Gamma$ ca be defined as [43]:

$$\|\boldsymbol{x}\|_p = \left(\sum_{\gamma=1}^{\Gamma} |x_\gamma|^p \right)^{\frac{1}{p}} \quad \text{, for } 1 \leq p < +\infty \,, \tag{A.1}$$

$$\|\boldsymbol{x}\|_\infty = \max_{1 \leq p \leq \Gamma} |x_\gamma| \quad \text{, for } p = +\infty \,. \tag{A.2}$$

In case of $\mathbb{F} = \mathbb{R}$, the normalized vector space is denoted as *Euclidian*, whereas in complex case $\mathbb{F} = \mathbb{C}$ *unitary* [43].

A.1.2 Unitary vector spaces

A vector space $\mathbb{V}$ over the field $\mathbb{C}$ can be denoted as *Pre-Hilbert space* if a *inner product* or *scalar product* is defined: $\langle f, g \rangle$ [43].

If further the *scalar product* is related to a *norm*, then the *Pre-Hilbert space* becomes not only *normalized* but also *unitary* [43]. Supplementary to properties like e.g. *associative*, *commutative*, *distributive law*, the following features can be added [43]:

- Triangle inequality: $\|f + g\|^2 \leq (\|f\| + \|g\|)^2$
- Cauchy-Schwarz inequality: $|(f, g)| \leq \sqrt{\langle f, f \rangle} \sqrt{\langle g, g \rangle}$
- Parallelogram equality: $\|f + g\|^2 + \|f - g\|^2 = 2 \left(\|f\|^2 + \|g\|^2 \right)$
- Continuity of scalar product: If f and g are steady, then $\langle f, g \rangle$ is also steady.

A.1.3 Hilbert spaces

A vector space $\mathbb{H}$ is *complete* if any *Cauchy sequence* converges [43]. If this vector space is additionally *unitary*, then $\mathbb{H}$ is a so-called *Hilbert (vector) space* [43].

A.1.4 Orthogonal and orthonormal vector spaces

The elements $f_1, \ldots, f_n$ shall be in *Hilbert space* $\mathbb{H}$ and *pairwise orthogonal*. Amongst others, this yields several remarkable properties [43]:

- Pythagorean theorem: $\| \sum_{i=1}^{n} f_i \|^2 = \sum_{i=1}^{n} \| f_i \|^2$.

- Projection theorem: The Hilbert space shall consist of two orthogonal parts $\mathbb{H} = \mathbb{H}_0 + \mathbb{H}_1$, $\mathbb{H}_0 \perp \mathbb{H}_1$. Each element $f \in \mathbb{H}$ can be separated into $f = f' + f''$ with $f' \in \mathbb{H}_0$, $f'' \in \mathbb{H}_1$.

- Approximation problem: $f \in \mathbb{H}$, $g \in \mathbb{H}_0$, $\| f - g \| \to inf$. This problem is uniquely solvable by the projection of f onto $\mathbb{H}_0$, f'. Thereby $f' = g$ provides the value with smallest distance $\mathbb{H}_0$ to f.

A *set* of elements $f_1, \ldots, f_i, f_j, \ldots, f_n \in \mathbb{H}$ are denoted as *orthogonal system*, when their *pairwise inner products* are *Kronecker delta* like [43]:

$$\langle f_i, f_j \rangle = \delta_{ij} = \begin{cases} 1, & \text{for } i = j \, , \\ 0, & \text{for } i \neq j \, . \end{cases} \tag{A.3}$$

If additionally each norm is $\| f_i \| = 1$, $\forall i \in [1; n]$, then the system is even *orthonormal* [43].

A.2 Fresnel integrals

There are several *Fresnel integrals* which deliver odd functions [32]:

$$C(x) = \int_0^x \cos\left(\frac{\pi}{2} t^2 \right) dt \quad , \ S(x) = \int_0^x \sin\left(\frac{\pi}{2} t^2 \right) dt \quad , \tag{A.4}$$

$$F(x) = C(x) - j\, S(x) = \int_0^x \mathrm{e}^{-j \frac{\pi}{2} t^2} \, dt \quad . \tag{A.5}$$

For $x \to +\infty$, the asymptotic values are [32]:

$$C(x) = S(x) \to \frac{1}{2} \,,\, F(x) \to \frac{1-j}{2} \text{ for } x \to +\infty \,. \tag{A.6}$$

For large positive values of x, the following approximations are feasible:

$$C(x) \approx \frac{1}{2} + \frac{1}{\pi x} \sin\left(\frac{\pi}{2} x^2\right) \quad,\quad S(x) \approx \frac{1}{2} - \frac{1}{\pi x} \cos\left(\frac{\pi}{2} x^2\right) \quad, \tag{A.7}$$

$$F(x) \approx \frac{1-j}{2} + \frac{j}{\pi x} \, e^{-j\frac{\pi}{2} x^2} \quad. \tag{A.8}$$

For small values of x: $F(x) \approx x$. Furthermore: $F(0) = 0$, $\frac{dF}{dx}(0) = 1$.

There is further a group of so-called *type 2 Fresnel integrals*, defined as [32]:

$$C_2(x) = \int_0^x \frac{\cos t}{\sqrt{2\pi\,t}} \, dt \quad,\quad S_2(x) = \int_0^x \frac{\sin t}{\sqrt{2\pi\,t}} \, dt \quad, \tag{A.9}$$

$$F_2(x) = C_2(x) - j\,S_2(x) = \int_0^x \frac{e^{-j\,t}}{\sqrt{2\pi\,t}} \, dt \quad. \tag{A.10}$$

For $x \geq 0$, the substitution $t = \frac{\pi}{2} x^2$ gives:

$$C(x) = C_2\left(t = \frac{\pi}{2} x^2\right) \,,\, S(x) = S_2\left(\frac{\pi}{2} x^2\right) \,,\, F(x) = F_2\left(\frac{\pi}{2} x^2\right) \,. \tag{A.11}$$

However, for $x < 0$: $F(x) = -F(-x) = -F_2(\frac{\pi}{2} x^2)$

Special types of Fresnel integrals In section 10, derivates of afore mentioned Fresnel integrals were used for describing the radiation pattern of horn antennas, such like [32]:

$$F_0(\chi,\varsigma) = \int_{-1}^{+1} e^{+j\,\pi\chi\,\xi} \cdot e^{-j\frac{\pi}{2}\varsigma\,\xi^2} \, d\xi =$$

$$\tag{A.12}$$

$$= \ldots = \frac{1}{\varsigma} e^{+j\frac{\pi\chi^2}{2\varsigma^2}} \left[F\left(\frac{\chi}{\varsigma} + \varsigma\right) - F\left(\frac{\chi}{\varsigma} - \varsigma\right) \right] \,.$$

Or [32]:

$$F_1(\chi, \varsigma) = \int\limits_{-1}^{+1} \cos\left(\frac{\pi}{2}\xi\right) \cdot e^{+j\,\pi\chi\xi} \cdot e^{-j\,\frac{\pi}{2}\varsigma\,\xi^2}\, d\xi =$$

$$= \ldots = \frac{1}{2}\left[F_0\left(\chi + {}^1\!/_2,\, \varsigma\right) + F_0\left(\chi - {}^1\!/_2,\, \varsigma\right)\right] \quad . \tag{A.13}$$

A.3 Method of stationary phase

An integral over a product of two functions, one with a rather slow change in amplitude and phase and another one with a rapidly oscillating behaviour, can be approximated by the *method of stationary phase* [61], [71], [32].

The basic idea behind this is, that rapidly oscillating parts contain a lot of positive and negative half-waves which more or less cancel out themselves and deliver nearly no contribution to the value of the integral. In general such an integral can be described as follows [61], [71], [32]:

$$I = \int\limits_{-\infty}^{+\infty} f(x)\, e^{+j\,\Phi(x)}\, dx \quad . \tag{A.14}$$

The phase $\Phi(x)$ is approximated by a Taylor series expansion which is aborted after the quadratic term [32]:

$$\Phi(x) \approx \Phi(x_0) + \frac{\partial \Phi}{\partial x}(x_0)\,(x - x_0) + \frac{1}{2}\frac{\partial^2 \Phi}{\partial x^2}(x_0)\,(x - x_0)^2 =$$

$$= \Phi(x_0) + \Phi'(x_0) + \frac{1}{2}\,\Phi''(x_0)\,(x - x_0)^2 \quad . \tag{A.15}$$

The point of interest is the flat point at $\Phi' = 0$, the *stationary point* x_*. There, equation (A.14) is equal to [32]:

$$I = \int\limits_{-\infty}^{+\infty} f(x_*)\, e^{+j\,\Phi(x_*)} \cdot e^{+j\,\frac{1}{2}\Phi''(x_*)\,(x-x_*)^2}\, dx =$$

$$= f(x_*)\, e^{+j\,\Phi(x_*)} \cdot \int\limits_{-\infty}^{+\infty} e^{+j\,\frac{1}{2}\Phi''(x_*)\,(x-x_*)^2}\, dx \quad . \tag{A.16}$$

The next step is, to substitute $\Phi''(x_*)\,(x-x_*)^2$ by $\pi\chi^2$. With $dx = \sqrt{\frac{\pi}{\Phi''(x_*)}}\,d\chi$, this gives for the remaining integral into the complex conjugated Fresnel integral [32]:

$$
\int\limits_{-\infty}^{+\infty} \mathrm{e}^{+j\,\frac{1}{2}\,\Phi''(x_*)\,(x-x_*)^2}\,dx = \sqrt{\frac{\pi}{\Phi''(x_*)}}\,\int\limits_{-\infty}^{+\infty} \mathrm{e}^{+j\,\frac{\pi}{2}\,\chi^2}\,d\chi =
$$
$$
= \dots = \sqrt{\frac{\pi}{\Phi''(x_*)}}\cdot\sqrt{2j}. \tag{A.17}
$$

The integral from equation (A.14) can therefore be expressed by [32]:

$$
I = \int\limits_{-\infty}^{+\infty} f(x)\,\mathrm{e}^{+j\,\Phi(x)}\,dx \approx \sqrt{\frac{2\pi j}{\Phi''(x_*)}}\cdot f(x_*)\,\mathrm{e}^{+j\,\Phi(x_*)} \quad . \tag{A.18}
$$

A.4 Linear coordinate transformation

A linear coordinate transformation, like e. g. a rotation, is a common problem for Fourier transformation [34]:

$$
f\,(\boldsymbol{A}\,\boldsymbol{x}) \;\circ\!\!-\!\!\bullet\; \frac{1}{|\det\boldsymbol{A}|}\,F\,(\boldsymbol{B}\,\boldsymbol{k}) \quad\text{, with } \boldsymbol{B} = \left(\boldsymbol{A}^{-1}\right)^{\mathrm{T}} \quad . \tag{A.19}
$$

B System theory on electromagnetics

B.1 Electromagnetic fields from vector potentials

Since the divergence of the magnetic flux density B is zero, thus B is *solenoid*, it can be derived from a *magnetic vector potential* A [31], [32], [36]:

$$\nabla \cdot B(r,t) = 0 \quad , \nabla \cdot \nabla \times A(r,t) = 0 \quad ,$$
$$\Rightarrow \quad B(r,t) = \nabla \times A(r,t) \quad . \tag{B.1}$$

B.2 The rectangular hollow waveguide

The propagation of a general electromagnetic fields within a rectangular hollow waveguide can be described by a magnetic and an electric vector potential, whereby a component in propagation direction (here z-direction) is sufficient [58] (Fig. B.1).

$$A_m = A_{m,z}\, e_z \quad , \quad A = A_z\, e_z \quad . \tag{B.2}$$

The first possible field mode is denoted by $TE_{mn} = TE_{10}$ (former H_{10}) and represents the standard mode for operation [58]. Here $A = 0$. The

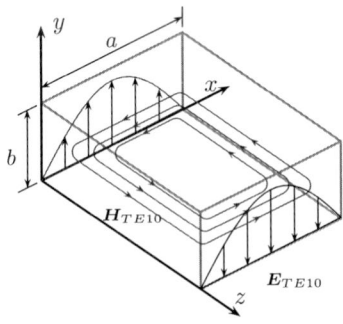

Fig. B.1 Electric and magnetic field distribution in rectangular hollow waveguide at TE_{10} (standard) mode.

consideration shall be restricted to TE_{mn} modes here. The corresponding fields $\boldsymbol{E}$ and $\boldsymbol{H}$ can be calculated by [31],[32], [36]:

$$\boldsymbol{E} = -\frac{1}{\epsilon} \, \nabla \times \boldsymbol{A}_m \, , \tag{B.3}$$

$$\boldsymbol{H} = -j\omega \, \boldsymbol{A}_m + \frac{1}{j\omega \, \epsilon\mu} \, \nabla \left(\nabla \cdot \boldsymbol{A}_m \right) \, . \tag{B.4}$$

The standard operational mode of a rectangular hollow waveguide is a so-called TE-mode (former H-mode). This wave type can be expressed mathematically by an electric vector potential consisting of only a component in propagation direction, e.g. $\boldsymbol{A}_m = A_m(x,y) \cdot \mathrm{e}^{-j\beta\,z} \cdot \boldsymbol{e}_z$ [32], [36], [37], [58] etc. The boundary condition is a *Neumann boundary condition* [32], [36], [37], [58] etc. The standard operational mode of a rectangular hollow waveguide is depicted as TE_{10} (former H_{10}) (compare to appendix B.2). Adapted to the coordinate system in Fig. B.2, this gives [32], [58]:

$$\begin{aligned} \boldsymbol{E}_{TE10}(\boldsymbol{r}) &= -\frac{A_0}{\epsilon} \, k_x \cdot \cos\left(k_x \, x\right) \, \boldsymbol{e}_y \cdot \mathrm{e}^{-j\,\beta z} = \\ &= E_0 \cdot \cos\left(k_x \, x\right) \, \boldsymbol{e}_y \cdot \mathrm{e}^{-j\,\beta z} \quad , \text{ with } k_x = \frac{\pi}{a} \, , \end{aligned} \tag{B.5}$$

$$\begin{aligned} \boldsymbol{H}_{TE10}(\boldsymbol{r}) &= \frac{A_0}{j\omega \, \epsilon\mu} \, k_x \cdot \left[j\beta \cos\left(k_x \, x\right) \, \boldsymbol{e}_x - k_x \sin\left(k_x \, x\right) \, \boldsymbol{e}_z \right] \cdot \mathrm{e}^{-j\,\beta z} \, , \\ &= -E_0 \cdot \left[\frac{1}{Z_{TE10}} \cos\left(k_x \, x\right) \, \boldsymbol{e}_x - \frac{k_x}{j\omega\mu} \sin\left(k_x \, x\right) \, \boldsymbol{e}_z \right] \cdot \mathrm{e}^{-j\,\beta z} \, , \end{aligned} \tag{B.6}$$

$$\text{with } Z_{TE10} = \frac{Z_{F0}}{\sqrt{1 - \left(\frac{\lambda_0}{2a}\right)^2}} = \frac{120\pi \, \Omega}{\sqrt{1 - \left(\frac{\lambda_0}{2a}\right)^2}} \quad .$$

B.3 Radiation from apertures

On apertures, the tangential field components can be considered as equivalent surface current densities at $\boldsymbol{r}_a$:

$$\boldsymbol{J}_s = \boldsymbol{n} \times \boldsymbol{H}(\boldsymbol{r}_a) \quad , \text{ with } \boldsymbol{n} \text{ as normal vector on aperture.} \tag{B.7}$$

In terms of duality, the equivalent magnetic surface current density $\boldsymbol{J}_{m,s}$ is [32]:

$$\boldsymbol{J}_{m,s} = -\boldsymbol{n} \times \boldsymbol{E}(\boldsymbol{r}_a) \quad , \text{ with } \boldsymbol{n} \text{ as normal vector on aperture.} \tag{B.8}$$

The radiation vectors for far-field condition can be determined by a two dimensional Fourier transformation over the aperture (here: x-y-plane) [32]:

$$\boldsymbol{F}(\boldsymbol{r}) = \boldsymbol{n} \times \int\limits_{x,y} \boldsymbol{H} \, \mathrm{e}^{+j\,(k_x\,x + k_y\,y)} \, dx\,dy = \boldsymbol{n} \times \boldsymbol{f}(\boldsymbol{r}) \quad , \tag{B.9}$$

$$\boldsymbol{F}_m(\boldsymbol{r}) = -\boldsymbol{n} \times \int\limits_{x,y} \boldsymbol{E} \, \mathrm{e}^{+j\,(k_x\,x + k_y\,y)} \, dx\,dy = -\boldsymbol{n} \times \boldsymbol{f}_m(\boldsymbol{r}) \quad . \tag{B.10}$$

B.4 Horn antennas

Horn antennas are so-called aperture antennas [31], [32] (Fig. B.2). Usually the feed waveguide is designed to be operated by a single waveguide mode in order to avoid dispersion effects. This depends highly on the inner dimensions of the waveguide and its (dielectric) filling [32], [36], [37] etc., appenidx B.2. It shall be assumed that the boundaries of the waveguide are made of perfect electric conductors (PEC), thus the tangential components of the electric field are zero. The antenna is fed with a TE$_{10}$-wave.

Now, due to the larger cross-section of the flaring (linearly from a, b to A, B), more and more field modes can not only be excited but are also capable to propagate at different distances from the feed waveguide. Since those field modes differ in their wavenumbers β, this causes dispersion, thus a broadening of the transmit pulse. However, for a first examination of the horn antennas, it shall be assumed that no energy couples from the standard TE$_{10}$-mode into other field modes during the transition through the flaring. For the radiated fields, the tangential field components at the aperture plane are of interest. It can be seen that the field components propagating to the edges of the aperture underlie a longer path length than those who end at the middle of the aperture.

The transversal plane from the waveguide is bended driven by the condition of being perpendicular onto the inner side of the flaring. The surface of special interest is tangent to the aperture in the origin, with the following condition: $r_b = R_b$. From each point on the surface $\boldsymbol{s}$, a path is considered perpendicular to $\boldsymbol{s}$ until it pierces the plane $z = 0$. The length of such a path shall be denoted by $\tilde{\delta}$. The difference in path length $\tilde{\delta}(x,y)$, thus the

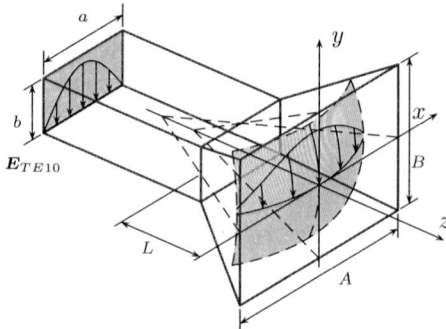

Fig. B.2 Pyramidal horn antenna

distance from the bended surface with $r_b = R_b$ to the aperture at $z = 0$, can be obtained by approximately [32]:

$$\tilde{\delta}(x, y) \approx R_a \cdot \frac{\alpha_a^2}{2} + R_b \cdot \frac{\alpha_b^2}{2} \approx$$
$$\approx \frac{x^2}{2\,R_a} + \frac{y^2}{2\,R_b} = \delta_a + \delta_b = \delta(x, y) \quad . \tag{B.11}$$

This approximation yields into two separate contributions depending only on one variable x or y. Hence, the problem can be examined separately in

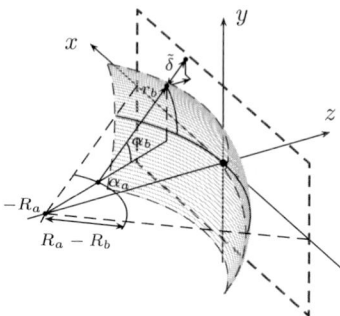

Fig. B.3 Phase fronts of a pyramidal horn antenna (rear view).

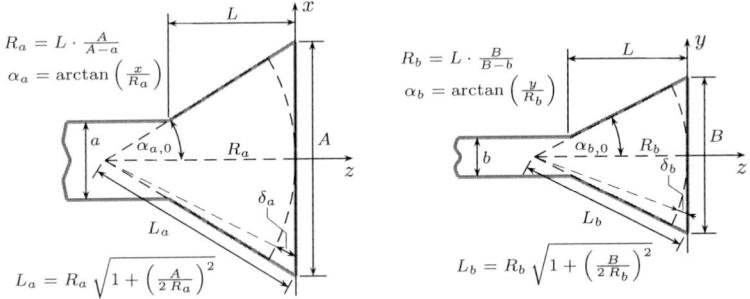

Fig. B.4 Pyramidal horn antenna: H-plane (left), E-plane (right)

E- and H-plane. In the light of all those approximations, the wavenumber k_x is also approximated by π/A at the aperture [32]. With the phase shifts, the tangential components of the fields can then be described by:

$$
\begin{aligned}
\boldsymbol{E}_{ap}(x,y) &= E_0 \cdot \cos\left(\frac{\pi}{A}\, x\right) \boldsymbol{e}_y \cdot \mathrm{e}^{-jk_0\,[\delta_a(x)+\delta_b(y)]} \\
&= E_0 \cdot \cos\left(\frac{\pi}{A}\, x\right) \boldsymbol{e}_y \cdot \mathrm{e}^{-jk_0\,\frac{x^2}{2R_a}} \cdot \mathrm{e}^{-jk_0\,\frac{y^2}{2R_b}} \ .
\end{aligned}
\tag{B.12}
$$

$$
\boldsymbol{H}_{ap}(x,y) = -\frac{E_0}{Z_{TE10}} \cdot \cos\left(\frac{\pi}{A}\, x\right) \boldsymbol{e}_x \cdot \mathrm{e}^{-jk_0\,\frac{x^2}{2R_a}} \cdot \mathrm{e}^{-jk_0\,\frac{y^2}{2R_b}} \ .
\tag{B.13}
$$

B.4.1 Radiated fields from horn antennas

If the equations (B.12), (B.13) are compared to each other, the following condition can be derived [32]:

$$
\boldsymbol{H}_{ap} = \frac{1}{Z_{TE10}}\, \boldsymbol{e}_z \times \boldsymbol{E}_{ap} \quad , \text{Huygens source.}
\tag{B.14}
$$

This is also known as *Huygens source* [32], thus the complete field distribution can be described by either the $\boldsymbol{E}$- or $\boldsymbol{H}$-field. The other one can be calculated by a cross product and a transverse impedance, depending on the excited field mode [32]. The radiated $\boldsymbol{E}$-field can be determined by [32]:

$$E_\vartheta = jk\,\frac{e^{-jk\,r}}{4\pi\,r} \cdot (1 + \cos\vartheta) \cdot (f_{m,x}\cos\varphi + f_{m,y}\sin\varphi) \quad , \tag{B.15}$$

$$E_\varphi = jk\,\frac{e^{-jk\,r}}{4\pi\,r} \cdot (1 + \cos\vartheta) \cdot (f_{m,y}\cos\varphi - f_{m,x}\sin\varphi) \quad , \tag{B.16}$$

$$\boldsymbol{f}_m = \iint\limits_{x\,y} \boldsymbol{E}(x,y) \cdot e^{+j(k_x\,x + k_y\,y)}\,dx\,dy \quad . \tag{B.17}$$

The terms $F_0(\nu_y, \sigma_b)$ and $F_1(\nu_x, \sigma_a)$ are also known as *Fresnel diffraction integrals* [32]. The radiated $\boldsymbol{E}$-field is then:

$$\begin{aligned}
\boldsymbol{E}(\vartheta,\varphi) = +\,jk\,\frac{e^{-jk\,r}}{4\pi\,r} \cdot (1 + \cos\vartheta) \cdot E_0\cdot \\
\cdot\,\frac{AB}{4} \cdot F_1(\nu_x,\sigma_a) \cdot F_0(\nu_y,\sigma_b) \cdot (\sin\varphi\,\boldsymbol{e}_\vartheta + \cos\varphi\,\boldsymbol{e}_\varphi) \quad .
\end{aligned} \tag{B.18}$$

The magnetic field is then analogously:

$$\begin{aligned}
\boldsymbol{H}(\vartheta,\varphi) = -\,j\frac{k}{Z_{TE10}}\,\frac{e^{-jk\,r}}{4\pi\,r} \cdot (1 + \cos\vartheta) \cdot E_0\cdot \\
\cdot\,\frac{AB}{4} \cdot F_1(\nu_x,\sigma_a) \cdot F_0(\nu_y,\sigma_b) \cdot (\cos\varphi\,\boldsymbol{e}_\vartheta - \sin\varphi\,\boldsymbol{e}_\varphi) \quad .
\end{aligned} \tag{B.19}$$

Please note, that the parameters ν_x and ν_y can be expressed in spherical coordinates [32]:

$$\begin{aligned}
\nu_x(\vartheta,\varphi) = k_x\,\frac{A}{2\pi} = \frac{2\pi}{\lambda}\,\frac{A}{2\pi}\,\boldsymbol{e}_x \cdot \boldsymbol{e}_r = \frac{A}{\lambda}\,\sin\vartheta\cos\varphi \quad , \\
\nu_y(\vartheta,\varphi) = k_y\,\frac{B}{2\pi} = \frac{2\pi}{\lambda}\,\frac{A}{2\pi}\,\boldsymbol{e}_y \cdot \boldsymbol{e}_r = \frac{B}{\lambda}\,\sin\vartheta\sin\varphi \quad .
\end{aligned} \tag{B.20}$$

B.4.2 Directivity and footprint

For a directivity analysis the complex *Poynting vector* $\boldsymbol{T}$ is first determined from equations (B.12) and (B.13)

$$\boldsymbol{T} = \frac{1}{2}\,\boldsymbol{E} \times \boldsymbol{H}^* = \frac{1}{2} \cdot \frac{|E_0|^2}{Z_{TE10}} \cdot \cos^2\left(\frac{\pi\,x}{A}\right) \cdot \boldsymbol{e}_z \quad . \tag{B.21}$$

Then T is calculated from equations (B.18) and (B.19) for the closer analysis of directivity. Since the field components had already been defined for far-field, this *Poynting vector* becomes real and represents the effective power density S_m:

$$
T = \frac{1}{2} E \times H^* =
$$
$$
= \frac{1}{2} \frac{|E_0|^2}{Z_{TE10}} \left(\frac{k}{4\pi\,r}\right)^2 (1 + \cos\vartheta)^2 \left(\frac{AB}{4}\right)^2 \cdot \qquad\text{(B.22)}
$$
$$
\cdot\, |F_1(\nu_x, \sigma_a)|^2\, |F_0(\nu_y, \sigma_b)|^2 \cdot e_z = S_m \quad .
$$

By inserting equation (B.18) in equation (B.22) and a resort of the latter, it can be obtained:

$$
T = \underbrace{P_{rad} \cdot \frac{1}{4\pi\,r^2}}_{\text{isotrop. radiator } p_{iso}} \cdot \underbrace{\left(\frac{1 + \cos\vartheta}{2}\right)^2}_{\text{obliquity factor } c_{ob}^2} \cdot
$$
$$
\cdot\, \underbrace{\frac{4\pi}{\lambda^2}}_{\frac{1}{A_{iso}}} \cdot \underbrace{AB}_{A_{ap}} \cdot \underbrace{\frac{1}{8} |F_1(\nu_x, \sigma_a)|^2\, |F_0(\nu_y, \sigma_b)|^2}_{\text{aperture efficiency } \eta_{ap}} \cdot e_z = \qquad\text{(B.23)}
$$
$$
= p_{iso} \cdot \underbrace{c_{ob}^2 \cdot \frac{A_{ap}}{A_{iso}} \cdot \eta_{ap}}_{\text{directivity } D(\vartheta, \varphi)} = p_{iso} \cdot D(\vartheta, \varphi) \quad .
$$

The directivity $D(\vartheta, \varphi)$ is then [32]:

$$
D(\vartheta, \varphi) = \left(\frac{1 + \cos\vartheta}{2}\right)^2 \cdot \frac{4\pi}{\lambda^2} \cdot AB \cdot \frac{1}{8} |F_1(\nu_x, \sigma_a)|^2\, |F_0(\nu_y, \sigma_b)|^2 \quad . \quad\text{(B.24)}
$$

This is not equal to the gain of the antenna, since the fields reaching the aperture ready for being radiated were taken into account, thus the radiated power not the input power at the antenna feed. Hence for a correct gain, the losses of the feeding waveguide and horn flaring due to non-infinite conductivity must be considered additionally [31].

C Signal theory

C.1 The residual video phase method

The *residual video phase* (RVP) method is explained explicitly in [64] and applied in [65]. The RVP effect is that two targets with constant radial distance appear with larger separation at larger distances [64]. This is especially a problem for systems with large *round-trip delays*. In section 4.4.3, the mixer output signal was modelled as:

$$s_{mx}(t) \approx \sum_{q=1}^{Q} U_q \cdot e^{+j\,\varphi_q(t)} \cdot e^{-j\,\pi K \tau_{0q}^2} \ . \tag{C.1}$$

Depending on conventions, the complex term $\exp\left(-j\,\pi K\,\tau_{0q}^2\right)$ is already depicted as *residual video phase* (RVP) [64] or alternatively, the remaining phase distortion after the attempt of compensating $\exp\left(-j\,\pi K\,\tau_{0q}^2\right)$ [65]. In section 4.4.3, *Fourier* transformation was applied for range compression. However, the term $\exp\left(-j\,\pi K t^2\right)$ is invariant with respect to *Fourier* transformation over t:

$$
\begin{aligned}
s_{rc}(f) &= \mathcal{F}\left\{ \sum_{q=1}^{Q} U_q \cdot e^{+j\,\varphi_q(t)} \cdot e^{-j\,\pi K \tau_{0q}^2} \right\} = \\
&= \sum_{q=1}^{Q} \mathcal{F}\left\{ U_q \cdot e^{+j\,\varphi_q(t)} \right\} \cdot e^{-j\,\pi K \tau_{0q}^2} = \\
&= \sum_{q=1}^{Q} \mathcal{F}\left\{ U_q \cdot e^{+j\,\varphi_q(t)} \right\} \cdot e^{-j\,\pi \frac{f_{0q}^2}{K}} \ .
\end{aligned}
\tag{C.2}
$$

In section 4.4.3, the relation between *round trip delay* and frequency was demonstrated: $f = K\tau$, which was used in equation (C.2). The kernel of RVP method is now to multiply the spectrum in equation (C.2) the complex correction term $\exp\left(+j\,\pi/K\,f^2\right)$ and to transfer the result back to time domain in order to obtain a corrected signal [64], [65]:

$$s_{mx,corr}(t) = \mathcal{F}^{-1}\left\{s_{rc}(f)\cdot e^{+j\frac{\pi}{K}f^2}\right\} =$$

$$= \mathcal{F}^{-1}\left\{\sum_{q=1}^{Q}\mathcal{F}\left\{U_q\cdot e^{+j\,\varphi_q(t)}\right\}\cdot e^{+j\frac{\pi}{K}\left[f^2-(K\,\tau_{0q})^2\right]}\right\}. \quad (C.3)$$

Due remaining effects like e. g. a finite window length, thus *Gibb's phenomenon* [56] etc., still a little phase term $\exp\left(+j\,\delta\varphi_{rvp,q}\right)$ remains [64], [65].

$$s_{mx,corr}(t) \approx \sum_{q=1}^{Q} U_q\cdot e^{+j\,\varphi_q(t)}\cdot e^{+j\,\delta\varphi_{rvp,q}}. \quad (C.4)$$

C.2 Zero-padding

The discrete *Fourier* transformation (DFT) can be described as [56]:

$$X[k] = \frac{1}{\sqrt{N}}\sum_{n=0}^{N-1}x[n]\cdot e^{-j\,\frac{2\pi}{N}\,k\,n}. \quad (C.5)$$

Which is nothing else but a discrete correlation of a non-zero time series $x[n]$ with orthonormal basis functions $\psi_k[n]$ at discrete frequencies $\omega_k = 2\pi\,k/N$:

$$\psi_k[n] = \frac{1}{\sqrt{N}}\cdot e^{+j\,2\pi\,\frac{k}{N}\,n}. \quad (C.6)$$

If the non-zero $x[n]$ is elongated by a vector containing only zeros, this is denoted as *zero padding* [61]. Formally no new information is brought into the signal analysis, however, N is increased, thus the frequency grid ω_k gets narrowed. Therefore, $x[n]$ is correlated with intermediate basis functions providing complex values for intermediate frequency points, which is a form of interpolation [56], [61]. Of course, this is also valid for the inverse DFT [56], [61].

C.3 Window functions

The optimal window function providing the best compromise between minimum mainlobe broadening and required sidelobe suppression, is a *Dolph-*

Chebyshev filter/window/weighting [47], [56], [59], [67]. However, this filter cannot be implemented properly [47], [56], [59], [67]. Feasible approximations for a *Dolph-Chebyshev window* are *Taylor windows* [47], [56], [59] or *Hamming windows* [56], [67].

The *Kaiser window* exhibits the advantage of defining the stopband attenuation by a single shape parameter [47], [56], [59], [67].

C.3.1 Taylor and Hamming window

The *Taylor window* for amplitude weighting in time or spectral domain is defined by a chain of coefficients weighting trigonometric functions. The coefficients corresponding to a wanted sidelobe suppression are listed in tables [59], [67]. Here we concentrate on 40 dB suppression, which leads to seven coefficients [59]:

$$w_{tay}(t) = 1 + 2 \sum_{\nu=1}^{7} \gamma_\nu \cdot \sin\left(2\pi\,\nu\frac{t}{T}\right) . \tag{C.7}$$

The corresponding coefficients for $t \in [0; T]$ are listed in Tab. C.1 [59].

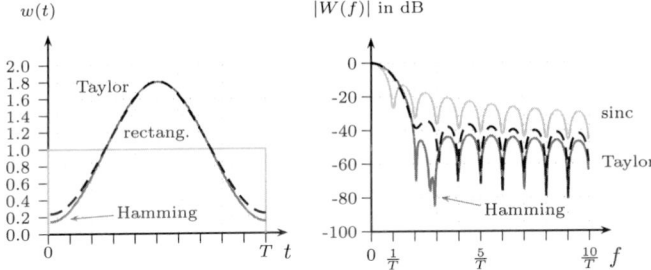

Fig. C.1 *Taylor window* with 40 dB sidelobe suppression and *Hamming window*, in time domain (left), and frequency domain (right) [56], [59].

The *Hamming window* features properties very similar to *Taylor window* [56], [67]. However, the mathematical formulation for $t \in [0; T]$ is a little simpler [56]:

$$w_{ham}(t) = \alpha \left[0.54 - 0.46 \cos\left(2\pi\frac{t}{T}\right)\right] , \text{ here } \alpha = 1.8 . \tag{C.8}$$

γ_1	γ_2	γ_3	γ_4
-0.387560	$0.954603 \cdot 10^{-2}$	$-0.470359 \cdot 10^{-2}$	$0.135350 \cdot 10^{-2}$

γ_5	γ_6	γ_7
$-0.332979 \cdot 10^{-4}$	$-0.357716 \cdot 10^{-3}$	$0.290474 \cdot 10^{-3}$

Tab. C.1 Coefficients for Taylor window with 40 dB sidelobe suppression [47], [59].

The factor α is usually not necessary. But for a similar performance to a 40 dB *Taylor window*, α must be set to 1.8. Please note, for for $t \in \left[-\frac{T}{2}; +\frac{T}{2} \right]$, the signs for all coefficients must be inverted.

C.3.2 The Kaiser window

A well-known and approved window function in signal processing is the *kaiser window function* [56], [61], [160].

$$
w_{T,\beta}(t) = \begin{cases} \dfrac{Y_0\left[\beta \sqrt{1 - \left(\frac{2t}{T} \right)^2} \right]}{Y_0(\beta)}, & -\dfrac{T}{2} \le t \le +\dfrac{T}{2}, \\ 0, & \text{else.} \end{cases} \tag{C.9}
$$

Or equivalently:

$$
w_{T,\beta}(t) = \begin{cases} \dfrac{Y_0\left[\beta \sqrt{1 - \left(\frac{2t-T}{T} \right)^2} \right]}{Y_0(\beta)}, & 0 \le t \le T, \\ 0, & \text{else.} \end{cases} \tag{C.10}
$$

Hereby, Y_0 is a *modified Bessel function* (type 1) of zero order [43], [56], [61]. While T is the filter length, β is the *shape parameter* [56] (it does not really meet the termini *form factor* or *aspect ratio*). For $\beta = 0$ the kaiser window becomes rectangular [56]. Kaiser proofed that the attenuation in stopband *att* is only determined by β [56]. Further, Kaiser proposed empirical

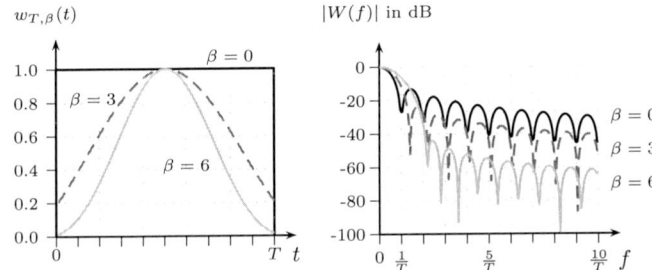

Fig. C.2 Kaiser window in time domain (left), and frequency domain (right) [56], [61], [160].

formulas to determine β by specifying the stop-band attenuation att in dB [61], [56], [160]:

$$
\beta = \begin{cases}
0.1102 \cdot (att - 8.7) \,, & att > 50\text{dB} \,, \\
0.5842 \cdot (att - 21)^{0.4} + 0.07886 \cdot (att - 21) \,, & 21\text{dB} \leq att \leq 50\text{dB} \,, \\
0 \,, & att < 21\text{dB}
\end{cases}
$$

$$(\text{C.11})$$

D Array processing fundamentals

D.1 Steering weights from Laurent series

The start is equation 3.33:

$$w_{st,n} = \frac{1}{2\pi j} \oint_{\partial C} \frac{AF(z)}{z^{n+1}}\, dz = \frac{1}{2\pi j} \oint_{\partial C} AF(z) \cdot z^{-(n+1)}\, dz \quad . \quad (D.1)$$

The parameter ∂C stands for a closed curve within the *domain of definition* of $AF(z)$. Now, z is again substituted by $z = e^{j\,\Psi}$. The integration increment, dz, must then be substituted by $j\,e^{j\,\Psi}\,d\Psi$. Thereby Ψ is completely real, thus $\Im\{\Psi\} = 0$. This means that the closed path ∂C can be chosen arbitrarily in the upper or lower half space. The other way round, it can be stated that the path ∂C can be degraded to line integral on the real axis. Hence, according to *residual theorem* [43], [44], the closed path ∂C is substituted by the integration interval $[-\infty\,;\,+\infty]$. However, due to the periodicity of the exponential function, $AF(\Psi)$ is periodic in Ψ-domain by steps of 2π. Thus, it is sufficient to chose the path ∂C within the *principal value* of $[-\pi\,;\,+\pi]$. Combining this and manipulating equation (D.1), it can be obtained:

$$
\begin{aligned}
w_{st,n} &= \frac{1}{2\pi j} \oint_{\partial C} AF(\Psi)\,e^{-j\,\Psi\,(n+1)}\,j\,e^{j\,\Psi}\,d\Psi = \\[2mm]
&= \frac{1}{2\pi} \int_{-\infty}^{+\infty} AF(\Psi) \cdot e^{-j\,\Psi\,n}\,d\Psi = \frac{1}{2\pi} \int_{-\pi}^{+\pi} AF(\Psi) \cdot e^{-j\,\Psi\,n}\,d\Psi \quad .
\end{aligned}
\tag{D.2}
$$

In section 3.2, the array factor was stated to be the *spatial Fourier series* with the steering weights as coefficients. Now, equation (D.2) can directly be recognized as the analysis formula for its *Fourier series coefficients* [43].

In standard case, thus a rectangular window along the coordinate ξ, the array factor is a si-function. The back-transformation is then:

$$w_{st,n} \approx \frac{1}{2\pi} \int\limits_{-\pi}^{+\pi} \frac{\sin\left(\Gamma/2\,\Psi\right)}{\Gamma/2\,\Psi}\, \mathrm{e}^{-j\,\Psi n}\, d\Psi$$

$$w_{st,n} = \frac{1}{2\pi} \int\limits_{-\pi}^{+\pi} \left(\sum_{\xi=0}^{\Gamma-1} \frac{1}{\Gamma} \mathrm{e}^{+j\,\Psi\xi} \right) \mathrm{e}^{-j\,\Psi n}\, d\Psi =$$

$$= \frac{1}{2\pi\,\Gamma} \sum_{n=0}^{\Gamma-1} \int\limits_{-\pi}^{+\pi} \mathrm{e}^{+j\,(\xi-n)\,\Psi}\, d\Psi = \frac{1}{2\pi\,\Gamma} \sum_{\xi=0}^{\Gamma-1} 2\pi\,\delta\left(\xi - n\right) =$$

$$= \frac{1}{\Gamma} \sum_{n=0}^{\Gamma-1} \delta\left(\xi - n\right) \text{ , q.\,e.\,d.}$$

(D.3)

Especially interesting for the zero redundancy array considerations are si-function with an added constant term C:

$$w_{st,n} = \frac{1}{2\pi} \int\limits_{-\pi}^{+\pi} \left[C + \sum_{\xi=0}^{\Gamma-1} \frac{1}{\Gamma} \mathrm{e}^{+j\,\Psi\xi} \right] \mathrm{e}^{-j\,\Psi n}\, d\Psi = \ldots =$$

$$= \frac{1}{2\pi}\, C \cdot \int\limits_{-\pi}^{+\pi} \mathrm{e}^{-j\,\Psi n}\, d\Psi + \frac{1}{\Gamma} \sum_{n=0}^{\Gamma-1} \delta\left(\xi - n\right) = \ldots =$$

$$= C \cdot \frac{\sin(\pi\,n)}{\pi\,n} + \frac{1}{\Gamma} \sum_{n=0}^{\Gamma-1} \delta\left(n\right) = C \cdot \mathrm{sinc}(n) + \frac{1}{\Gamma} \sum_{n=0}^{\Gamma-1} \delta\left(\xi - n\right) =$$

$$= C\,\delta(n) + \frac{1}{\Gamma} \sum_{n=0}^{\Gamma-1} \delta\left(\xi - n\right) \text{ .}$$

(D.4)

D.2 Zero redundancy arrays

This section shall introduce the idea of proof for zero redundancy arrays as proposed in [25] and exploited in [26].

A flat, thus constant spatial sensivity function (spatial auto-correlation of array weights $w_{st,i}$) would lead to a si-shaped radiation pattern with negative values for power density which states a conflict (see Fig. D.1) [25]: si$\left(\Gamma/2\,\Psi\right)$. A possible remedy is a constant added value in Ψ-domain

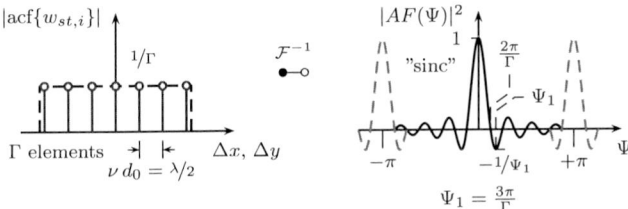

Fig. D.1 Flat spatial auto-correlation (*spatial frequency spectrum*) would mean a si-shaped radiation pattern with physically not plausible negative values for power density.

[25]. The absolute value of a si- or sinc-function declines with $1/x$ over x [43]. Thereby, the tangential points can be found at $x = (n + 1/2)\,\pi$ with $n \in \mathbb{N}$ for the standard si-function [43]. Scaled to $x = \Psi\,\Gamma/2$, the tangential points can be found at $\Psi = (n + 1/2)\,2\pi\,1/\Gamma$. The very first one is then at $\Psi_1 = 3\pi/\Gamma$. The corresponding function value is $-\Gamma/3\pi$, due to the $1/x$-property. However, the first minimum of an si-function is at

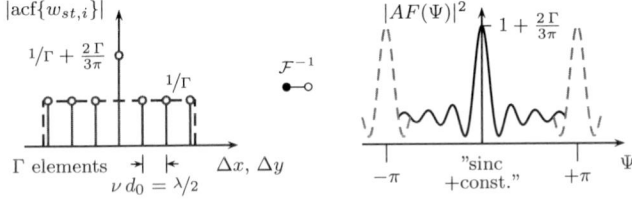

Fig. D.2 Flat spatial auto-correlation (spatial sensivity) with a constant additive term gives a si-shaped radiation pattern shifted to positive values only, for power density.

$x \approx (n + 1/2)\,\pi - \frac{1}{(n + 1/2)\,\pi}$ [43]. Hence, in order to secure that the result is only positive, Bracewell added $2 \cdot 3\pi/\Gamma$ to the si-function [25]. This results in a radiation pattern of si $(\Gamma/2\,\Psi) + 2 \cdot 3\pi/\Gamma$. Therefore, the auto-correlation function gets an additional value $2 \cdot 3\pi/\Gamma\,\delta(n)$ [25] (see Fig. D.2). Now, the task is to find the corresponding weights $w_{st,n}$ in order to form the auto-correlation function in Fig. D.2. Bracewell found four possible array configurations and has postulated that their are no more arrays possible which consist of more than four elements (see Fig. 6.4) [25].

D.3 Array processing algoithms

D.3.1 Capon's minimum variance estimator

The start is the output of the conventional beamformer, however with a linear constraint [45], [89]:

$$\min a_\nu^{\mathrm{H}}(\Psi_\nu)\, R_n\, a_\nu(\Psi_\nu) \quad \text{subject to} \quad a_\nu^{\mathrm{H}}(\Psi_\nu)\, a_q(\Psi_q) = 1 \ . \tag{D.5}$$

This constraint optimization suppresses contributions from unwanted directions [45]. With help of *Lagrange method*, equation (D.5) can be converted into an unconstrained optimization problem :

$$\mathcal{L} = a_\nu^{\mathrm{H}}\, R_n\, a_\nu + \lambda \left(a_\nu^{\mathrm{H}}\, a_q - 1\right) \ , \text{with } \lambda \in \mathbb{R} \ . \tag{D.6}$$

The output power of Capon's minimum variance estimator is [45], [89]:

$$P_c(\Psi_\nu) = \ldots = \frac{1}{a_\nu^{\mathrm{H}}\, R_n^{-1}\, a_\nu} \ . \tag{D.7}$$

A short derivation can be found in [89]. Synonyms for this algorithm are e. g. *Linearly constrained minimum-variance beamformer* (LCMV), or *minimum variance distortionless response spectrum* (MVDR) [89]. In general, it can be stated that the resolution capability of Capon's method is superior, compared to conventional beamformer, however, the fundamental signal level is much lower [45].

D.3.2 Linear estimation, auto-regressive models or maximum entropy method

The fundamental idea behind this method is a *prediction error filter*[45], [87], [89]. One array element's signal is predicted by knowledge of all others and compared to the actually measured signal. The error is taken as cost function and minimized [45], [87], [89]. Therefore, the array signal process is modelled as *moving average* process [45], [87], [89]. The filter coefficients b_i can be found with help of covariance matrix [45], [87], [89]:

$$R_n \begin{bmatrix} b_{\Gamma-1} \\ \vdots \\ b_1 \\ 1 \end{bmatrix} = \begin{bmatrix} 0 \\ \vdots \\ 0 \\ \delta_0 \end{bmatrix} \ . \tag{D.8}$$

The coefficients b_i form a transfer function $H(\Psi)$ [45], [89]. The inverse process is an *auto-regressive* (all pole) process which is fed by the above mentioned error [45], [89]. The processing output is a quasi spectral representation with peaks at the corresponding directions [45], [89]:

$$S_x(\Psi) = \frac{\delta_0}{|H(\Psi)|^2} \ . \tag{D.9}$$

A different ansatz leading to a similar result is the maximum entropy method introduced by Burg [92]. In order to obtain a fine spectral resolution, the signal gets extended beyond its limits in time domain, with respect to achieve maximum signal entropy [92]. The condition for maximum entropy for Gaussian processes in turn can be imposed more easily in spectral domain [92]. Plenty of literature has been published on this topic. The reader shall be referred to short selection: [92], [94], [120], [164], [165], [166], [167], [168].

D.3.3 Eigenanalysis methods

The eigenvalue shift, as well as noise sub-space orthogonal to signal sub-space was introduced in section 5.3.3. This knowledge shall be applied in the following [45].

MUSIC

A well-known method using eigenvector decomposition of R_n is MUSIC: MUltiple SIgnal Classification [45]. There, basically an eigenvector decomposition is applied on R_n. For the signal processing, the eigenvectors $a_{e,i}$ ($i \in [\Gamma_s + 1, \Gamma]$) belonging to *noise sub-space* $\mathcal{N}$ are exploited. Equation (5.34) has demonstrated that the *noise sub-space* $\mathcal{N}$ eigenvectors are orthogonal to R. Therefore, the directional vectors $a_\nu(\Psi_\nu)$ (out of the so-called dictionary) are tested to be orthogonal to *noise sub-space* $\mathcal{N}$ [45], [87]:

$$P(\Psi_\nu) = \frac{1}{\sum_{i=\Gamma_s+1}^{\Gamma} \left|a_{e,i}^{\mathrm{H}} a_\nu\right|^2} \ . \tag{D.10}$$

Since the correct directional vectors a_q are orthogonal to the *noise sub-space* $\mathcal{N}$, the dominator in equation (D.10) becomes minimal, thus the complete term maximum in case of $a_q = a_\nu$. However, at least one noise eigenvector is needed, thus only a maximum of $\Gamma - 1$ linear independent directional vectors can be identified [45], [87]. MUSIC can be applied to general array setups and is not restricted to ULAs [93], nor it is a-priori restricted to *uncorrelated signal scenarios* [45].

ESPRIT

ESPRIT stands for **E**stimation of **S**ignal **P**arameters via **R**otational **I**nvariance **T**echniques [45], [93] and is applied to *coherent signal scenario* [45], [93]. In explicit words, this method exploits the *Vandermode* character of A_q [45], [48], [93]. Such a type of matrix (or its transposed version) can be written as [48]:

$$
\begin{bmatrix}
1 & z_1 & z_2 & \cdots & z_q & \cdots & z_Q \\
1^2 & z_1^2 & z_2^2 & \cdots & z_q^2 & \cdots & z_Q^2 \\
\vdots & \vdots & \vdots & & \vdots & & \vdots \\
1^{\Gamma-1} & z_1^{\Gamma-1} & z_2^{\Gamma-1} & \cdots & z_q^{\Gamma-1} & \cdots & z_Q^{\Gamma-1}
\end{bmatrix} .
\tag{D.11}
$$

The array shall consist of two equal array setups, which can bear an arbitrary antenna configuration, however, each element must have equivalent element within the other setup, in order to form functional couples [93]. The common sense behind is now, that each couple exhibits the same spatial shift [93]. In case of the simpler case of a ULA, the first partial array is enumberated from element number 1 to $(\Gamma - 1)$, whereas the second one starts at element number 2 and terminates at number Γ. The first partial array shall be labelled by l, the second one r. According to equation (5.3) and with respect to the directional vectors, it can be stated [45], [93]:

$$
a_{q,r} = \mathrm{e}^{-j\,\Psi_q}\, a_{q,l} \,,\; s_{q,r} = \mathrm{e}^{-j\,\Psi_q}\, s_{q,l} \;\Rightarrow\; A_{q,r} = A_{q,l}\, C \,,
$$
$$
\text{with } C = \mathrm{diag}\left\{\ldots, \mathrm{e}^{-j\,\Psi_q}, \ldots\right\} .
\tag{D.12}
$$

Both sub-arrays provide similar signals, but shifted by one element phase difference [45], [93]:

$$
s_l(t) = A_{q,l}\, u(t) + n_l(t) \quad,
\tag{D.13}
$$

$$
s_r(t) = A_{q,r}\, u(t) + n_r(t) = A_{q,l}\, C\, u(t) + n_r(t) \quad.
\tag{D.14}
$$

Equation (5.5) must be imposed twice, hence, two covariance matrices are formed [45], [93]:

$$
\begin{aligned}
R_{ll,n} = \mathcal{E}\left\{s_{p,l}\, s_{p,l}^{\mathrm{H}}\right\} &= R_{ll} + \mathrm{diag}\left\{\sigma_1^2, \ldots, \sigma_{\Gamma-1}^2\right\} = \\
&= A_{q,l}\, \varsigma\varsigma^{\mathrm{H}}\, A_{q,l}^{\mathrm{H}} + \mathrm{diag}\left\{\sigma_1^2, \ldots, \sigma_{\Gamma-1}^2\right\} ,
\end{aligned}
\tag{D.15}
$$

$$\boldsymbol{R}_{lr,n} = \mathcal{E}\left\{\boldsymbol{s}_{p,l}\,\boldsymbol{s}_{p,r}^{\mathrm{H}}\right\} = \boldsymbol{R}_{lr} + \mathrm{diag}\left\{\sigma_1^2,\,\ldots,\,\sigma_{\Gamma-1}^2\right\} =$$
$$= \boldsymbol{A}_{q,l}\,\varsigma\varsigma^{\mathrm{H}}\,\boldsymbol{C}^{\mathrm{H}}\,\boldsymbol{A}_{q,l}^{\mathrm{H}} + \mathrm{diag}\left\{\sigma_2^2,\,\ldots,\,\sigma_{\Gamma}^2\right\}\;. \tag{D.16}$$

In case of equal noise variances per channel $\sigma_i^2 = \sigma_n^2$, two noise-free covariance matrices can be formed [45], [93]:

$$\boldsymbol{R}_{ll} = \boldsymbol{R}_{ll,n} - \sigma_n^2\cdot\mathbf{1}_{\Gamma-1}\;,\;\; \boldsymbol{R}_{lr} = \boldsymbol{R}_{lr,n} - \sigma_n^2\,\mathbf{1}_{\Gamma-1}\;. \tag{D.17}$$

Both sub-arrays provide similar signals which differ by the phase shift of one element. Therefore $\boldsymbol{R}_{ll,n}$ and $\boldsymbol{R}_{lr,n}$ can be exploited for a *generalized eigenvalue problem* [43], [45], [93]:

$$\boldsymbol{R}_{ll}\,\boldsymbol{x} = \lambda\,\boldsymbol{R}_{lr}\,\boldsymbol{x}\quad,$$
$$\left(\boldsymbol{R}_{ll} - \lambda\,\boldsymbol{R}_{lr}\right)\boldsymbol{x} = \boldsymbol{0}\quad, \tag{D.18}$$
$$\boldsymbol{A}_{q,l}\,\varsigma\varsigma^{\mathrm{H}}\left(\boldsymbol{E}_Q - \lambda\,\boldsymbol{C}^{\mathrm{H}}\right)\boldsymbol{A}_{q,l}^{\mathrm{H}}\,\boldsymbol{x} = \boldsymbol{0}\quad.$$

The general eigenvalues are determined by the characteristic polynomial [43]:

$$\det\left(\boldsymbol{R}_{ll} - \lambda\,\boldsymbol{R}_{lr}\right) = 0\;\Leftrightarrow\;\det\left(\boldsymbol{E}_Q - \lambda\,\boldsymbol{C}^{\mathrm{H}}\right) = 0\quad. \tag{D.19}$$

A closer look on the outcome of general eigenvalues in equation (D.19) discloses their correspondence to phase shifts between two neighbour elements [45], [93]:

$$\lambda_q = \mathrm{e}^{-j\,\Psi_q\cdot 1}\;,\;\text{for}\;q\in[1,\,\Gamma_s]\;. \tag{D.20}$$

GEESE

GEESE stands for **GE**neralized **E**igenvalues utilizing **s**ignal **S**ubspace **E**igenvectors [45]. Again an eigenvalue decomposition is applied on $\boldsymbol{R}_n$. Those eigenvectors are aligned in the matrix $\boldsymbol{A}_e$ [45]. Section 5.3.3 demonstrated, that the directional vectors $\boldsymbol{a}_q$ span the same *signal sub-space* $\mathcal{S}$ [45]. Now in GEESE method the eigenvectors $\boldsymbol{a}_e$ are expressed by Γ_s linear independent directional vectors $\boldsymbol{a}_q$ [45]:

$$\boldsymbol{a}_e = \sum_{i=1}^{\Gamma_s} c_i\cdot\boldsymbol{a}_q\;\Rightarrow\;\boldsymbol{A}_e = \boldsymbol{A}_q\,\boldsymbol{\Lambda}\;\text{with}\;\boldsymbol{C}\in\mathbb{J}^{\Gamma_s\times\Gamma_s}\;. \tag{D.21}$$

Similar to ESPRIT, two sub-matrices are now formed: The first one, $A_{e,l}$ with the row 1 to $\Gamma - 1$ of A_e, the second one, $A_{e,r}$, with the rows 2 to Γ [45]:

$$A_{e,l} = A_{q,l} \, \Lambda \quad , \tag{D.22}$$

$$A_{e,r} = A_{q,r} \, \Lambda = A_{q,l} \, C \, \Lambda \quad . \tag{D.23}$$

The matrix C is the same diagonal matrix like for ESPRIT algorithm and contains Γ_s phasors between two neighbour array elements [45]. Again, a *generalized eigenvalue problem* can be formulated [45]:

$$A_{e,l} \, x = \lambda \, A_{e,r} \, x \quad ,$$

$$(A_{e,l} - \lambda \, A_{e,r}) \, x = 0 \quad , \tag{D.24}$$

$$A_{e,l} \, (E_Q - \lambda \, C) \, \Lambda \, x = 0 \quad .$$

The general eigenvalues can be determined again by the characteristic polynomial [43], [45]:

$$\det (A_{e,l} - \lambda \, A_{e,r}) = 0 \; \Leftrightarrow \; \det (E_Q - \lambda \, C) = 0 \quad . \tag{D.25}$$

This time, the general eigenvalues correspond to the complex conjugated phasors belonging to the Γ_s directions since C is not in its *hermitian* form [45]:

$$\lambda_q = \mathrm{e}^{+j \, \Psi_q \cdot 1} \; , \; \text{for} \; q \in [1, \, \Gamma_s] \quad . \tag{D.26}$$

List of Figures

E List of variables

List of variables

variable	(value +) unit	declaration
$\boldsymbol{A}$	$1\,\mathrm{Vs}/\mathrm{m}$	magnetic vector potential
A_{ap}	$1\,\mathrm{m}^2$	area size of aperture
acf		auto-correlation function
$\boldsymbol{A}_m$	$1\,\mathrm{As}/\mathrm{m}$	electric vector potential
$a(t)$	$1\,\mathrm{V},\ 1\,\mathrm{A},\ 1\,\sqrt{\mathrm{W}}$	signal envelope function
AF		array factor
$\boldsymbol{a}_{\nu,q}$		directional vectors , atoms
$\boldsymbol{A}_{\nu,q}$		matrix of directional vectors
$\boldsymbol{A}_\nu,\ \boldsymbol{A}_{\nu,u}$		dictionaries for compressed sensing
α	$1\,^\circ,\ 1\,\mathrm{rad},\ \mathrm{none}$	double use for angle or complex weight
$\boldsymbol{b}_{n,i}$		eigenvectors
$\boldsymbol{B}$	$1\,\mathrm{T}{=}1\,\mathrm{Vs}/\mathrm{m}^2$	magnetic flux density (appendix)
B_0	$1\,\mathrm{Hz}$	inherent (noise) bandwidth
$\boldsymbol{B}_c$		spreading matrix
$\boldsymbol{B}_s$		shift matrix
β	$1\,^\circ,\ 1\,\mathrm{rad},\ \mathrm{none}$	double use for angle or complex weight
c	$1\,\mathrm{m}/\mathrm{s}$	speed of light in general medium
c_0	$1\,\mathrm{m}/\mathrm{s}$	speed of light in vacuum
$\mathbb{C}$		field of complex numbers
$\boldsymbol{C}$		rotational variance matrix
$\chi(\tau, f_d)$		ambiguity function
$\boldsymbol{d}$	$1\,\mathrm{m}$	distance vector
d_n	$1\,\mathrm{m}$	entry in distance vector
D	$1\,\mathrm{m}$	general dimension e. g. antenna diameter
$D(\vartheta, \varphi)$		directivity of antenna
$\boldsymbol{D}$		matrix in spatial smoothing
$\delta(\xi)$		Dirac delta function

Tab. E.1 List of variables in the thesis - continued

variable	(value +) unit	declaration
δ_{ij}		Kronecker delta function
δ_i^2	$1\,\mathrm{m}$	remaining curvature of phase fronts
$\boldsymbol{E}$	$1\,\mathrm{V/m}$	electric field
$\boldsymbol{e}_i$		unit vector along coordinate i
$\epsilon = \epsilon_r\,\epsilon_0$	$1\,\mathrm{As/Vm}$	general permittivity
ϵ_0	$= 8.854 \cdot 10^{-12}\,\mathrm{As/Vm}$	vacuum permittivity
ϵ_r		relative permittivity
f	$1\,\mathrm{Hz}{=}1\,\mathrm{1/s}$	frequency
$\boldsymbol{F}(\boldsymbol{r})$	$1\,\mathrm{Am}$	radiation vector
$\mathcal{F}\{\}$		Fourier transformator
$\mathcal{F}_0, \mathcal{F}_1$		Fresnel integrals
ΔF	$1\,\mathrm{Hz}$	modulation bandwidth
$\mathbb{F}$		field (functional analysis)
f_{IF}	$1\,\mathrm{Hz}$	intermediate frequency
$f_{r,q}$	$1\,\mathrm{Hz}$	distance frequency of object q
$f_{d,q}$	$1\,\mathrm{Hz}$	Doppler frequency of object q
φ	$1\,\mathrm{rad}, 1\,^\circ$	(azimuth) angle or phase term
$\phi_c(\xi)$		discrete correlation function (lag ξ)
ϕ_{ij}		cross-correlation function
$\Phi(t)$	$1\,\mathrm{rad}$	phase
$\boldsymbol{\Phi}(\vartheta, \varphi, \delta^2, t)$		signal matrix
$g(\boldsymbol{r})$	$1\,\mathrm{1/m}$	Green's function
$G(\boldsymbol{k})$		Fourier transform of Green's function
$G(\vartheta, \varphi)$		antenna gain
γ_r		array redundancy, figure or merit
γ_{mr}	$\gamma_{mr} = \gamma_r - 1$	pure array redundancy
$\Gamma, \Gamma_d, \Gamma_u$		enumerating parameter, virtual elements
$\boldsymbol{H}$	$1\,\mathrm{A/m}$	magnetic field
$\mathcal{H}\{\}$		Hilbert transformator
$\mathbb{H}$		Hilbert vector space
$\Im\{\}$		imaginary part of complex term
$\boldsymbol{J}$	$1\,\mathrm{A/m^2}$	current density
j		imaginary unit
$k = \omega/c$	$1\,\mathrm{1/m}$	wavenumber
K	$1\,\mathrm{1/s^2}$	FMCW modulation slope/ ramp
$\mathbb{K}$		field (functional analysis)

Tab. E.1 List of variables in the thesis - continued

variable	(value +) unit	declaration
κ		running index
λ	$1\,\mathrm{m}$	wavelength
$\lambda_1 \ldots \lambda_\Gamma$		eigenvalues of matrices
$\boldsymbol{\Lambda}$		diagonal matrix of (general) eigenvalues
$\boldsymbol{\Lambda}_{delta}$		signal matrix
M		enumerating parameter
$\mu = \mu_r\,\mu_0$	$1\,\mathrm{Vs}/\mathrm{Am}$	general permeability
(μ)		turn/ stage in iteration
μ_0	$= 4\pi \cdot 10^{-7}\,\mathrm{Vs}/\mathrm{Am}$	vacuum permeability
μ_r		relative permeability
n_i	$= \sqrt{\epsilon_{r,i}\mu_{r,i}} = \frac{c_0}{c_i}$	refraction index of medium i
N		enumerating parameter
ν		enumerating parameter
$o(\boldsymbol{r})$	$1\,\mathrm{m}$	object function
$\mathcal{O}(n \log n)$		order function
ω	$1\,{}^1\!/\mathrm{s}$	angular frequency
P	$1\,\mathrm{W}$	power
π	≈ 3.14159	natural constant
PCR		pulse compression ratio
ψ	$1\,\mathrm{V}$	scalar electric potential
ψ_m	$1\,\mathrm{A}$	scalar magnetic potential
$\psi_{\nu,\eta}$		basis function, atom
Ψ	$1\,\mathrm{rad}$	digital wavenumber
psf		point-spread function
Q		quality factor of e. g. filters
q	$q \in [1; Q]$	running index, for e. g. objects
$q(\boldsymbol{r},t)$	$1\,\mathrm{V},\,1\,\mathrm{A},\,1\,\sqrt{\mathrm{W}}$	source function Helmholtz equation
$Q(\boldsymbol{k},\omega)$	$1\,\mathrm{V}/\mathrm{Hz},\,1\,\mathrm{A}/\mathrm{Hz},\,1\,\sqrt{\mathrm{W}}/\mathrm{Hz}$	Fourier transform of source function q
$\boldsymbol{r}$	$1\,\mathrm{m}$	vector of coordinates
$r(\xi)$		covariance value
$r_{\alpha,\beta}$		entry in covariance matrix
$\boldsymbol{R},\,\boldsymbol{R}_n,\,\boldsymbol{R}_u$		covariance matrices
$\mathbb{R}$		field of real numbers
$\Re\{\}$		real part of complex term
res		residual

Tab. E.1 List of variables in the thesis - continued

variable	(value +) unit	declaration
RVP		residual video phase, quadratic phase term
ρ_{ij}		correlation coefficient
ϱ	$1\,\mathrm{As/m^3}$	electric charge density
ϱ_m	$1\,\mathrm{Vs/m^3}$	equivalent magnetic charge density
rcs	$1\,\mathrm{m^2}$, $1\,\mathrm{dBm^2}$	radar cross section
$s(t)$	$1\,\mathrm{V}$, $1\,\mathrm{A}$, $1\,\sqrt{\mathrm{W}}$	signal
$\boldsymbol{s}$	$1\,\mathrm{V}$, $1\,\mathrm{A}$, $1\,\sqrt{\mathrm{W}}$	signal vector
$S(f)$	$1\,\mathrm{Vs}$, $1\,\mathrm{As}$, $1\,\sqrt{\mathrm{W}}\mathrm{s}$	signal spectrum
σ_q	$1\,\mathrm{1/m}$	complex reflectivity factor
σ_n	$1\,\mathrm{V}$, $1\,\mathrm{A}$, $1\,\sqrt{\mathrm{W}}$	noise standard deviation
σ_n^2	$1\,\mathrm{V^2}$, $1\,\mathrm{A^2}$, $1\,\mathrm{W}$	noise variance
ς		complex envelope function
SNR		signal-to-noise ratio
$\mathrm{si}(x)$	$\frac{\sin x}{x}$	non-normalized *sinus cardinalis*
$\mathrm{sinc}(x)$	$\frac{\sin(\pi x)}{\pi x}$	normalized *sinus cardinalis*
t	$1\,\mathrm{s}$	variable for time
T	$1\,\mathrm{s}$	period
τ	$1\,\mathrm{s}$	propagation time, round-trip delay
T_c	$1\,\mathrm{s}$	chirp/ sweep length/ duration
TBP		time-bandwidth product
T_p	$1\,\mathrm{s}$	pulse length/ duration
T_s	$1\,\mathrm{s}$	sampling period
ϑ	$1\,\mathrm{rad}$, $1\,^\circ$	(polar) angle
$u(t)$	$1\,\mathrm{V}$, $1\,\mathrm{A}$, $1\,\sqrt{\mathrm{W}}$	general signal
$\boldsymbol{u}$		lateral directional vector
$\mathcal{U}(t)$	$1\,\mathrm{V}$, $1\,\mathrm{A}$, $1\,\sqrt{\mathrm{W}}$	signal amplitude
$U_q(t)$	$1\,\mathrm{V^2}$, $1\,\mathrm{A^2}$, $1\,\mathrm{W}$	squared signal amplitude
$U(\boldsymbol{k},\omega)$	$1\,\mathrm{V/Hz}$, $1\,\mathrm{A/Hz}$, $1\,\sqrt{\mathrm{W}}/\mathrm{Hz}$	Fourier transform of $u(t)$
Υ		accumulated complex expression
$\mathcal{U}(t)$		envelope function
V	$1\,\mathrm{m^3}$	volume (for integration purposes)
$\mathbb{V}$		vector space
v	$1\,\mathrm{m/s}$	velocity
$\boldsymbol{v}$		lateral directional vector
$\boldsymbol{w}$		directional vector

Tab. E.1 List of variables in the thesis - continued

variable	(value +) unit	declaration
$\mathbb{W}$		vector space
$\boldsymbol{W}$		matrix in cyclic shifting
w_n		entry in direction vector
$\boldsymbol{w}_{st}$		steering vector
$w_{st,n}$		steering weight
W_{vr}	$1\,\mathrm{rad}$, $1\,°$	visible region of arrays
ξ		correlation lag, integration variable
$\boldsymbol{Y}_c$		coefficient position matrix
$\boldsymbol{y}_{nm}$	$1\,\mathrm{m}$	absolute positions of virtual elements
$\boldsymbol{y}_{c,nm}$		virtual elements' position vector
$\boldsymbol{y}_{Tx}$	$1\,\mathrm{m}$	absolute y-position of Tx elements
$\boldsymbol{y}_{Rx}$	$1\,\mathrm{m}$	absolute y-position of Rx elements
$\Delta\boldsymbol{y}_{nm}$	$1\,\mathrm{m}$	distances between virtual elements
z	$= \mathrm{e}^{+j\Psi}$	complex number for z-transformation
ζ	$1\,\sqrt{\mathrm{m}}$	complex reflection factor

Tab. E.1 List of variables in thesis

Bibliography

[1] LI, Jian ; STOICA, Petre: *MIMO Radar Signal Processing*. John Wiley and Sons, Inc., 2009

[2] LEHMANN, N.H. ; FISHLER, E. ; HAIMOVICH, A.M. ; BLUM, R.S. ; CHIZHIK, D. ; CIMINI, L.J. ; VALENZUELA, R.A.: Evaluation of Transmit Diversity in MIMO-Radar Direction Finding. In: *Signal Processing, IEEE Transactions on* 55 (2007), May, Nr. 5, S. 2215–2225. http://dx.doi.org/10.1109/TSP.2007.893220. – DOI 10.1109/TSP.2007.893220. – ISSN 1053–587X

[3] HAIMOVICH, A.M. ; BLUM, R.S. ; CIMINI, L.J.: MIMO-Radar with widely separated antennas. In: *IEEE Signal Processing Magazine* (2008), January

[4] FISHLER, E. ; HAIMOVICH, A. ; BLUM, R.S. ; CIMINI, L.J. ; CHIZHIK, D. ; VALENZUELA, R.A.: Spatial diversity in radars-models and detection performance. In: *Signal Processing, IEEE Transactions on* 54 (2006), March, Nr. 3, S. 823–838. http://dx.doi.org/10.1109/TSP.2005.862813. – DOI 10.1109/TSP.2005.862813. – ISSN 1053–587X

[5] FISHLER, E. ; HAIMOVICH, A. ; BLUM, R. ; CIMINI, R. ; CHIZHIK, D. ; VALENZUELA, R.: Performance of MIMO radar systems: advantages of angular diversity. In: *Signals, Systems and Computers, 2004. Conference Record of the Thirty-Eighth Asilomar Conference on* Bd. 1, 2004, S. 305–309 Vol.1

[6] SHEIKHI, A. ; ZAMANI, A.: Coherent Detection for MIMO Radars. In: *Radar Conference, 2007 IEEE*, 2007. – ISSN 1097–5659, S. 302–307

[7] LI, Jian ; XU, Luzhou ; STOICA, Petre ; FORSYTHE, K.W. ; BLISS, D.W.: Range Compression and Waveform Optimization for MIMO Radar: A Cramer-Rao Bound Based Study. In: *Signal Processing, IEEE Transactions on* 56 (2008), Jan, Nr. 1, S. 218–232. http://dx.doi.org/10.1109/TSP.2007.901653. – DOI 10.1109/TSP.2007.901653. – ISSN 1053–587X

[8] MECCA, V.F. ; KROLIK, J.L. ; ROBEY, F.C.: Beamspace slow-time MIMO radar for multipath clutter mitigation. In: *Acoustics, Speech and Signal Processing, 2008. ICASSP 2008. IEEE International Conference on*, 2008. – ISSN 1520–6149, S. 2313–2316

[9] HASSANIEN, A. ; VOROBYOV, S.A.: Transmit/receive beamforming for MIMO radar with colocated antennas. In: *Acoustics, Speech and Signal Processing, 2009. ICASSP 2009. IEEE International Conference on*, 2009. – ISSN 1520–6149, S. 2089–2092

[10] WEISS, M.: Multistatic Surveillance and Reconaissance: Sensor, Signals and Data Fusion. In: *RTO Nato Lecture, NATO* (2009). – Wachtberg, Germany

[11] LI, Jian ; STOICA, Petre: MIMO Radar with Colocated Antennas. In: *Signal Processing Magazine, IEEE* 24 (2007), Sept, Nr. 5, S. 106–114. http://dx.doi.org/10.1109/MSP.2007.904812. – DOI 10.1109/MSP.2007.904812. – ISSN 1053–5888

[12] ZHANG, J.J. ; PAPANDREOU-SUPPAPPOLA, A.: MIMO Radar with Frequency Diversity. In: *Waveform Diversity and Design Conference, 2009 International*, 2009, S. 208–212

[13] BLISS, D.W. ; FORSYTHE, K.W.: Multiple-input multiple-output (MIMO) radar and imaging: degrees of freedom and resolution. In: *Signals, Systems and Computers, 2004. Conference Record of the Thirty-Seventh Asilomar Conference on* Bd. 1, 2003, S. 54–59 Vol.1

[14] DE MAIO, A. ; LOPS, M.: Design Principles of MIMO Radar Detectors. In: *Aerospace and Electronic Systems, IEEE Transactions on* 43 (2007), July, Nr. 3, S. 886–898. http://dx.doi.org/10.1109/TAES.2007.4383581. – DOI 10.1109/TAES.2007.4383581. – ISSN 0018–9251

[15] YOU, Qu J. ; JIANYUN, Zhang ; CHUNQUAN, Liu: The Ambiguity Function of MIMO Radar. In: *Microwave, Antenna, Propagation and EMC Technologies for Wireless Communications, 2007 International Symposium on*, 2007, S. 265–268

[16] SAN ANTONIO, G. ; FUHRMANN, D.R. ; ROBEY, F.C.: MIMO Radar Ambiguity Functions. In: *Selected Topics in Signal Processing, IEEE Journal of* 1 (2007), June, Nr. 1, S. 167–177. http://dx.doi.org/10.1109/JSTSP.2007.897058. – DOI 10.1109/JSTSP.2007.897058. – ISSN 1932–4553

[17] CHEN, Chun-Yang ; VAIDYANATHAN, P.P.: Properties of the MIMO radar ambiguity function. In: *Acoustics, Speech and Signal Processing, 2008. ICASSP 2008. IEEE International Conference on*, 2008. – ISSN 1520–6149, S. 2309–2312

[18] ABRAMOVICH, Y.I. ; FRAZER, Gordon J.: Bounds on the Volume and Height Distributions for the MIMO Radar Ambiguity Function. In: *Signal Processing Letters, IEEE* 15 (2008), S. 505–508. http://dx.doi.org/10.1109/LSP.2008.922514. – DOI 10.1109/LSP.2008.922514. – ISSN 1070–9908

[19] POURVOYEUR, K. ; FEGER, R. ; SCHUSTER, S. ; STELZER, A. ; MAURER, L.: Ramp sequence analysis to resolve multi target scenarios for a 77-GHz FMCW radar sensor. In: *Information Fusion, 2008 11th International Conference on*, 2008, S. 1 –7

[20] FEGER, R. ; WAGNER, C. ; SCHUSTER, S. ; SCHEIBLHOFER, S. ; JAGER, H. ; STELZER, A.: A 77-GHz FMCW MIMO Radar Based on an SiGe Single-Chip Transceiver. In: *Microwave Theory and Techniques, IEEE Transactions on* 57 (2009), May, Nr. 5, S. 1020–1035. http://dx.doi.org/10.1109/TMTT.2009.2017254. – DOI 10.1109/TMTT.2009.2017254. – ISSN 0018–9480

[21] ENDER, J.H.G. ; KLARE, J.: System architectures and algorithms for radar imaging by MIMO-SAR. In: *Radar Conference, 2009 IEEE*, 2009. – ISSN 1097–5659, S. 1–6

[22] WEISS, M. ; PETERS, O. ; ENDER, J.: A three dimensional SAR system on an UAV. In: *Geoscience and Remote Sensing Symposium, 2007. IGARSS 2007. IEEE International*, 2007, S. 5315–5318

[23] ENDER, J.H.G. ; GIERULL, C.H. ; CERUTTI-MAORI, D.: Improved Space-Based Moving Target Indication via Alternate Transmission and Receiver Switching. In: *Geoscience and Remote Sensing, IEEE Transactions on* 46 (2008), Dec, Nr. 12, S. 3960–3974. http://dx.doi.org/10.1109/TGRS.2008.2002266. – DOI 10.1109/TGRS.2008.2002266. – ISSN 0196–2892

[24] SKOLNIK, Merrill I.: *Introduction to Radar Systems*. 3. McGraw-Hill Books, Inc., 2001

[25] Astrophysics Radio Astronomy Techniques. In: BRACEWELL, R. N.: *Miscellaneous - Verschiedenes*. Bd. 54: *Encyclopedia of Physics - Handbuch der Physik*. Springer Verlag, Berlin, 1962, S. 42–129

[26] MOFFET, A.T.: Minimum-Redundancy Linear Arrays IEEE, Transaction on Antennas and Propagation, March 1968

[27] DONG, Jian ; LI, QingXia ; HE, Fangmin ; NI, Wei ; ZHU, Yaoting: Co-array properties of minimum redundancy linear arrays with minimum sidelobe level. In: *Antennas, Propagation and EM Theory, 2008. ISAPE 2008. 8th International Symposium on*, 2008, S. 74–77

[28] CHEN, C.Y. ; VAIDYANATHAN, P.P.: Minimum Redundancy MIMO Radars IEEE, International Symposium on Circuits and Systems IS-CAS, 2008

[29] SCHMID, C.M. ; FEGER, R. ; WAGNER, C. ; STELZER, A.: Design of a linear non-uniform antenna array for a 77-GHz MIMO FMCW radar. In: *Wireless Sensing, Local Positioning, and RFID, 2009. IMWS 2009. IEEE MTT-S International Microwave Workshop on*, 2009, S. 1–4

[30] DETLEFSEN, Jürgen ; SIART, Uwe: *Grundlagen der Hochfrequenztechnik*. 3. Oldenbourg Verlag München, 2009

[31] BALANIS, Constantine A.: *Antenna Theory*. John Wiley and Sons Inc., 1997

[32] ORFANIDIS, Sophocles J.: *Electromagnetic Waves and Antennas*. Rutgers University, 2008. – download: http://www.ece.rutgers.edu/ orfanidi/ewa/

[33] HOVANESSIAN, S.A.: *Radar System Design and Analysis*. Artech House, Inc., 1984

[34] BAMLER, R.: *Mehrdimensionale lineare Systeme*. Springer Verlag, 1989

[35] DETLEFSEN, Jürgen: *Computational Methods in Electromagnetics*. lecture notes, 2006. – Technische Universität München

[36] LEHNER, Günther: *Elektromagnetische Feldtheorie für Ingenieure und Physiker*. 5th edition. Springer Verlag, 2005

[37] POZAR, David M.: *Microwave Engineering*. John Wiley, 1998

[38] GERTHSEN, Christian ; MESCHEDE, Dieter (Hrsg.): *Gerthsen Physik*. 21. Springer Verlag, 2001

[39] BERTL, Sebastian: *Abbildung mit Millimeterwellen zur Personenkontrolle*, Technische Universität München, Diss., 2009

[40] JONES, D.S.: *Methods in Electromagnetic Wave Propagation*. Oxford University Press, 1979

[41] DALLINGER, Alexander: *Zirkulares Synthetisches Apertur Radar zur Personenkontrolle mit Millimeterwellen*, Technische Universität München, Diss., 2007

[42] KLAUSING, Helmut ; HOLPP, Wolfgang: *Radar mit realer und synthetischer Apertur*. Oldenbourg Verlag München, 2000

[43] BRONSTEIN ; SEMENDJAJEW: *Taschenbuch der Mathematik*. Harri Deutsch Verlag, 2001

[44] MEYBERG, K. ; VACHENAUER, P.: *Höhere Mathematik 2*. 3. Springer Verlag, 1999

[45] PILLAI, S. U.: *Array Signal Processing*. Springer Verlag, 1989

[46] DETLEFSEN, Jürgen: *Radio Navigation and Location*. lecture notes, 2009. – Technische Universität München

[47] PEEBLES, Peyton Z.: *Radar Principles*. John Wiley and Sons, Inc., 1998

[48] PETERSEN, Kaare B. ; PEDERSEN, Michael S.: *The Matrix Cookbook*. http://matrixcookbook.com, November 2008

[49] KIRSCHNER, Andreas ; BERTL, Sebastian ; GUETLEIN, Johanna ; DETLEFSEN, Juergen: Comparison and tests of different virtual arrays for MIMO radar applications. In: *International Radar Symposium (IRS), 2011 Proceedings International*, 2011, S. 697–702. – Leipzig, Germany

[50] HÄNSLER, Eberhard: *Statistische Signale*. Springer Verlag, 1991

[51] HAGENAUER, Joachim: *Nachrichtentechnik 1*. lecture notes, 2006. – Technische Universität München

[52] KRIEGER, G. ; YOUNIS, M. ; HUBER, S. ; BORDONI, F. ; PATYUCHENKO, A. ; KIM, J. ; LASKOWSKI, P. ; VILLANO, M. ; ROMMEL, T. ; LOPEZ-DEKKER, P. ; MOREIRA, A.: MIMO-SAR and the orthogonality confusion. In: *Geoscience and Remote Sensing Symposium (IGARSS), 2012 IEEE International*, 2012. – ISSN 2153–6996, S. 1533–1536

[53] RIGOLL, G.: *Signaldarstellung.* lecture notes, September 2005. – Technische Universität München

[54] HAGENAUER, Joachim: *Nachrichtentechnik 2.* lecture notes, 2005/2006. – Technische Universität München

[55] LEVANON, Nadav ; MOZESON, Eli: *Radar Signals.* John Wiley and Sons, Inc., 2004

[56] OPPENHEIM, Alan V. ; SCHAEFER, Ronald W.: *Zeitdiskrete Signalverarbeitung.* Oldenourg Verlag, 1999. – 3

[57] LÜKE, Hans D.: *Signalübertragung - Grundlagen der digitalen und analogen Nachrichtenübertragungssysteme.* 6th. Springer Verlag, 1995

[58] DETLEFSEN, Jürgen: *Mikrowellensystemtechnik.* lecture notes, 2009. Technische Universität München

[59] SKOLNIK, Merrill I.: *Radar Handbook.* McGraw-Hill Books, Inc., 1970

[60] WEHNER, Donald R.: *High-Resolution Radar.* 2nd. Artech House, 1995

[61] CUMMING, Ian G. ; WONG, Frank H.: *Digital Processing of Synthetic Aperture Radar Data.* Artech House, 2005

[62] GRIFFITHS, H.D.: New ideas in FM radar. In: *Electronics Communication Engineering Journal 2* (1990), Oct, Nr. 5, S. 185–194. – ISSN 0954–0695

[63] STOVE, A.G.: Linear FMCW radar techniques. In: *Radar and Signal Processing, IEE Proceedings F 139* (1992), Oct, Nr. 5, S. 343–350. – ISSN 0956–375X

[64] CARRARA, Walter G. ; GOODMAN, Ron S. ; MAJEWSKI, Ronald M.: *Spotlight Synthetic Aperture Radar.* Artech House, Boston, 1995. – Signal Processing Algorithms

[65] META, A ; HOOGEBOOM, P. ; LIGTHART, L.P.: Range Non-linearities Correction in FMCW SAR. In: *Geoscience and Remote Sensing Symposium, 2006. IGARSS 2006. IEEE International Conference on,* 2006, S. 403–406

[66] BARTON, David K. ; LEONOV, Sergey A.: *Radar Technology Encyclopedia.* Artech House, 1998

[67] LUDLOFF, Albrecht: *Praxiswissen Radar und Radarsignalverarbeitung.* 3rd. Vieweg Verlag, 2002

[68] ROHLING, H. ; MEINECKE, M.-M.: Waveform design principles for automotive radar systems. In: *Radar, 2001 CIE International Conference on, Proceedings*, 2001, S. 1–4

[69] ROHLING, H. ; MOLLER, C.: Radar waveform for automotive radar systems and applications. In: *Radar Conference, 2008. RADAR '08. IEEE*, 2008. – ISSN 1097–5659, S. 1–4

[70] MEYBERG, K. ; VACHENAUER, P.: *Höhere Mathematik 1.* 6. Springer Verlag, 2003

[71] SOUMEKH, Mehrdad: *Synthetic Aperture Radar Signal Processing.* John Wiley and Sons, Inc., 1999

[72] SCHELKSHORN, Simon: *Multisensorielle Positionsbestimmung aus Dopplersignalen*, Technische Universität München, Diss., 2007

[73] VARSHNEY, L.R. ; THOMAS, D.: Sidelobe reduction for matched filter range processing. In: *Radar Conference, 2003. Proceedings of the 2003 IEEE*, 2003, S. 446–451

[74] ZHAO, Shi ; XIN, He J.: Study of Side Lobes Suppression for Using Pulse Compression in Weather Radar. In: *Image and Signal Processing, 2009. CISP '09. 2nd International Congress on*, 2009, S. 1–3

[75] ASHE, J.M. ; NEVIN, R.L. ; MURROW, D.J. ; URKOWITZ, H. ; BUCCI, N.J. ; NESPOR, J.D.: Range sidelobe suppression of expanded/compressed pulses with droop. In: *Radar Conference, 1994., Record of the 1994 IEEE National*, 1994, S. 116–122

[76] FENG, Luo ; LITING, Ruan ; SHUNJUN, Wu ; QIANSHENG, Zhao ; ZHIQIANG, Zhang: Design of modified spectrum filter based on Mismatched Window for NLFM signal. In: *Synthetic Aperture Radar, 2009. APSAR 2009. 2nd Asian-Pacific Conference on*, 2009, S. 274–277

[77] VIZITIU, I.-C. ; ENACHE, F. ; POPESCU, F.: Sidelobe reduction in pulse-compression radar using the stationary phase technique: An extended comparative study. In: *Optimization of Electrical and Electronic Equipment (OPTIM), 2014 International Conference on*, 2014, S. 898–901

[78] COLLINS, T. ; ATKINS, P.: Nonlinear frequency modulation chirps for active sonar. In: *Radar, Sonar and Navigation, IEE Proceedings* 146 (1999), Dec, Nr. 6, S. 312–316. http://dx.doi.org/10.1049/ip-rsn:19990754. – DOI 10.1049/ip–rsn:19990754. – ISSN 1350–2395

[79] BOUKEFFA, S. ; JIANG, Y. ; JIANG, T.: Sidelobe reduction with nonlinear frequency modulated waveforms. In: *Signal Processing and its Applications (CSPA), 2011 IEEE 7th International Colloquium on*, 2011, S. 399–403

[80] JOHNSTON, J.A. ; FAIRHEAD, A.C.: Waveform design and doppler sensitivity analysis for nonlinear FM chirp pulses. In: *Communications, Radar and Signal Processing, IEE Proceedings F* 133 (1986), April, Nr. 2, S. 163–175. http://dx.doi.org/10.1049/ip-f-1:1986027. – DOI 10.1049/ip–f–1:19860027. – ISSN 0143–7070

[81] GUETLEIN, J. ; BERTL, S. ; KIRSCHNER, A. ; DETLEFSEN, J.: Switching scheme for a FMCW-MIMO radar on a moving platform. In: *Radar Conference (EuRAD), 2012 9th European*, 2012, S. 91–94. – Amsterdam, Netherlands

[82] GUETLEIN, Johanna ; KIRSCHNER, Andreas ; DETLEFSEN, Juergen: Motion compensation for a TDM FMCW MIMO radar system. In: *Radar Conference (EuRAD), 2013 10th European*, 2013, S. 37–40. – Nuremberg, Germany

[83] GUETLEIN, Johanna ; KIRSCHNER, Andreas ; DETLEFSEN, Juergen: Modulation scheme for a TDM FMCW MIMO radar in the presence of motion. In: *Radar Symposium (IRS), 2013 14th International* Bd. 1, 2013, S. 77–82. – Dresden, Germany

[84] GUETLEIN-HOLZER, Johanna ; KIRSCHNER, Andreas ; SPECK, Christian ; DETLEFSEN, Juergen: Comparison of Motion Compensation Methods applied to a TDM FMCW MIMO Radar System IDE, Radar 2014. – Lille, France

[85] CHEN, Chun-Yang ; VAIDYANATHAN, P.P.: Beamforming issues in modern MIMO Radars with Doppler. In: *Signals, Systems and Computers, 2006. ACSSC '06. Fortieth Asilomar Conference on*, 2006. – ISSN 1058–6393, S. 41–45

[86] ROHLING, H.: Ordered statistic CFAR technique - an overview. In: *Radar Symposium (IRS), 2011 Proceedings International*, 2011, S. 631–638

[87] PILLAI, S. ; HABER, F. ; BAR-NESS, Y.: A new approach to array geometry for improved spatial spectrum estimation. In: *Acoustics, Speech, and Signal Processing, IEEE International Conference on ICASSP '85.* Bd. 10, 1985, S. 1816–1819

[88] PILLAI, S.U. ; HABER, F.: Statistical analysis of a high resolution spatial spectrum estimator utilizing an augmented covariance matrix. In: *Acoustics, Speech and Signal Processing, IEEE Transactions on* 35 (1987), Nov, Nr. 11, S. 1517–1523. http://dx.doi.org/10.1109/TASSP.1987.1165068. – DOI 10.1109/TASSP.1987.1165068. – ISSN 0096–3518

[89] HAYKIN, Simon: *Adaptive Filter Theory.* Prentice Hall, 2002

[90] BURG, Klemens ; HAF, Herbert ; WILLE, Friedrich: *Höhere Mathematik für Ingenieure.* Teubner Verlag, Stuttgart, 1987 (Band 2 Lineare Algebra). – Professors at Kassel University

[91] KALMAN, R. E.: A New Approach to Linear Filtering and Prediction Problems. In: *Transactions of the ASME* (1960), Nr. 82, Series D, S. 35–45. – Journal of Basic Engineering

[92] BURG, John P.: *Maximum Entropy Spectral Analysis*, Stanford University, Diss., 1975. – Department of Geophysics

[93] ROY, R. ; KAILATH, T.: ESPRIT-estimation of signal parameters via rotational invariance techniques. In: *Acoustics, Speech and Signal Processing, IEEE Transactions on* 37 (1989), Jul, Nr. 7, S. 984–995. http://dx.doi.org/10.1109/29.32276. – DOI 10.1109/29.32276. – ISSN 0096–3518

[94] LANG, S. ; DUCKWORTH, G. ; MCCLELLAN, J.: Array design for MEM and MLM array processing. In: *Acoustics, Speech, and Signal Processing, IEEE International Conference on ICASSP '81.* Bd. 6, 1981, S. 145–148

[95] VAN TREES, Harry L.: *Optimum Array Processing.* Bd. Part IV of Detection, Estimation, and Modulation Theory. John Wiley and Sons, 2002

[96] JORGENSON, M.B. ; FATTOUCHE, M. ; NICHOLS, S.T.: Applications of minimum redundancy arrays in adaptive beamforming. In: *Microwaves, Antennas and Propagation, IEE Proceedings H* 138 (1991), Oct, Nr. 5, S. 441–447. – ISSN 0950–107X

[97] GOLOMB, S.W. ; TAYLOR, H.: Two-dimensional synchronization patterns for minimum ambiguity. In: *Information Theory, IEEE Transactions on* 28 (1982), Jul, Nr. 4, S. 600–604. http://dx.doi.org/10.1109/TIT.1982.1056526. – DOI 10.1109/TIT.1982.1056526. – ISSN 0018–9448

[98] BOMER, L. ; ANTWEILER, M.: Two-dimensional binary arrays with constant sidelobe in their PACF. In: *Acoustics, Speech, and Signal Processing, 1989. ICASSP-89., 1989 International Conference on*, 1989. – ISSN 1520–6149, S. 2768–2771 vol.4

[99] NOZAKI, Shinya ; FUJIOKA, Shinsuke ; UEDA, Tatsuki ; CHEN, Yen wei ; NAMIHIRA, Y.: A Heuristic Decoding Method for Coded Images of Random Uniformly Redundant Array. In: *Intelligent Information Hiding and Multimedia Signal Processing, 2009. IIH-MSP '09. Fifth International Conference on*, 2009, S. 771–774

[100] FENIMORE, E. E. ; CANNON, T. M.: Coded aperture imaging with uniformly redundant arrays. In: *Applied Optics* 17 (1978), February, Nr. 3

[101] In: GUNSON, J. ; POLYCHRONOPULOS, B.: *Optimum Design of a Coded Mask X-Ray Telescope For Rocket Applications.* Bd. 177. Royal Astronomical Society, 1975, S. 485–497. – provided by NASA Astrophysical Data System

[102] BUSBOOM, A. ; ELDERS-BOLL, H. ; DIETER SCHOTTEN, H.: Combinatorial design of near-optimum masks for coded aperture imaging. In: *Acoustics, Speech, and Signal Processing, 1997. ICASSP-97., 1997 IEEE International Conference on* Bd. 4, 1997. – ISSN 1520–6149, S. 2817–2820 vol.4

[103] BLOOM, G.S. ; GOLOMB, S.W.: Applications of numbered undirected graphs. In: *Proceedings of the IEEE* 65 (1977), April, Nr. 4, S. 562–570. http://dx.doi.org/10.1109/PROC.1977.10517. – DOI 10.1109/PROC.1977.10517. – ISSN 0018–9219

[104] LANGE, K. (Hrsg.) ; K.-H., Löcherer (Hrsg.): *Taschenbuch der Hochfrequenztechnik.* 5. Springer Verlag, 1992. – original authors: Meinke, Gundlach

[105] WOODS, J.W. ; EKSTROM, M.P. ; PALMIERI, T.M. ; TWOGOOD, R.E.: Best Linear Decoding of Random Mask Images. In: *Nuclear Science, IEEE Transactions on* 22 (1975), Feb, Nr. 1, S.

379–383. http://dx.doi.org/10.1109/TNS.1975.4327666. – DOI 10.1109/TNS.1975.4327666. – ISSN 0018–9499

[106] WEISS, H.: Nonredundant Point Distribution for Coded Aperture Imaging with Application to Three-Dimensional On-Line X-Ray Information Retrieving. In: *Computers, IEEE Transactions on* C-24 (1975), April, Nr. 4, S. 391–394. http://dx.doi.org/10.1109/T-C.1975.224232. – DOI 10.1109/T–C.1975.224232. – ISSN 0018–9340

[107] GOTTESMAN, S.R. ; SCHNEID, E.J.: PNP - A New Class of Coded Aperture Arrays. In: *Nuclear Science, IEEE Transactions on* 33 (1986), Feb, Nr. 1, S. 745–749. http://dx.doi.org/10.1109/TNS.1986.4337206. – DOI 10.1109/TNS.1986.4337206. – ISSN 0018–9499

[108] CANNON, T.M. ; FENIMORE, E.E.: A Class of near-Perfect Coded Apertures. In: *Nuclear Science, IEEE Transactions on* 25 (1978), Feb, Nr. 1, S. 184–188. http://dx.doi.org/10.1109/TNS.1978.4329301. – DOI 10.1109/TNS.1978.4329301. – ISSN 0018–9499

[109] CHANG, L.T. ; MACDONALD, B. ; PEREZ-MENDEZ, V. ; SHIRAISHI, L.: Coded Aperture Imaging of Gamma-Rays Using Multiple Pinhole Arrays and Multiwire Proportional Chamber Detector. In: *Nuclear Science, IEEE Transactions on* 22 (1975), Feb, Nr. 1, S. 374–378. http://dx.doi.org/10.1109/TNS.1975.4327665. – DOI 10.1109/TNS.1975.4327665. – ISSN 0018–9499

[110] PRICE, LeRoy R.: An Improved Coded-Aperture System for Emission Computed Tomography (ECT). In: *Nuclear Science, IEEE Transactions on* 26 (1979), April, Nr. 2, S. 2794–2796. http://dx.doi.org/10.1109/TNS.1979.4330538. – DOI 10.1109/TNS.1979.4330538. – ISSN 0018–9499

[111] ROSENFELD, Dov ; MACOVSKI, A.: Time Modulated Apertures for Tomography in Nuclear Medicine. In: *Nuclear Science, IEEE Transactions on* 24 (1977), Feb, Nr. 1, S. 570–576. http://dx.doi.org/10.1109/TNS.1977.4328742. – DOI 10.1109/TNS.1977.4328742. – ISSN 0018–9499

[112] COOK, W.R. ; FINGER, M. ; PRINCE, T.A. ; STONE, E.C.: Gamma-Ray Imaging with a Rotating Hexagonal Uniformly Redundant Array. In: *Nuclear Science, IEEE Transactions on* 31 (1984), Feb, Nr. 1, S. 771–775. http://dx.doi.org/10.1109/TNS.1984.4333364. – DOI 10.1109/TNS.1984.4333364. – ISSN 0018–9499

[113] MCCONNELL, M.L. ; FORREST, D.J. ; CHUPP, E.L. ; DUNPHY, P.P.: A Coded Aperture Gamma Ray Telescope. In: *Nuclear Science, IEEE Transactions on* 29 (1982), Feb, Nr. 1, S. 155–159. http://dx.doi.org/10.1109/TNS.1982.4335818. – DOI 10.1109/TNS.1982.4335818. – ISSN 0018–9499

[114] NOZAKI, Shinya ; FUJIOKA, S. ; CHEN, Yen wei: A Heuristic Decoding Method for Coded Images of Uniformly Redundant Array. In: *Intelligent Systems Design and Applications, 2008. ISDA '08. Eighth International Conference on* Bd. 3, 2008, S. 132–135

[115] CHEN, S.S. ; DONOHO, D.L. ; SAUNDERS, M.A.: Atomic Decomposition by Basis Pursuit. Stanford, CA 94305, USA : http://www-stat.stanford.edu/ donoho/reports.html, 1995. – American Mathematical Society

[116] MALLAT, Stéphane G. ; ZHANG, Zhifeng: Matching Pursuit With Time-Frequency Dictionaries IEEE, Transaction on Signal Processing, December 1993

[117] BOSSERT, Martin: *Einführung in die Nachrichtentechnik.* Oldenbourg Verlag, 2012

[118] DONOHO, D.L. ; HUO, X.: Uncertainty Principles and Ideal Atomic Decomposition IEEE, Transactions on Information Theory, November 2001

[119] CHEN, Shaobing ; DONOHO, David: Basis Pursuit. Stanford, CA 94305, USA : http://www-stat.stanford.edu/ donoho/reports.html, 1994. – American Mathematical Society

[120] PRESS, William H. ; FLANNERY, Brian P. ; TEUKOLSKY, Saul A. ; VETTERLING, William T.: *Numerical Recipes in C.* Cambridge University Press, 1988. – The Art of Scientific Computing

[121] MANCERA, L. ; PORTILLA, J.: L0-Norm-Based Sparse Representation Through Alternate Projections. In: *Image Processing, 2006 IEEE International Conference on*, 2006. – ISSN 1522–4880, S. 2089–2092

[122] LEI, Zhu ; CHUNTING, Qiu: Application of compressed sensing theory to radar signal processing. In: *Computer Science and Information Technology (ICCSIT), 2010 3rd IEEE International Conference on* Bd. 6, 2010, S. 315–318

[123] ROSSI, M. ; HAIMOVICH, A.M. ; ELDAR, Y.C.: Global methods for compressive sensing in MIMO radar with distributed sensors. In: *Signals, Systems and Computers (ASILOMAR), 2011 Conference Record of the Forty Fifth Asilomar Conference on*, 2011. – ISSN 1058–6393, S. 1506–1510

[124] WIPF, David ; RAO, Bhaskar: L0-norm minimization for basis selection. In: *Advances in Neural Information Processing Systems 17*, 2005, S. 1513–1520

[125] STROHMER, T. ; FRIEDLANDER, Benjamin: Compressed sensing for MIMO radar - algorithms and performance. In: *Signals, Systems and Computers, 2009 Conference Record of the Forty-Third Asilomar Conference on*, 2009. – ISSN 1058–6393, S. 464–468

[126] CHEN, Chun-Yang ; VAIDYANATHAN, P.P.: Compressed sensing in MIMO radar. In: *Signals, Systems and Computers, 2008 42nd Asilomar Conference on*, 2008. – ISSN 1058–6393, S. 41–44

[127] DONOHO, David L.: For Most Large Underdetermined Systems of Equations, the Minimal l1-norm Near-Solution Approximates the Sparsest Near-Solution. In: *Department of Statistics Stanford University* (2004), August

[128] BOYD, Stephen ; VANDENBERGHE, Lieven: *Convex Optimization*. 7. Cambridge University Press, 2004. – corrections 2009

[129] COIFMAN, R.R. ; WICKERHAUSER, M.V.: Entropy-based algorithms for best basis selection. In: *Information Theory, IEEE Transactions on* 38 (1992), March, Nr. 2, S. 713–718. http://dx.doi.org/10. 1109/18.119732. – DOI 10.1109/18.119732. – ISSN 0018–9448

[130] DONOHO, David L.: CART and Best-Ortho-Basis: A Connection. In: *The annals of Statistics* Bd. 25, Stanford University and University of California, Berkley, 1997, S. 1870–1911

[131] GRAEB, Helmut: *Optimization Methods for Circuit Design*. lecture notes, 2010. – Technische Universität München

[132] LUSTIG, S. ; DONOHO, D.L. ; SANTOS, J.M. ; PAULY, J.M.: Compressed Sensing MRI. Stanford, CA 94305, USA : http://www-stat.stanford.edu/ donoho/reports.html, 2007. – American Mathematical Society

[133] HERMAN, M. ; STROHMER, T.: Compressed sensing radar. In: *Acoustics, Speech and Signal Processing, 2008. ICASSP 2008. IEEE International Conference on*, 2008. – ISSN 1520–6149, S. 1509–1512

[134] MEES, Wim ; HEREMANS, Roel: Surveillance in an Urban Environment using Mobile Sensors SPIE Security and Defence, 2012. – Edinburgh, Scotland

[135] DILL, Stephan ; PEICHL, Markus ; RUDOLF, Daniel: SUMIRAD - A near real-time MMW radiometer imaging system for threat detection in an urban environment SPIE Security and Defence, 2012. – Edinburgh, Scotland

[136] DILL, Stephan ; PEICHL, Markus ; RUDOLF, Daniel: SUMIRAD - A close to real time MMW radiometer imaging system. In: *Geoscience and Remote Sensing Symposium (IGARSS), 2012 IEEE International*, 2012. – ISSN 2153–6996, S. 4797–4800. – Munich, Germany

[137] DILL, Stephan ; PEICHL, Markus ; RUDOLF, Daniel: SUMIRAD - A fast imaging MMW radiometer for security and safety applications. In: *Radar Conference (EuRAD), 2013 European*, 2013, S. 323–326. – Nuremberg, Germany

[138] MEES, Wim ; HEREMANS, Roel: Multisensor data fusion for IED threat detection SPIE Security and Defence, 2012. – Edinburgh, Scotland

[139] HAGELEN, M. ; BRIESE, G. ; ESSEN, H. ; BERTUCH, T. ; KNOTT, P. ; TESSMANN, A.: A millimetrewave landing aid approach for helicopters under brown-out conditions. In: *Radar Conference, 2008. RADAR '08. IEEE*, 2008. – ISSN 1097–5659, S. 1–4

[140] PAGELS, A. ; HAGELEN, M. ; BRIESE, G. ; TESSMANN, A.: Helicopter Assisted Landing System - Millimeter-Wave against Brown-Out. In: *German Microwave Conference, 2009*, 2009, S. 1–3

[141] SCHLECHTWEG, Michael ; LEUTHER, Arnulf ; TESSMANN, Axel ; HÜLSMANN, Axel ; AMBACHER, Oliver ; HÄGELEN, Manfred ; BRIESE, Gunnar ; ESSEN, Helmut: Advanced Millimeter-Wave Radar Modules for Helicopter Landing Aid Conference Future Security, 2009. – Karlsruhe, Germany

[142] MAAS, Stephen A.: *The RF and Microwave Circuit Design Cookbook.* Artech House, 1998

[143] GIANNINI, F. ; LEUZZI, G.: *Nonlinear Microwave Circuit Design.* John Wiley and Sons, Ltd., 2004

[144] JARRY, Pierre ; BENEAT, Jacques: *Advanced Design Techniques and Realizations of Microwave and RF Filters.* John Wiley and Sons, Inc., 2008

[145] GRIFFITHS, H.D.: The effect of phase and amplitude errors in FM radar. In: *High Time-Bandwidth Product Waveforms in Radar and Sonar, IEE Colloquium on,* 1991, S. 9/1–9/5

[146] RUSSER, Peter: *Hochfrequenztechnik 1.* Lehrstuhl für Hochfrequenztechnik TU München : lecture notes, 1998. – 8th edition

[147] FUCHS, J. ; WARD, K.D. ; TULIN, M.P. ; YORK, R.A: Simple techniques to correct for VCO nonlinearities in short range FMCW radars. In: *Microwave Symposium Digest, 1996., IEEE MTT-S International* Bd. 2, 1996. – ISSN 0149–645X, S. 1175–1178 vol.2

[148] VOSSIEK, M. ; HEIDE, P. ; NALEZINSKI, M. ; MAGORI, V.: Novel FMCW radar system concept with adaptive compensation of phase errors. In: *Microwave Conference, 1996. 26th European* Bd. 1, 1996, S. 135–139

[149] VOSSIEK, M. ; KERSSENBROCK, T.V. ; HEIDE, P.: Signal Processing Methods for Millimetrewave FMCW Radar with High Distance and Doppler Resolution. In: *Microwave Conference, 1997. 27th European* Bd. 2, 1997, S. 1127–1132

[150] SCHEIBLHOFER, S. ; SCHUSTER, S. ; STELZER, A.: High-Speed FMCW Radar Frequency Synthesizer With DDS Based Linearization. In: *Microwave and Wireless Components Letters, IEEE* 17 (2007), May, Nr. 5, S. 397–399. http://dx.doi.org/10.1109/LMWC.2007. 895732. – DOI 10.1109/LMWC.2007.895732. – ISSN 1531–1309

[151] PIPER, S.O.: Homodyne FMCW radar range resolution effects with sinusoidal nonlinearities in the frequency sweep. In: *Radar Conference, 1995., Record of the IEEE 1995 International,* 1995, S. 563–567

[152] PIPER, S.O.: FMCW range resolution for MMW seeker applications. In: *Southeastcon '90. Proceedings., IEEE,* 1990, S. 156–160 vol.1

[153] ZHU, Zhaodu ; YU, Weidong ; ZHANG, Xinggan ; QIU, Xiaohui: A correction method for distortions in FM-CW imaging system. In: *Aerospace and Electronics Conference, 1996. NAECON 1996., Proceedings of the IEEE 1996 National* Bd. 1, 1996. – ISSN 0547–3578, S. 323–326 vol.1

[154] ANGHEL, A. ; VASILE, G. ; CACOVEANU, R. ; IOANA, C. ; CIOCHINA, S.: Nonlinearity correction algorithm for wideband FMCW radars. In: *Signal Processing Conference (EUSIPCO), 2013 Proceedings of the 21st European*, 2013, S. 1–5

[155] GENERAL MICROWAVE (Hrsg.): *Broad Band VCOs - V60 series*. General Microwave, 2009. – datasheet

[156] HÄMMERLIN, Günther ; HOFFMANN, Karl-Heinz: *Numerische Mathematik*. Springer Verlag, 1994. – up to 2nd edition volume 7 of Grundwissen Mathematik series

[157] AHLBERG, J.H. ; NILSON, E.N. ; WALSH, J.L.: *The Theory of Splines and Their Applications*. Academic Press, 1967

[158] HERRMANN, Norbert: *Höhere Mathematik*. 2. Oldenbourg Verlag, 2007

[159] CHUI, Charles K.: *Multivariate Splines*. Society for Industrial and Applied Mathematics SIAM, Philadelphia, Pennsylvania, 1988. – Texas A and M University

[160] MATHWORKS, Inc.: *Matlab documentation*. product help, 2011. – versions R2011b, R2012a

[161] SCHOENBERG, I.J.: *Cardinal Spline Interpolation*. Society for Industrial and Applied Mathematics SIAM, Philadelphia, Pennsylvania, 1973. – University of Wisconsin-Madison

[162] SCHÜSSLER, H.W.: *Digitale Signalverarbeitung*. Bd. 1. 3. Springer Verlag, 1992

[163] PARK, Jung D. ; KIM, Wan J. ; LEE, Chang W.: A novel method for beat frequency error correction for a low cost FMCW radar using VCO sweep characteristics. In: *Radar Conference, 2005. EURAD 2005. European*, 2005, S. 371–374

[164] PILLAI, S.U.: Generalization of maximum entropy spectrum extension method. In: *Acoustics, Speech, and Signal Processing, 1990. ICASSP-90., 1990 International Conference on*, 1990. – ISSN 1520–6149, S. 2619–2622 vol.5

[165] PILLAI, S.U. ; SHIM, T.I. ; BENTEFTIFA, M.H.: A new spectrum extension method that maximizes the multistep minimum prediction error-generalization of the maximum entropy concept. In: *Signal Processing, IEEE Transactions on* 40 (1992), Jan, Nr. 1, S. 142–158. http://dx.doi.org/10.1109/78.157189. – DOI 10.1109/78.157189. – ISSN 1053–587X

[166] ROBINSON, E.A.: A historical perspective of spectrum estimation. In: *Proceedings of the IEEE* 70 (1982), Sept, Nr. 9, S. 885–907. http://dx.doi.org/10.1109/PROC.1982.12423. – DOI 10.1109/PROC.1982.12423. – ISSN 0018–9219

[167] KAY, S.M.: Maximum entropy spectral estimation using the analytical signal. In: *Acoustics, Speech and Signal Processing, IEEE Transactions on* 26 (1978), Oct, Nr. 5, S. 467–469. http://dx.doi.org/10.1109/TASSP.1978.1163127. – DOI 10.1109/TASSP.1978.1163127. – ISSN 0096–3518

[168] LACOUME, J.-L. ; GHARBI, M. ; LATOMBE, C. ; NICOLAS, J.: Close frequency resolution by maximum entropy spectral estimators. In: *Acoustics, Speech and Signal Processing, IEEE Transactions on* 32 (1984), Oct, Nr. 5, S. 977–984. http://dx.doi.org/10.1109/TASSP.1984.1164436. – DOI 10.1109/TASSP.1984.1164436. – ISSN 0096–3518